T0256383

ELASTIC WAVES
IN
ANISOTROPIC
LAMINATES

G.R. Liu
Z.C. Xi

CRC PRESS

Boca Raton   London   New York   Washington, D.C.

Library of Congress Cataloging-in-Publication Data

Catalog record is available from the Library of Congress

Visit the CRC Press Web site at www.crcpress.com

© 2002 by CRC Press LLC

No claim to original U.S. Government works
International Standard Book Number 0-8493-1070-9

Preface

With many outstanding properties superior to traditional materials, advanced composite materials have found wide application in various engineering sectors. In most applications, composite structures are subjected to a variety of dynamic loadings. A study of wave propagation in such structures is very helpful for understanding their dynamic characteristics and failure mechanism. Moreover, ultrasonic-based non-destructive evaluation (NDE) plays more and more important roles in the determination of material properties and the detection of defects (cracks and flaws) in composites. An analysis of wave behavior is prerequisite in applying effectively and innovatively NDE techniques using ultrasonic and elastic waves.

Analysis of elastic waves in anisotropic media is much more complicated than that for isotropic media. Composite structures in engineering practices are made of a stack of plies, each of which is reinforced by fibers. The ply orientation, ply material, stacking sequence, and the number of layers vary with the different requirements of design. This results in an anisotropic and inhomogeneous nature in the material properties for the laminate. The governing equations for the anisotropic laminates are coupled with each other, and general solutions to these equations are usually very difficult to obtain by conventional methods of analysis. With the rapid development of computers, numerical methods of full domain discretization, such as finite element method (FEM) and finite difference method (FDM), have evolved as flexible and powerful computational tools for a wide range of engineering practical problems. These numerical methods are very often both extravagant and time-consuming in the treatment of waves propagating in anisotropic laminates. The results obtained using entirely numerical methods are usually in the form of a vast volume of data, from which important phenomena and characteristics of wave propagation are extremely difficult to extract and reveal in an explicit manner. This creates obstacles between the physics of wave propagation phenomenon and the analysts (students, researchers, and engineers) who are striving to learn and to understand these important phenomena. Methods of analytical nature with minimum domain discretization are therefore preferred for studying and investigating wave phenomena in composite laminates, as they can provide insightful results revealing important characteristics. This book provides a set of high-performance analytical-numerical methods for elastic wave analysis of anisotropic layered structures. At present, despite certain existing books on wave propagation in anisotropic media, there is no text available that is devoted entirely to modern analytical-numerical methods on this subject.

The material covered in this text is developed from the research work of the first author and co-workers in the past 15 years. The salient feature of this work is that the methods presented are novel and efficient in the treatment of wave propagation in anisotropic layered media. A comprehensive introduction, theoretical development, formulation and applications of these methods are provided. Key techniques discussed are

1. Layered media are discretized only in the thickness direction, so as to take advantage of the FEM to handle difficulties caused by the inhomogeneity of the media in the thickness direction. The use of the FEM leads to a set of dimension-reduced partial differential equations (not algebraic equations).

2. Analytic techniques, such as the Fourier transform technique and modal superimposition method, are used to deal with the dimension-reduced partial differential equations.

3. Strip element method (SEM) is introduced to deal with laminates with delaminations and flaws. Clear advantages of the SEM are demonstrated through examples by comparing the results obtained by the SEM and the FEM. SEM provides also the Green's function for anisotropic laminates. Green's function is essential for the BEM, and it is difficult to obtain for anisotropic media using conventional means.

4. Techniques for problems in both time and frequency domains are treated in great detail. Complex path method is introduced for wavenumber-space domain transformation.

5. A set of six characteristic wave surfaces is introduced to clearly visualize the characteristics of waves propagating in anisotropic layered media.

6. Application of these techniques to smart materials, plates, and shells composed of functionally graded and piezoelectric materials is discussed.

7. Methods for inverse problems related to wave propagation are presented, including material property characterization, impact-loading identification, and crack and flaw detection for anisotropic laminates using elastic waves.

8. Methods proposed for inverse problems are conjugate gradient methods, genetic algorithms, and conventional curve fitting techniques. These methods are effectively combined with forward solvers of wave propagation problems.

The general philosophy governing the writing of the book is to make all the topics insightful but simple, informative but interesting, and theoretical but practical. It attempts to show the reader that the topic of wave propagation and its inverse problems are actually not that difficult.

The book is written primarily for senior university students, postgraduate students, and engineers in civil, mechanical, geographical, aeronautical engineering and engineering mechanics. The book is written in an easy-to-understand manner, using simple and common terminology, so that anyone with an *elementary* knowledge of matrix algebra, the Fourier transform, complex variables, and the linear theory of elasticity should be able to understand the contents fairly easily.

A picture tells a thousand words. Numerous drawings and charts are used to describe important concepts, theories, and results. This is very important for readers coming from non-engineering backgrounds.

A large number of examples are included in the text. The examples are mostly generated using the methods discussed. Software codes of some of these methods are available on the website http://www.nus.edu.sg/ACES. Instruction on the usage of these software codes is also provided there.

The chapters are written in a relatively independent manner. After reading the first chapter, readers can jump to any chapter for the problem or method of his or her interest. Cross-references are provided to link relevant information or materials in other chapters. Readers are advised to read Chapter 1 before reading other chapters. For advanced readers, the material in Chapter 1 might be familiar and too simple, but quickly going through the material could be helpful in the use of terminologies used in the book. The following table gives a concise summary of the content of the book.

Chapters	Topics	Methods Introduced
1	Preliminaries and fundamentals of waves in solids; P wave in a bar	Common terminologies, Analytic procedure, Fourier transform, Cauchy's theorem
2	Waves in functionally graded materials (FGM), P wave in inhomogeneous media	Exact method, Confluent hypergeometric functions, Adaptive integrals
3	P wave, SV wave, SH wave, Characteristics of Lamb waves in laminates, Dispersion of wave velocity, Anisotropy of wave velocity, Wave energy distribution in laminates	Exact method, classic formulation, Method of matrix transfer
4	Harmonic wave in laminates, Transient wave in laminates	Exact method, Matrix formulation, Complex path method, Exponential window method, adaptive integral schemes
5	Characteristics of Lamb waves in laminates, Dispersion relation, Group velocity, six wave surfaces	Layer element method, Matrix formulation, Rayleigh quotient
6	Characteristics of waves in laminated bars	Finite strip element method
7	Characteristics of waves in laminated bars, Edge wave	Semi-exact method
8	Transient wave in laminates	Hybrid numerical method (HNM), Integral schemes for oscillatory integrands

9	Transient wave in FGM plates	HNM, Layer element method
10	Transient wave in Piezoelectric FGM plates, Mechanical excitation, Electric excitation	HNM, Layer element method
11	Wave in solids, Rayleigh wave	Strip element method (SEM)
12	Wave scattering by cracks in laminates, Crack characterization	SEM
13	Wave scattering by flaw in laminates and sandwich laminates, SH wave, Flaw characterization	SEM
14	Bending deformation of laminated plates, Bending wave in laminated plates	SEM
15	Characteristics of helical waves in laminated cylinders, Dispersion relation, Group velocity, six wave surfaces	Layer element method, Rayleigh quotient,
16	Wave scattering by cracks in laminated cylinders, Crack characterization	SEM
17	Inverse reconstruction of impact loading	HNM, Conjugate gradient method
18	Inverse characterization of material property of laminates	HNM, Genetic algorithms

A chapter-by-chapter description of the book is given below.

Chapter 1 provides fundamentals and terminologies for wave propagation in elastic media. One-dimensional longitudinal waves are analyzed in both frequency and time domains. This chapter familiarizes the reader with standard procedures in dealing with wave propagation problems.

Chapter 2 deals with harmonic and transient longitudinal waves in plates made of functionally grade materials. An idealized one-dimensional problem is studied. We formulate an inhomogeneous element method and apply it to a functionally grade plate subjected to harmonic or transient excitations. Waves propagating across the plate are revealed numerically. A novel integral technique is introduced for evaluating the confluent hypergeometric functions with complex valued argument.

Chapter 3 inquires into Lamb wave propagation in anisotropic laminates by an exact method based on classic formulation. The method of matrix transfer is employed. The dispersion and anisotropy of phase velocities for fundamental modes are computed and discussed in detail. The energy distributions in the thickness direction of the laminate are calculated for various modes of Lamb waves.

Chapter 4 formulates an exact method for analyzing elastic waves (both harmonic and transient) propagating in laminates. The dynamic equilibrium equation in the wave number domain is developed by the Fourier transform technique. The solutions in the spatial domain are obtained with the inverse Fourier transform technique. A complex path technique for evaluating the inverse Fourier integration is introduced. The transient solutions are found by applying the Fourier superposition method. The exponential window method is introduced in order to avoid the singularity of integration. Examples are presented to demonstrate the effectiveness and efficiency of the method.

Chapter 5 studies the frequency and group velocity dispersions and characteristic wave surfaces in laminated composite plates. The layer element method is introduced in detail. Dispersion curves for composite laminates are computed. A formula for group velocity is established using the Rayleigh's quotient. Group velocity spectra for various composite laminates are calculated. Six characteristic wave surfaces are defined, formulated, and visualized for composite laminates. These characteristic surfaces can illustrate efficiently the qualitative properties of Lamb waves propagating in anisotropic laminates.

Chapter 6 analyzes harmonic waves in anisotropic laminated bars. The finite strip element method is formulated and used to investigate wave modes in anisotropic laminated bars. The bar is divided into layer elements in the thickness direction and series expansion is employed in the width direction.

Chapter 7 formulates a semi-exact method for waves in laminated bars. The bar is divided into layer elements in the thickness direction, and the displacement is approximated using shape function only in the thickness direction. The solution in width direction is solved in an exact manner. Wave modes in bars are computed using the semi-exact method and plotted in a number of graphs. Edge waves on the edge of anisotropic laminates are also investigated.

Chapter 8 describes the hybrid numerical method (HNM) used for wave propagation in composite laminates. The laminate is divided into layer elements, and the displacement in the thickness direction of the laminate is approximated using shape functions. The system equations are derived by the principle of virtual work. The transient waves are simulated using the Fourier transform in conjunction with the modal analysis method. Quadrature methods for carrying out the inverse Fourier transform are also introduced.

Chapter 9 applies the layer element method and HNM introduced in the preceding chapters to analyze dispersion and transient waves in plates made of functionally grade materials. Strong surface waves are observed in a number of example problems.

Chapter 10 analyzes waves in plates made of functionally graded piezoelectric materials (FGPM). Detailed formulation for waves in such plates is provided. Piezoelectrical-mechanical coupled system equation is derived using variational principle. Characteristics of waves are investigated, and displacement and electrostatic potential responses are computed for mechanical and electric excitations. Strong surface waves are observed on the softer surface of the plate.

Chapter 11 formulates a strip element method (SEM) for analyzing waves in linearly elastic solids. A number of nonreflection boundary conditions are discussed in detail. The efficiency of the SEM is proved through the comparison with results from the literature for a variety of classical wave propagation problems.

Chapter 12 presents the SEM for dealing with wave scattering by a crack in composite laminates. A laminate with cracks is divided into subdomains to which the SEM is applicable. Procedures of assembling the SEM equations of all the subdomains are detailed. Numerical examples are given for wave scattering by both horizontal and vertical cracks in laminates. Characterization of cracks in composite laminates using the SEM is also studied in detail.

Chapter 13 investigates the scattering of harmonic waves in an anisotropic laminate containing a flaw. A simple rectangular flaw is considered and formulated using SEM. Results indicate that it is possible to detect the presence of a flaw, determine its position, and estimate its size by examining the scattering of the wave response. Characterization of defects in sandwich plates is also studied using SH waves. SEM formulation for SH waves is presented. Based on the wave field scattered by defects, methods for characterizing the size and location of the defects are described.

Chapter 14 applies the strip element method to deal with a bending deformation and bending waves in anisotropic laminates. Detailed SEM formulae are provided. Example problems for plates with various boundary conditions are analyzed.

Chapter 15 formulates a layer element method for the dispersion and wave surfaces in laminated composite cylinders. The helical wave in anisotropic hollow cylinders is analyzed. Numerical examples include frequency and group velocity spectra, as well as a set of six characteristic wave surfaces.

Chapter 16 discusses the waves scattered by a crack in laminated composite cylinders. The SEM described in Chapters 11 and 12 is generalized to the case for cracked laminated composite cylinders. The formulation for axisymmetric guided waves is derived. Numerical results are given for wave scattering by axial and radial cracks in both hollow cylinders and circular cylindrical shells.

Chapter 17 introduces techniques for inverse identification of impact loadings on the surface of laminates using elastic waves simulated by HNM. Two- and three-dimensional problems are studied. Both time function and spatial distribution function are reconstructed for a number of example problems.

Chapter 18 introduces techniques for inverse characterization of material properties of laminates using elastic waves simulated by HNM. Two-dimensional problems are studied. A genetic algorithm is employed for minimizing the error function in the determination of the elastic constants of laminates.

Authors

G. R. Liu, Ph.D., received his doctorate from Tohoku University, Japan in 1991. He was a postdoctoral fellow at Northwestern University, U.S. He is currently the Director of the Centre for Advanced Computations in Engineering Science (ACES, http://www.nus.edu.sg/ACES/), National University of Singapore. Dr. Liu is an SMA Fellow at the Singapore-MIT Alliance Program. He is also an associate professor at the Department of Mechanical Engineering, National University of Singapore. He authored more than 200 technical publications, including 120 international journal papers. He is also the author of three books. Dr. Liu is the recipient of the Outstanding University Researchers Award (1998) for his development of the Strip Element Method. He is also a recipient of the Defence Technology Prize (national award, 1999) for his contribution to development of underwater shock technology at Singapore. He won the Silver Award at CrayQuest 2000 (nationwide competition in 2000) for his development of meshless methods. His research interests include computational mechanics, element free methods, vibration and wave propagation in composites, mechanics of composites and smart materials, inverse problems and numerical analysis.

Z. C. Xi, Ph.D., received his B.Eng. in the Department of Engineering Mechanics, East China Technical University of Water Resource, Nanjing, China in 1985, the M.Eng. at the Research Institute of Engineering Mechanics, Dalian University of Technology, Dalian, China in 1988, and the Ph.D. in the Department of Mechanical Engineering, The Hong Kong Polytechnic University in 1997. He joined the Research Institute of Engineering Mechanics, Dalian University of Technology in 1987 as an assistant professor to work on numerical analyses in mechanics of advanced composite materials. In 1997, he was promoted to associate professor and headed the Division of Structural Mechanics. Since 1998, Dr. Xi has been in the Department of Mechanical Engineering, National University of Singapore to work on wave propagation in laminated composite plates and shells containing cracks. His research led to the Outstanding Young Researcher Award of Liaoning Provincial Government, China in 1992 for his contributions to hydrothermal effects in aircraft made of composite materials. He also was awarded the Taipei Trading Centre Scholarship Prize of The Hong Kong Polytechnic University in 1996 for his contributions in the area of nonlinear vibration of fluid-filled laminated composite shells of revolution. His current research interests include wave propagation in cracked composite plates and shells, fluid-composite structure interaction, structural dynamics of advanced composite materials, and thermal stresses. Dr. Xi is the author/co-author of 30 refereed journal papers.

Contents

1

Fundamentals of Waves in Elastic Solids

1.1 Introduction

Waves are a disturbance propagating in a medium, which can be air, water, or solid. Unlike acoustic waves in air, waves in solids are inaudible to human ears. Unlike surface waves in water, waves in solids are invisible to human eyes. Waves in solids are, however, real, physical, and very important to engineering applications. Mathematical and numerical tools are required to analyze and simulate the phenomena of waves in solids. These tools can help us to construct virtual views of waves in our mind. This chapter attempts to introduce some of the important basic concepts, methodology, and procedures through the analysis of a simple example of waves in an elastic bar. We first explain some common and important terminology used in this book.

Solids are stressed when they are subjected to external *forces* or *loads*. The *stresses* are, in general, not uniform as the forces usually vary with coordinates. The stresses lead to *strains*, which can be observed as *deformation* or *displacement*. Solid mechanics and structural mechanics deal with the relationships between stresses and strains, displacements and forces, and stresses (strains) and forces for given boundary conditions imposed on solids. These relationships are essential in simulating and analyzing the mechanical status including wave motion in solids and structures made of solids.

Forces can be static and/or dynamic. *Statics* deals with the mechanics of solids and structures subject to static loads. Solids will experience dynamic motion under the action of dynamic forces that vary with time. The stress, strain, and the displacement caused by dynamic forces will also be functions of time, and principles and theories of *dynamics* must apply. The dynamic motion is often observed and noted as *vibration* or *wave motion*. It is not easy to draw exactly a clear line between vibration and wave motion, but, in general, we can say that a wave is a localized vibration and a vibration is a motion of waves with very long wavelength. When talking about waves, one usually pays special attention to the motion or propagation of a localized mechanical disturbance. When talking about vibration, one is concerned more with the global motion of the entire structure. Mathematically, both vibration and wave motion are governed by the same dynamic equation of motion, which is derived basically using Newton's Law. This book discusses these principles, theories, and methods dealing with *wave motion* in solids, with special focus on laminated structures.

The time function of dynamic forces can be divided broadly into two large categories. One is the *harmonic* force that varies with time in a sinusoidal fashion, and the other is the *transient* force that varies with time in an arbitrary manner.

Depending on the material property, solids can be *elastic* meaning that the deformation in the solids disappears completely after it is unloaded. There are also solids that are *plastic* meaning that the deformation in the solids cannot be fully recovered when it is unloaded. In fact, most of the materials are plastic if they are stressed beyond a certain limit. Many materials are elastic if they are stressed below the limit called *yield stress*. *Elasticity* deals with solids and structures of elastic materials while *plasticity* deals with those of plastic materials. This book deals mainly with solids of elastic materials. Waves propagating in elastic material are termed *elastic wave*. One of the major applications of elastic waves is in the area of nondestructive evaluation. In such an application, one always keeps the stress level as low as possible, and within the elastic range. Otherwise, it could be destructive. Therefore, the topics in this book are fully applicable in areas relating to nondestructive evaluation.

Materials can be *anisotropic*, meaning that the material property varies with direction. Deformation in an anisotropic material caused by a force applied in a direction may be different from that caused by the same force applied in another direction. Composite materials, especially fiber-reinforced composites, are often strongly anisotropic. A number of material constants have to be used to define the material property of anisotropic materials. Many engineering materials can be assumed isotropic whose material property is direction independent. Isotropic materials are a special case of anisotropic materials. There are only two independent elastic material constants for isotropic materials, often known as Young's modulus and Poisson's ratio. This book deals with primarily anisotropic materials. All the formulations are, however, applicable to isotropic materials as a special case. Waves in an anisotropic material exhibit anisotropic characteristics, meaning that their properties (such as velocity) are direction dependent.

Composites are often inhomogeneous in microscopic scale (micro-meters). In fiber-reinforced materials, the diameters of the fibers are micrometers. Micromechanics details with issues related to the interfaces of fibers and matrices. Macroscopically, we are concerned more about inhomogeneity in a larger scale that is the laminar level (minimeters), because many composite structures are made of laminae of fiber-reinforced materials. A lamina is treated as a (macroscopically) homogeneous material. This book deals with wave phenomena on a macroscopic level with a particular focus on the effects of laminated structures. The terminology is often confusing, such as "laminate" and "laminated plate." Physically, a laminate and a laminated plate are the same. However, in mechanics they are treated differently in this book. A laminate is mechanically governed by solid mechanics while a laminated plate is governed by a theory of laminated plates.

Boundary conditions are another important consideration in wave propagation problems, as waves will reflect on the boundary. Interfaces of solids, such as laminates, are another important factor affecting wave propagation. The presence of boundaries and interfaces cause waves to reflect and/or refract resulting in *dispersion*, meaning that the velocity depends on the frequency of excitation and shape of the wave changes during the propagation.

The deformation in solids and structures can be large and small depending on the external forces. Large deformations often lead to both geometrical and material nolinearity that should be treated by nonlinear mechanics. This book assumes also that the deformation or displacement in the solid is very small so that the displacement and strain relation can be linear. Therefore, this book deals with waves that are *linear elastic waves*.

Free wave motion refers to wave motion in media free of external excitation; one is concerned with only the characteristics or features of harmonic waves in a given configuration of the media. We want to study what *could* be happening in the media under its *natural* status, rather than what *will* be happening under a specified loading condition. In particular, one needs to find the velocity, natural frequency, and wave modes in relation to wavelength or wavenumber. In contrast, forced wave motion refers to waves in media excited by an external excitation. One is interested in what *will* be happening under a specified loading condition and is concerned more about the response of the media to a particular type of external excitation. The excitation could be harmonic or transient. Frequency response refers to the response of the media to a harmonic excitation, and the results are presented in the form of frequency spectrum of displacement (or velocity or acceleration) response. Analysis of waves generated by harmonic excitation is also called *frequency analysis* and is also referred to as wave analysis in frequency domain. Transient response refers to the response of the media to a transient excitation, and the results are presented in the form of time history of displacement (or velocity or acceleration) response. Analysis of transient waves is naturally called *transient analysis*. It is also referred to as wave analysis in time domain.

The complexity of wave propagation problems depends very much on the complexity of the geometry of the domain where the waves are propagating. However, the general concept and procedure are very much the same, although one may have to use numerical means to compute the results for complex cases. In order to illustrate the concepts and procedures in an easy-to-understand manner, the following introduces methods and procedures through solving a one-dimensional wave propagation problem: longitudinal waves in bars.

1.2   Formulation of Longitudinal Wave in a Bar

1.1.1   Statement of the Problem

Consider a uniform and isotropic *thin* bar or rod, whose lateral dimension is much smaller than its longitudinal, as shown in Figure 1.1. Bars can be found in many engineering applications, such as pillars, columns, and piles in structural systems. Bars are very often used in experiments for determining dynamic properties of materials, such as the famous Hopkinson bar test which makes use of waves propagating in bars. The shape of the cross-section of the bar is immaterial as long as it is thin. The bar is subjected to a uniform *traction, p(x)*, at the cross-section of the bar at point x and in the x (axial) direction. Traction is defined in this book as a uniformly distributed force. Traction is very much similar to pressure, except that it is allowed to be negative. Our purpose is to analyze waves propagating in the bar excited by the dynamic force that is totaled $p(x)A$ at point x, where A is the cross-sectional area of the bar.

As the traction p is applied uniformly on a thin bar only in the x direction, the displacement u in the x direction will be dominant. The problem can be therefore considered as one dimensional, meaning that the field variables (displacement, stress, strain, etc.) are a function only of x, and independent

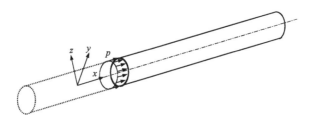

FIGURE 1.1
Thin bar subjected to axial dynamic force.

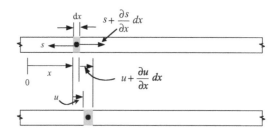

FIGURE 1.2
Motion of a representative cell in a bar.

of y and z. The governing equation for the one-dimensional wave motion problem can be derived as follows.

1.1.2 Strain-Displacement Relation

We consider now a representative cell (shaded portion) isolated from a uniform, isotropic and elastic bar, whose mechanical properties do not vary with coordinates and direction. First, we introduce the linear strain-displacement relation

$$\varepsilon = \frac{\partial u}{\partial x} \tag{1.1}$$

where ε is the strain in the material and u is the displacement at point x in the bar.

1.1.3 Constitutive Equation

Using Hooke's law of linear elastic material of the bar, the linear stress-strain relation can be written as

$$\sigma = E\varepsilon \tag{1.2}$$

where σ is the stress in the material, and E is Young's modulus of the material of the bar. Hooke's law for higher dimensional problems is often termed as a *constitutive equation*. Substituting Eq. (1.1) into (1.2), we have

$$\frac{\partial u}{\partial x} = \frac{\sigma}{E} = \frac{s}{AE} \tag{1.3}$$

where $s = A\sigma$ is the total axial *internal* force acting on the cross-section, which becomes visible (external) when the cell is isolated from the bar. By differentiating with respect to x, Eq. (1.2) becomes

$$AE\frac{\partial^2 u}{\partial x^2} = \frac{\partial s}{\partial x} \tag{1.4}$$

1.1.4 Equation of Motion for Free Wave Motion

The motion of the representative cell has to be governed by Newton's law, which states that the summation of all the unbalanced forces is equal to the product of the mass and acceleration of the cell, i.e.,

$$\frac{\partial s}{\partial x} dx = \rho A dx \frac{\partial^2 u}{\partial t^2} \tag{1.5}$$

where ρ is the mass density of the material of the cell. Using Eq. (1.4), the foregoing equation becomes

$$\frac{\partial^2 u}{\partial t^2} = \left(\frac{E}{\rho}\right)\frac{\partial^2 u}{\partial x^2} \tag{1.6}$$

or

$$\frac{\partial^2 u}{\partial x^2} = \frac{1}{c^2}\frac{\partial^2 u}{\partial t^2} \tag{1.7}$$

where

$$c = \sqrt{\frac{E}{\rho}} \tag{1.8}$$

Equation (1.7) is the so-called *wave equation*, which governs the free wave motion in the bar.

1.3   Free Wave Motion in Infinite Bars

The classical form of solution for Eq. (1.7) can be written as

$$u = F_1(x - ct) + F_2(x + ct) \tag{1.9}$$

where F_1 and F_2 are arbitrary functions that represent the shape of the wave, with function F_1 representing the shape of waves propagating in the positive x direction and F_2 representing the shape of waves in the negative x direction. This solution can be verified simply by substituting Eq. (1.9) into (1.7). Equation (1.9) shows that c is the velocity of the shape of waves propagating in the bar. It is therefore called the *velocity* of the wave propagation along the bar. Equation (1.8) shows that the velocity of the wave depends only on the material properties, Young's modulus and mass density, and is independent of the frequency of excitation. Waves with constant velocity are said to be nondispersive. We will find in the later chapters that the velocity is frequency (or wavelength) dependent for *dispersive waves*. In general, functions F_1 and F_2 are not necessarily the same, but they propagate at the same velocity, and the shape is always kept the same during the propagation in the bar, as illustrated in Figure 1.3. The shapes of dispersive waves and nondispersive waves in higher dimensional media change during propagation. Equation (1.9) is often called D'Alembert's solution. The following procedure determines explicitly the solution of functions F_1 and F_2.

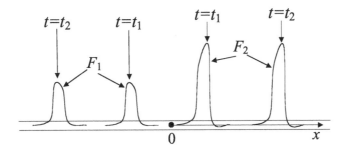

FIGURE 1.3
Waves propagating in a bar.

We consider first the harmonic motion of waves that could be generated by a harmonic force. The harmonic force is also termed *sinusoidal* force, meaning that the force varies with time harmonically or sinusoidally. Mathematically, it can be expressed in one of the following sine, cosine, and exponential forms of excitation.

$$p(x, t) = P(x) \sin(\omega t) \tag{1.10}$$

$$p(x, t) = P(x) \cos(\omega t) \tag{1.11}$$

$$p(x, t) = P(x) \exp(-i\omega t) \tag{1.12}$$

where $i = \sqrt{-1}$. In Eqs. (1.10), (1.11), and (1.12), P is, in general, a given function of x, and ω is the *angular frequency* of the force. The angular frequency relates to the *frequency, f*, as

$$\omega = 2\pi f \tag{1.13}$$

The three expressions given in Eqs. (1.10), (1.11), and (1.12) are different in form but can all, in general, represent a harmonic force. Equation (1.12) is used relatively more often, as it is most convenient in analytically deriving solutions for wave propagation problems. Moreover, Eq. (1.12) can be rewritten as

$$\begin{aligned} p(x, t) &= P(x) \exp(-i\omega t) = P(x)(\cos \omega t - i \sin \omega t) \\ &= P(x) \cos \omega t - iP(x) \sin \omega t \end{aligned} \tag{1.14}$$

which consists of cosine excitation as the real part and sine excitation as the imaginary part. The response of the system to an exponential excitation will also consist of real and imaginary parts. For linear systems, the real part of response corresponds to the cosine excitation, and the imaginary part of excitation corresponds to the sine excitation. Therefore, if the response to the exponential excitation is known, one knows responses to both cosine and sine excitations.

Under the action of harmonic force, the particles in the solid naturally undergo a harmonic motion. If the wave motion is caused by a harmonic excitation, the displacement u must also be harmonic. It can therefore be written in the form of

$$u(x, t) = U_c(x) \exp(-i\omega t) \qquad (1.15)$$

where U_c is a function of x to be determined later, and ω is the angular frequency of the wave. The substitution of the foregoing equation into Eq. (1.7) leads to

$$\frac{d^2 U_c}{dx^2} = \frac{-\omega^2}{c^2} U_c \qquad (1.16)$$

or

$$\frac{d^2 U_c}{dx^2} + k^2 U_c = 0 \qquad (1.17)$$

where

$$k = \frac{\omega}{c} \qquad (1.18)$$

is termed as *wavenumber*. Since c is constant, the wavenumber k is proportional to angular frequency ω for nondispersion waves, and ω-k curve will be a straight line with gradient of $1/c$. For nondispersive waves, ω and k are related in a much more complex manner. The relationship between ω and k is termed a dispersion relation, and ω-k curve is called a dispersion curve. Wavenumber relates to *wavelength* λ and frequency f as

$$\lambda = \frac{2\pi}{k} = \frac{2\pi c}{\omega} = \frac{c}{f} \qquad (1.19)$$

For convenience, we list in Table 1.1 simple and important fundamental relations.

Equation (1.17) is a *homogeneous* differential equation of the second order. Its solution can be easily found by first assuming

$$U_c(x) = C \exp(i\alpha x) \qquad (1.20)$$

TABLE 1.1

Fundamental Relations

	f	ω	T	λ	k	c
Frequency f	1	$\omega/2\pi$	$1/T$	c/λ	$ck/2\pi$	c/λ
Angular frequency ω	$2\pi f$	1	$2\pi/T$	$2\pi c/\lambda$	ck	$2\pi c/\lambda$
Period T	$1/f$	$2\pi/\omega$	1	λ/c	$2\pi/ck$	λ/c
Wavelength λ	c/f	$2\pi c/\omega$	cT	1	$2\pi/k$	c/f
Wavenumber k	$2\pi f/c$	ω/c	$2\pi/Tc$	$2\pi/\lambda$	1	ω/c
Velocity c	λf	ω/k	λ/T	λ/T	ω/k	1

where C is an arbitrary constant. Substituting Eq. (1.20) into Eq. (1.17), we have

$$-\alpha^2 + k^2 = 0 \tag{1.21}$$

This polynomial equation of α possesses two roots of

$$\alpha = \pm k \tag{1.22}$$

which means that U_c has two possible solutions in the form of Eq. (1.20). The solution for Eq. (1.17) is then obtained by superimposing these two possible solutions as follows.

$$U_c = C_1 \exp(ikx) + C_2 \exp(-ikx) \tag{1.23}$$

where C_1 and C_2 are arbitrary constants. This solution can be verified very easily by substituting Eq. (1.23) into Eq. (1.17). Since U_c is a function of coordinate x, it represents the shape of the wave. Substituting the forgoing equation into Eq. (1.15) leads to the complete solution for the harmonic wave motion in a bar.

$$u = C_1 \exp[ik(x - ct)] + C_2 \exp[-ik(x + ct)] \tag{1.24}$$

Notice that the previous equation has the same form of Eq. (1.9). The functions F_1 and F_2 are, in this case, experiential functions.

The solution given by Eq. (1.23) is often called a complementary solution for infinite bars. There are still two constants, C_1 and C_2, to be determined. Boundary conditions at the two ends of the bar have to be used to determine C_1 and C_2. In other words, Eq. (1.23) is the *solution* for all the possible boundary conditions. Also, note that Eq. (1.23) is obtained without applying an external force. The solution is therefore for free wave motion in infinite bars. Imposition of boundary conditions at the two ends will lead to wave modes and their corresponding natural frequencies for a constrained bar of finite length. This is further illustrated in the following section.

1.4   Free Wave Motion in a Finite Bar

Consider now a uniform bar of length l, Young's modulus E, and density ρ, as shown in Figure 1.4. As an example, assume that the bar is free of support at two ends. Now determine the natural frequencies and the shape of wave modes for harmonic waves propagating in the bar using boundary conditions. At the free ends of the bar, the stresses must be zero. Because the stress is obtained as $\sigma_x = E\frac{\partial u}{\partial x}$, the unit strain at the ends must be zero, which is translated to

$$\frac{\partial U_c}{\partial x} = 0 \quad \text{at } x = 0 \tag{1.25}$$

$$\frac{\partial U_c}{\partial x} = 0 \quad \text{at } x = l \tag{1.26}$$

Using Eq. (1.23), Eq. (1.25) can be written in detail as

$$\left(\frac{\partial U_c}{\partial x}\right)_{x=0} = ik[C_1 \exp(ikx) - C_2 \exp(-ikx)]|_{x=0} = 0 \tag{1.27}$$

which implies

$$C_1 = C_2 \tag{1.28}$$

Substituting Eq. (1.23) into Eq. (1.26), gives

$$\left(\frac{\partial U_c}{\partial x}\right)_{x=l} = ik[C_1 \exp(ikx) - C_2 \exp(-ikx)]_{x=l} = 0 \tag{1.29}$$

which leads to

$$C_1 \exp(ikl) - C_2 \exp(-ikl) = 0 \tag{1.30}$$

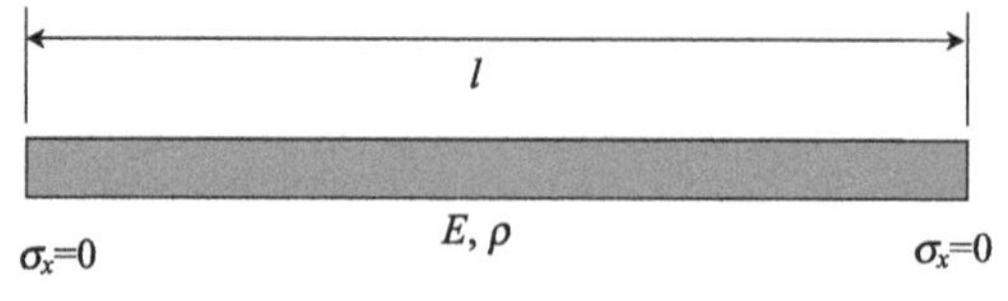

FIGURE 1.4
Uniform bar with free ends.

Solving Eqs. (1.28) and (1.30) simultaneously gives

$$\exp(ikl) - \exp(-ikl) = 0 \tag{1.31}$$

or

$$\sin(kl) = 0 \tag{1.32}$$

The roots of the previous equation are

$$k_n l = \frac{\omega_n l}{c} = \omega_n l \sqrt{\frac{\rho}{E}} = \pi, 2\pi, 3\pi, \dots, n\pi \tag{1.33}$$

The angular natural frequency of the nth wave mode of the bar is thus given by

$$\omega_n = k_n c = \frac{k_n l}{l} c = \frac{n\pi}{l} \sqrt{\frac{E}{\rho}}, \quad n = 1, 2, \dots, \infty \tag{1.34}$$

and the natural frequency of the nth mode is

$$f_n = \frac{\omega_n}{2\pi} = \frac{n}{2l} \sqrt{\frac{E}{\rho}}, \quad n = 1, 2, \dots, \infty \tag{1.35}$$

and the shape of the wave modes are

$$\sin \frac{\omega_n l}{c}, \quad n = 1, 2, \dots, \infty \tag{1.36}$$

Notice that there is an infinite number of natural frequencies and wave modes in an infinite bar with two free ends. It is a good practice for the reader to derive the wave modes and their corresponding natural frequencies of a bar fixed (clamped) at its two ends.

1.5 Forced Wave Motion in an Infinite Bar

Notice that in deriving Eq. (1.23), we did not specify exactly how the bar is excited or forced externally. *Particular solutions* refer to solutions particular to a specified external force or wave source. Now assume that the external traction is given by $p(x)$ as shown in Figure 1.5, which is a function of x

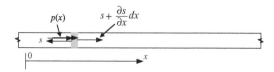

FIGURE 1.5
Motion of a cell in a bar excited by a force $p(x)$.

meaning that the force of $p(x)A$ is distributed in the x direction. In this case, Eq. (1.5) becomes

$$p(x)Adx + \frac{\partial s}{\partial x}dx = \rho Adx\frac{\partial^2 u}{\partial t^2} \qquad (1.37)$$

and the governing equation (wave equation) becomes

$$\frac{\partial^2 u}{\partial x^2} - \frac{1}{c^2}\frac{\partial^2 u}{\partial t^2} = \frac{-p(x)}{E} \qquad (1.38)$$

For harmonic excitation, the force should have the form of Eq. (1.12), and the displacement should have the form of Eq. (1.15). Substituting Eqs. (1.12) and (1.15) into the foregoing equation, we obtain

$$\frac{d^2 U_p}{dx^2} + \frac{\omega^2}{c^2}U_p = \frac{-P}{E} \qquad (1.39)$$

This is the governing equation for forced harmonic motion in a bar. To obtain the solution for the previous equation, we introduce one-dimensional Fourier transform (from spatial domain to wavenumber domain) defined by

$$\tilde{U}_p(k_x) = \int_{-\infty}^{+\infty} U_p(x)e^{ik_x x}dx \qquad (1.40)$$

where k_x is the variable of wavenumber with respect to x. Applying the Fourier transform to both sides of Eq. (1.39), we have

$$-k_x^2\tilde{U}_p + k^2\tilde{U}_p = \frac{-\tilde{P}}{E} \qquad (1.41)$$

where $\tilde{P}$ can be obtained for given P:

$$\tilde{P} = \int_{-\infty}^{+\infty} P(x)e^{ik_x x}dx \qquad (1.42)$$

Equation (1.41) is an algebraic equation for $\tilde{U}_p$ in the Fourier transform domain. Equation (1.41) can be solved easily to obtain

$$\tilde{U}_p(k_x) = \frac{\tilde{P}}{E(k_x^2 - k^2)} \tag{1.43}$$

To obtain the U_p, we need to perform an inverse transform.

$$U_p(x) = \frac{1}{2\pi} \int_{-\infty}^{+\infty} \tilde{U}_p(k_x) e^{-ik_x x} dk_x = \frac{1}{2\pi} \int_{-\infty}^{+\infty} \frac{\tilde{P} e^{-ik_x x}}{E(k_x^2 - k^2)} dk_x \tag{1.44}$$

Now assume the external force is a point force acting at $x = x_0$, which can be expressed mathematically by

$$P = \bar{P} \delta(x - x_0) \tag{1.45}$$

where $\bar{P}$ is a constant representing the amplitude of the force, and $\delta(x)$ is the Dirac delta function. Substituting the foregoing equation into Eq. (1.42) gives

$$\tilde{P} = \int_{-\infty}^{+\infty} P(x) e^{ik_x x} dx = \int_{-\infty}^{+\infty} \bar{P} \delta(x - x_0) e^{ik_x x} dx = \bar{P} e^{ik_x x_0} \tag{1.46}$$

Equation (1.44) now becomes

$$U_p(x) = \frac{\bar{P}}{2\pi E} \int_{-\infty}^{+\infty} \frac{e^{-ik_x(x - x_0)}}{(k_x^2 - k^2)} dk_x = \frac{\bar{P}}{2\pi E} \int_{-\infty}^{+\infty} \frac{e^{-ik_x(x - x_0)}}{(k_x + k)(k_x - k)} dk_x \tag{1.47}$$

The integration in Eq. (1.47) can be performed analytically using Cauchy's theorem. The integration of over $-\infty$ to $+\infty$ is replaced by the integration over a closed loop of a semicircle with a radius of ∞, as shown in Figure 1.6. When $x \geq x_0$, the lower semicircular loop must be chosen. When $x < x_0$, the upper semicircular loop must be chosen to ensure the integrand to be analytical on the circles, so that the integrations over the circles of infinite radius vanish. The integrand in Eq. (1.47) goes to infinity at two poles at $k_x = \pm k$. The integral is therefore singular and the result is not unique. Therefore, indentations over the poles need to be properly chosen to include or exclude the poles to ensure a physically meaningful result (satisfying the radiation condition or nonreflection condition). From Eq. (1.15), we know that the time dependence term is $\exp(-i\omega t)$, and $\exp[i(kx - \omega t)]$ represents the waves propagating in the positive x direction (rightward). At a point of $x \geq x_0$, we can observe only rightward waves, as the force (source) is applied at x_0. Therefore, the pole at $-k$ must be included, and that at k must be excluded

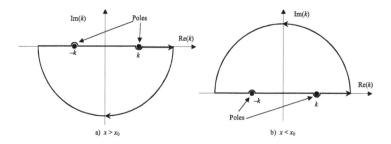

FIGURE 1.6
Contours and indentations for integration of Eq. (1.47).

(see Figure 1.6a). At a point of $x < x_0$, we can observe only leftward waves, which require a term of $\exp[i(kx + \omega t)]$. Therefore, the pole at k must also be included, and that at $-k$ must also be excluded (see Figure 1.6b). Finally, the solution of the integration should be

$$U_p(x) = \begin{cases} \dfrac{-i\bar{P}e^{ik(x-x_0)}}{2Ek}, & x \geq x_0 \\[3mm] \dfrac{i\bar{P}e^{-ik(x-x_0)}}{2Ek}, & x < x_0 \end{cases} \qquad \text{or} \qquad U_p(x) = \begin{cases} \dfrac{-iU_A e^{ik(x-x_0)}}{k}, & x \geq x_0 \\[3mm] \dfrac{-iU_A e^{-ik(x-x_0)}}{k}, & x < x_0 \end{cases}$$

$$(1.48)$$

where

$$U_A = \frac{\bar{P}}{2E} \tag{1.49}$$

Using Eq. (1.18), the foregoing equation can be alternatively written as

$$U_p(x, \omega) = \begin{cases} \dfrac{-iU_A c e^{i\omega(x-x_0)/c}}{\omega}, & x \geq x_0 \\[3mm] \dfrac{iU_A c e^{-i\omega(x-x_0)/c}}{\omega}, & x < x_0 \end{cases} \tag{1.50}$$

Equation (1.50) or (1.48) is the *particular solution* of the displacement to a point harmonic force acting on a bar at x_0. This function can be used as Green's function to calculate the solution of a force of an arbitrary distribution function of x.

1.6   Forced Wave Motion in a Finite Bar

The general solution U_g for a finite bar with boundary can be obtained by superimposing the complementary solution given in Eq. (1.23) and the particular solution given in Eq. (1.48), i.e.,

$$U_g(x) = U_c(x) + U_p(x) \tag{1.51}$$

The general solution can be used to provide solutions for a finite bar with boundaries at its ends. Now consider a bar with fixed-fixed boundary condition as shown in Figure 1.7. The boundary conditions at the left and right ends of the bar are expressed as

$$U_g\left(-\frac{l}{2}\right) = 0, \quad U_g\left(\frac{l}{2}\right) = 0 \tag{1.52}$$

Using Eqs. (1.52), (1.51), (1.50), and (1.23) leads to the following two equations:

$$C_1 \exp\left(-\frac{ikl}{2}\right) + C_2 \exp\left(\frac{ikl}{2}\right) + V_p^L = 0 \tag{1.53}$$

$$C_1 \exp\left(\frac{ikl}{2}\right) + C_2 \exp\left(-\frac{ikl}{2}\right) + V_p^R = 0 \tag{1.54}$$

where V_p^L and V_p^R are the displacement of the particular solution at the left and right boundaries that can be obtained using Eq. (1.48), i.e.,

$$V_p^L = U_p\left(x = -\frac{l}{2}\right) = \frac{iU_A e^{ikl/2}}{k}, \quad V_p^R = U_p\left(x = \frac{l}{2}\right) = \frac{-iU_A e^{ikl/2}}{k} \tag{1.55}$$

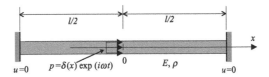

FIGURE 1.7
Uniform bar with fixed ends.

Solving Eqs. (1.53) and (1.54) simultaneously, we obtain

$$C_1 = -C_2 = \frac{-iU_A e^{ikl}}{k(1 - e^{ikl})} \tag{1.56}$$

Substituting Eq. (1.56) into Eq. (1.23) and then into Eq. (1.51), we have

$$U_g = \begin{cases} \dfrac{-iU_A}{k}\left[\dfrac{e^{ikx} - e^{ik(l-x)}}{1 - e^{ikl}}\right], & x \geq 0 \\[3mm] \dfrac{iU_A}{k}\left[\dfrac{e^{-ikx} - e^{ik(l+x)}}{1 - e^{ikl}}\right], & x < 0 \end{cases} \tag{1.57}$$

Using Eq. (1.18), the foregoing equation can be rewritten as

$$U_g(x, \omega) = \begin{cases} \dfrac{-iU_A c}{\omega}\left[\dfrac{e^{i\omega x/c} - e^{i\omega(l-x)/c}}{1 - e^{i\omega l/c}}\right], & x \geq 0 \\[3mm] \dfrac{iU_A c}{\omega}\left[\dfrac{e^{-i\omega x/c} - e^{i\omega(l+x)/c}}{1 - e^{i\omega l/c}}\right], & x < 0 \end{cases} \tag{1.58}$$

This is the solution for fixed-fixed bar subject to a point harmonic force at the center of the bar. The results are plotted in Figures 1.8 and 1.9.

For distributed forces of arbitrary function of x, the solution can be obtained by carrying out convolution using Eq. (1.57) as Green's function.

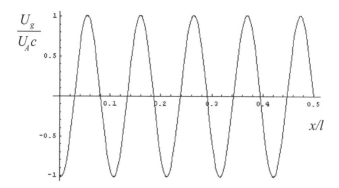

FIGURE 1.8
Displacement distribution of a fixed-fixed bar subject to a harmonic point force at the center of the bar ($\omega l/c = 60$).

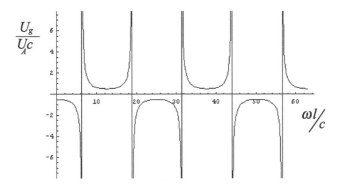

FIGURE 1.9
Response of a fixed-fixed bar subject to a harmonic point force at the center of the bar ($x = 0.25l$).

1.7 Transient Waves in an Infinite Bar

To obtain the transient wave field or the solution in the time domain, Fourier superposition will be applied, i.e.,

$$u(t) = \frac{1}{2\pi}\int_{-\infty}^{\infty} U_p(\omega)\hat{p}(\omega)e^{i\omega t}d\omega \tag{1.59}$$

where $U(\omega)$ is a solution in the frequency domain given in Eq. (1.50) for a unit force of $\bar{P} = 1$, while $\hat{p}(\omega)$ is the Fourier transform (from time domain to frequency domain) of the force function given by

$$\hat{p}(\omega) = \int_{-\infty}^{\infty} p(x, t)e^{-i\omega t}dt \tag{1.60}$$

Now assume the external force to be a point force of pulse of delta function applied at $t = 0$, which can be expressed mathematically by

$$p(x, t) = \bar{P}\delta(x - x_0)\delta(t) \tag{1.61}$$

We then have

$$\hat{p}(\omega) = \bar{P} \tag{1.62}$$

Substituting Eqs. (1.62) and (1.50) (with $\bar{P} = 1$) into Eq. (1.59), we have for $x \ge x_0$,

$$u(t) = \frac{1}{2\pi}\int_{-\infty}^{\infty}\frac{-ie^{-ik(x-x_0)}}{2Ek}\bar{P}e^{i\omega t}d\omega = \frac{1}{2\pi}\int_{-\infty}^{\infty}\frac{-ie^{-i\frac{\omega}{c}(x-x_0)}}{2E\frac{\omega}{c}}\bar{P}e^{i\omega t}d\omega \tag{1.63}$$

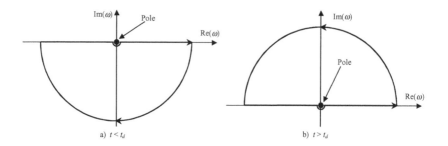

FIGURE 1.10
Contours and indentations for integration of Eq. (1.64).

Let $t_d = (x - x_0)/c$ be the time for the wave travel over the distance between x and x_0. The foregoing equation becomes

$$u(t) = \frac{1}{2\pi} \int_{-\infty}^{\infty} \frac{-ice^{-i\omega t_d}}{2E\omega} \bar{P} e^{i\omega t} d\omega = \frac{1}{2\pi} \frac{-i\bar{P}c}{2E} \int_{-\infty}^{\infty} \frac{e^{i\omega(t-t_d)}}{\omega} d\omega \qquad (1.64)$$

The integral in the above equation can be again performed by contour integration. Similar to the situation encountered in the integration of Eq. (1.47), the two possible contours have to be properly chosen, according to the physics of the problem. When $t < t_d$, the lower loop of contour has to be chosen, to ensure the integrand in Eq. (1.64) to be analytical and the integral exists (see Figure 1.10a). For the same reason, the upper loop of contour has to be chosen when $t > t_d$ (see Figure 1.10b). There is a pole in Eq. (1.64). For $t < t_d$, the pole must be excluded, as physically there should not be any response of displacement at $x > x_0$ when $t < t_d$. For $t > t_d$, the pole must be included because there must be some response of displacement under the action of force at $x = x_0$ even when $t > t_d$. For $t = t_d$, although mathematically the integral path can cut through the pole, we need not go any further, as there is no physical significance at $t = t_d$.

The results of the integral in Eq. (1.64) is finally obtained as

$$u(t) = \frac{1}{2\pi} \frac{-i\bar{P}c}{2E} \times \begin{cases} 2\pi i, & t > t_d \\ 0, & t < t_d \end{cases}$$

$$= \begin{cases} u_A, & t > t_d \\ 0, & t < t_d \end{cases} \qquad (1.65)$$

where u_A is the amplitude of the resultant displacement.

$$u_A = \frac{\bar{P}c}{2E} = \frac{\bar{P}}{2\rho c} \qquad (1.66)$$

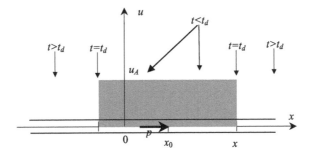

FIGURE 1.11
Wave shape at time $t = t_d$ in a bar excited by a point pulse at $x = x_0$.

This result is valid for $x > x_0$. Equation (1.65) gives the shape of the wave propagating rightward, under the action of a point force of pulse at $x = x_0$. Following exactly the same procedure, we can also obtain the solution for $x < x_0$ as

$$u(t) = \frac{1}{2\pi} \frac{i\bar{P}c}{2E} \times \begin{cases} -2\pi i, & t > t_d \\ 0, & t < t_d \end{cases}$$

$$= \begin{cases} u_A, & t > t_d \\ 0, & t < t_d \end{cases}$$

(1.67)

Readers are encouraged to confirm the results in the foregoing equation, which gives the shape of the wave propagating leftward, subject to a point force of pulse at $x = x_0$. The result is plotted in Figure 1.11. For distributed and/or arbitrary transient forces, the solution can be obtained by carrying out convolution using Eq. (1.65) or (1.67) as Green's function.

1.8 Remarks

Problems of longitudinal waves in bars have been solved using an analytical procedure. Solutions for a number of problems have been obtained in both frequency and time domain. For most complex wave propagation problems found in practice, it is usually very difficult to obtain a closed form of analytic solution, and numerical approaches are required to obtain an approximated solution. However, the solution procedure will be very similar to the procedure discussed in this chapter.

There are a number of other types of classic wave propagation problems, such as waves in strings, beams, plates, shells, and two- and three-dimensional solids. Details for solving these classic wave propagation problems are covered in excellent books by Achenbach (1973) and Graff (1991), among others. The following chapters of this book will focus waves in composite anisotropic laminates and laminated plates.

2

Waves in Plates of Functionally Graded Material

2.1 Introduction

To solve problems in the thermal-protection systems of spacecrafts, a new concept of materials, called functionally gradient material (FGM), has been proposed. Many techniques (e.g., Sasaki and Wang et al., 1989; Watanabe and Kawasaki et al., 1990; Takahashi and Itoh et al., 1990; etc.) have been developed for fabricating various FGM, in which the material properties change continuously in thickness. FGMs have found many applications not only in thermal protection systems for rocket and spacecraft engines but also in many other areas in engineering. Little work, however, is available in the literature for analyzing waves propagating in functionally graded plates.

This chapter introduces an exact method for analyzing waves, both harmonic and transient, in a FGM plate. The method was initially proposed by Liu and Han et al. in 1999, 2001a. In this method, the problem is idealized into a one-dimensional wave propagation problem by assuming that the excitation and field variables are independent of the coordinates in the plate plane. Compared to Chapter 1, the complexity of the problem of waves in FGM is due to the material variation with the coordinate. To solve the problem, the FGM plate is first divided into elements in which the material property is assumed as linear functions of the thickness coordinate to model the variation of material property of the FGM plate. Note that elements used here are merely for the purpose of simplifying the material property variation, and not for the displacement approximation. Hence, the method of wave analysis is still an exact approach in a mechanic sense. An analytical solution for the equation of motion governing the element is first derived. It is then solved for displacement and stresses in the frequency domain by employing boundary and continuity conditions for a FGM plate whose material property varies in an arbitrary manner. The response of the plate to a uniform surface traction (traction is defined as a distributed force) in the time domain is obtained using Fourier transform techniques. In addition, an integral technique is included in this chapter for evaluating the confluent hypergeometric functions with a complex valued argument, which arises from the wave propagation problem for plates of FGMs of linear property variation. Numerical examples are presented to demonstrate the efficiency of the present method for harmonic and transient waves in FGM plates.

2.2   Element of Linear Property Variation

Consider a plane harmonic wave propagating in an FGM plate of thickness H. The plane of the wave front is perpendicular to the z-axis, meaning that the waves are propagating in the z-direction. The FGM plate is divided into N elements in the thickness direction. The thickness of the nth element is denoted by H_n. In each element, the mass density ρ and the elastic constant c_{33} are assumed as linear functions of z (see Figure 2.1):

$$\rho = \rho_0(1 + b_\rho z), \quad c_{33} = c_{33}^0(1 + b_c z) \tag{2.1}$$

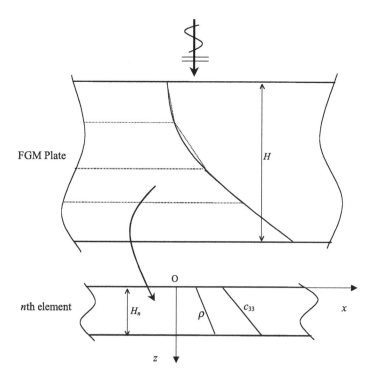

FIGURE 2.1
A plane wave propagating through a FGM plate in the thickness direction.

where b_p and b_c are gradient constants representing the change rate of material property along the thickness of the element. ρ_0 and c_{33}^0 are the mass density and the elastic constant of the material at the lower surface ($z = 0$) of the element.

Since the plane waves are propagating in the thickness (z) direction, the wave field is independent of x and y, and the problem is, in fact, one dimensional. The equation of motion governing the FGM can be expressed as (in the absence of body force)

$$\frac{\partial \sigma(z, t)}{\partial z} = \rho(z)\frac{\partial^2 w(z, t)}{\partial t^2} \tag{2.2}$$

where σ is the stress in the z direction that can be expressed as

$$\sigma = c_{33}(z)\varepsilon = c_{33}(z)\frac{\partial w}{\partial z} \tag{2.3}$$

where ε and w are, respectively, the strain and the displacement in the z direction. Substituting the foregoing equation into Eq. (2.2) gives

$$\frac{\partial}{\partial z}\left[c_{33}(z)\frac{\partial w}{\partial z}\right] = \rho(z)\frac{\partial^2 w}{\partial t^2} \tag{2.4}$$

Note that the c_{33} is a linear function of z. It is assumed that the displacement in z direction has the form of

$$w(z, t) = W(z)\exp(-i\omega t) \tag{2.5}$$

where ω is the angular frequency of the plane wave. Substituting Eqs. (2.5) and (2.1) into Eq. (2.4) leads to the following confluent hypergeometric differential equation.

$$(1 + b_c z)\frac{\partial^2 W}{\partial z^2} + b_c \frac{\partial W}{\partial z} + \frac{\rho_0 \omega^2}{c_{33}^0}(1 + b_p z)W = 0 \tag{2.6}$$

The solution of Eq. (2.6) can be obtained as follows.

For FGMs with $b_c \neq 0$, we have (Polyanin and Zaitsev, 1995)

$$W = \begin{cases} e^{hz}[A\phi(a, b, \xi) + B\varphi(a, b, \xi)] & b_p \neq 0 \\ AJ_0(\beta\sqrt{\zeta}) + BY_0(\beta\sqrt{\zeta}) & b_p = 0 \end{cases} \tag{2.7}$$

where A and B are integral constants to be determined by the boundary conditions on the surface of the element, and

$$h = D/(2b_c), \quad a = 0.5 + D/(4b_c^2) + \rho_0 \omega^2 b_c b_\rho/c_{33}^0, \quad b = 1$$

$$D^2 = -4\rho_0 \omega^2 b_c b_\rho/c_{33}^0 \quad \xi = -D(z + 1/b_c)/b_c \tag{2.8}$$

$$\beta = 2\sqrt{\rho_0 c_{33}^0}\,\omega \quad \zeta = (z + 1/b_c)/b_c$$

Functions ϕ and φ in Eq. (2.7) are the confluent hypergeometric functions (CHFs), while the functions J and Y are the Bessel functions of first and second kind, respectively. From Eq. (2.7), the normal stress can be obtained simply by differentiation, i.e.,

$$\sigma = \begin{cases} he^{hz}c_{33}^0(1 + b_c z)\{A[\phi(a, b, \xi) - 2\phi'(a, b, \xi)] \\ \quad + B[\varphi(a, b, \xi) - 2\varphi'(a, b, \xi)]\} & b_\rho \neq 0 \\ 0.5c_{33}^0(1 + b_c z)[AJ_0'(\beta\sqrt{\zeta}) + BY_0'(\beta\sqrt{\zeta})]\beta\xi^{-0.5}\dfrac{1}{b_c} & b_\rho = 0 \end{cases} \tag{2.9}$$

Solutions given by Eqs. (2.7) and (2.9) are applicable for all the N elements. Consequently, there are $2N$ constants to be determined by the boundary conditions at the surfaces of the plate and the continuity conditions at the interfaces between the elements.

2.3   Boundary and Continuity Conditions

Without the loss of generality, we assume that the plate is loaded on all the $(N-1)$ interfaces and the two surfaces by a harmonic traction with an angular frequency ω. The amplitude of the external traction vector can be written as

$$\mathbf{T}^T = \{T_1, T_2, T_3, \dots, T_N, T_{N+1}\} \tag{2.10}$$

where T_j is the amplitude of the external traction acting at the jth interface, $j = 1$ is for the lower surface, and $j = N + 1$ is for the upper surface of the plate. The boundary conditions and continuity conditions for the FGM plate can be written as follows.

On the lower surface of the plate

$$-\sigma_1^l = T_1 \tag{2.11}$$

At the interfaces

$$\sigma_n^u - \sigma_{n+1}^l = T_{n+1}, \qquad w_n^u = w_{n+1}^l \quad \text{for } 1 \le n \le (N-1) \tag{2.12}$$

On the upper surface of the plate

$$\sigma_N^u = T_{N+1} \tag{2.13}$$

The subscripts in Eqs. (2.11) to (2.13) stand for the element number, and the superscripts u and l stand for the upper and lower surfaces of the elements, respectively. Assembling all the elements together using Eqs. (2.11) to (2.13), we obtain the system equation of matrix form for the entire FGM plate as follows.

$$\mathbf{T} = \mathbf{KA} \tag{2.14}$$

where $\mathbf{A}$ is a constant vector for all the elements given by

$$\mathbf{A}^T = \{A_1, B_1, A_2, B_2, \dots, A_N, B_N\} \tag{2.15}$$

Matrix $\mathbf{K}$ in Eq. (2.14) is the stiffness matrix found to be

$$\mathbf{K} = \begin{bmatrix}
-K_{E11} & -K_{E12} & 0 & 0 & 0 & 0 & \cdot & 0 \\
T_{E11} & T_{E12} & -P_{E11} & -P_{E12} & 0 & 0 & \cdot & 0 \\
e_{11} & e_{12} & -K_{E21} & -K_{E22} & 0 & 0 & \cdot & 0 \\
0 & 0 & T_{E21} & T_{E22} & -P_{E21} & -P_{E22} & \cdot & 0 \\
0 & 0 & e_{21} & e_{22} & -K_{E31} & -K_{E32} & \cdot & 0 \\
\cdot & \cdot & \cdot & \cdot & \cdot & \cdot & \cdot & 0 \\
\cdot & \cdot & \cdot & \cdot & \cdot & \cdot & \cdot & 0 \\
\cdot & \cdot & \cdot & \cdot & \cdot & \cdot & \cdot & \cdot \\
0 & 0 & 0 & 0 & 0 & 0 & e_{N1} & e_{N2}
\end{bmatrix} \tag{2.16}$$

where

$$K_{En1} = h_n[\phi(a_n, b_n, -2h_n/b_{c(n)}) - 2\phi'(a_n, b_n, -2h_n/b_{c(n)})]c_{33(n)}^0 \tag{2.17}$$

$$K_{En2} = h_n[\varphi(a_n, b_n, -2h_n/b_{c(n)}) - 2\varphi'(a_n, b_n, -2h_n/b_{c(n)})]c_{33(n)}^0 \tag{2.18}$$

$$T_{En1} = e^{h_n H_n}\phi(a_n, b_n, -2h_n(H_n + 1/b_{c(n)})) \tag{2.19}$$

$$T_{En2} = e^{h_n H_n}\varphi(a_n, b_n, -2h_n(H_n + 1/b_{c(n)})) \tag{2.20}$$

$$P_{En1} = \phi(a_n, b_n, -2h_n/b_{c(n)}) \tag{2.21}$$

$$P_{En2} = \varphi(a_n, b_n, -2h_n/b_{c(n)}) \tag{2.22}$$

$$e_{n1} = e^{H_n h_n} h_n [\phi(a, b, -2h_n(H_n + 1/b_{c(n)}))$$
$$- 2\phi'(a_n, b_n, -2h_n(H_n + 1/b_{c(n)}))](1 + b_{c(n)}H_n)c_{33(n)}^0 \tag{2.23}$$

$$e_{n2} = e^{H_n h_n} h_n [\varphi(a, b, -2h_n(H_n + 1/b_{c(n)}))$$
$$- 2\varphi'(a_n, b_n, -2h_n(H_n + 1/b_{c(n)}))](1 + b_{c(n)}H_n)c_{33(n)}^0 \tag{2.24}$$

Equations (2.17) to (2.24) are obtained for $b_\rho \neq 0$. Similar equations can also be easily obtained for $b_\rho = 0$, and they are listed below.

$$K_{En1} = 0.5c_{33}^0 J_0'(\beta_n/b_{c(n)}) \tag{2.25}$$

$$K_{En2} = 0.5c_{33}^0 Y_0'(\beta_n/b_{c(n)}) \tag{2.26}$$

$$T_{En1} = J_0(\beta_n\sqrt{\zeta_n'}) \tag{2.27}$$

$$T_{En2} = Y_0(\beta_n\sqrt{\zeta_n'}) \tag{2.28}$$

$$P_{En1} = J_0(\beta_n/b_{c(n)}) \tag{2.29}$$

$$P_{En2} = Y_0(\beta_n/b_{c(n)}) \tag{2.30}$$

$$e_{n1} = 0.5c_{33(n)}^0(1 + b_{c(n)}H_n)\beta_n\zeta_n'^{-0.5}b_{c(n)}^{-1}J_0'(\beta_n\sqrt{\zeta_n'}) \tag{2.31}$$

$$e_{n2} = 0.5c_{33(n)}^0(1 + b_{c(n)}H_n)\beta_n\zeta_n'^{-0.5}b_{c(n)}^{-1}Y_0'(\beta_n\sqrt{\zeta_n'}) \tag{2.32}$$

In Eqs. (2.17) to (2.32),

$$\zeta_n' = (z + 1/b_c)/b_c \tag{2.33}$$

Solving Eq. (2.14), the constant vector **A** can be obtained. The displacement and stress in the frequency domain for each element can then be obtained using Eqs. (2.7) and (2.9), respectively.

2.4   Transient Response

Once the displacement in the frequency domain is obtained, the displacement in the time domain can be obtained using Fourier superposition, i.e.,

$$w_t(z, t) = \frac{1}{2\pi}\int_{-\infty}^{\infty} W(z, \omega)\tilde{P}(\omega)\exp(i\omega t)\,d\omega \tag{2.34}$$

where $\tilde{P}$ is the Fourier transform of the external traction $p(t)$ and is given by

$$\tilde{P}(\omega) = \int_0^{t_f} p(t)\exp(-i\omega t)\,dt \tag{2.35}$$

The traction is assumed to be uniformly applied on the surface of the plate, and it is independent of x and y. The time duration of the load is $(0, t_f)$. From Eq. (2.35), it is easily seen that

$$\tilde{P}(-\omega) = \tilde{P}^*(\omega) \tag{2.36}$$

where the asterisk denotes the complex conjugate. As we have seen in Chapter 1, the integrand in Eq. (2.34) will be singular at the poles on the integral axis. In Chapter 1 we managed to evaluate the integral analytically. However, the same cannot be done for FGM plates, as the integrand is very complex and cannot be written explicitly. An exponential window method (EWM) (see Section 4.6 for details) can be used to evaluate numerically the integrand in Eq. (2.34). In applying the EWM, we should make use of the fact that

$$W(z, -\omega) = W^*(z, \omega) \tag{2.37}$$

With the help of Eqs. (2.36) and (2.37), Eq. (2.34) can be changed to

$$w_t(z, t) = \frac{1}{\pi}\left[\int_0^\infty (W_R\tilde{P}_R - W_I\tilde{P}_I)\cos\omega t\,d\omega - \int_0^\infty (W_R\tilde{P}_I + W_I\tilde{P}_R)\sin\omega t\,d\omega\right] \tag{2.38}$$

where W_R and W_I are the real and imaginary parts of W, respectively. $\tilde{P}_R$ and $\tilde{P}_I$ are the real and imaginary parts of $\tilde{P}$, respectively. Using Eq. (2.38) is much more computationally efficient than using Eq. (2.34) because the computation needs to be done only in real numbers in Eq. (2.38). The stress in the time domain can be derived in exactly the same way as the displacement:

$$\sigma_t(z, t) = \frac{1}{\pi}\left[\int_0^\infty (S_R\tilde{P}_R - S_I\tilde{P}_I)\cos\omega t\,d\omega - \int_0^\infty (S_R\tilde{P}_I + S_R\tilde{P}_R)\sin\omega t\,d\omega\right] \tag{2.39}$$

where S_R and S_I are the real and imaginary parts of stress σ given by Eq. (2.9).

The integrals in Eqs. (2.34) and (2.39) can be evaluated by ordinary routines of numerical integration using equally spaced sampling points together with the EWM. To reduce the sampling points while achieving the desired accuracy, an adaptive integral scheme discussed in Section 4.5 can be employed.

2.5   Evaluation of Confluent Hypergeometric Function

The solution given in Eqs. (2.7) and (2.9) for harmonic waves in an element of linear property variation contains confluent hypergeometric functions (CHFs) with complex valued argument and orders. The evaluation of such CHFs must be carried out. There exist numerical methods (Slater, 1960; Erdelyi, et al., 1953; Abramowitz and Stegun, 1964) and efficient algorithms (Thompson, 1997) available for evaluating the CHFs. However, these methods can evaluate only CHFs with real parameters. Nardin et al. (1992a) proposed a method to evaluate CHFs of complex arguments with large magnitudes using a direct summation of Kummer's series. Here we introduce an integration scheme proposed by Liu and Han, et al., 2001, that makes use of the fact that CHFs can be given in the following integral form (Abramowitz and Stegun, 1964).

$$\phi(a, b, \eta) = \frac{\Gamma(b)}{\Gamma(a)\Gamma(b-a)} \int_0^1 e^{\eta\alpha} \alpha^{a-1}(1-\alpha)^{b-a-1} d\alpha \quad \text{real}(b) > \text{real}(a) > 0$$

$$(2.40)$$

$$\varphi(a, b, \eta) = \frac{1}{\Gamma(a)} \int_0^\infty e^{-\eta\alpha} \alpha^{a-1}(1+\alpha)^{b-a-1} d\alpha \quad \text{real}(a) > 0 \qquad (2.41)$$

$$\phi'(a, b, \eta) = \frac{a}{b}\phi(a+1, b+1, \eta), \quad \varphi'(a, b, \eta) = -a\varphi(a+1, b+1, \eta) \quad (2.42)$$

where Γ is the Gamma function, and it can be given in Euler's integral form of

$$\Gamma(\eta) = \int_0^\infty \alpha^{\eta-1} e^{-\alpha} d\alpha, \quad \text{real}(\eta) > 0 \qquad (2.43)$$

The task now is to evaluate numerically the integrals in Eqs. (2.40) and (2.43). As the integrand is singular, conventional numerical techniques cannot be used. In addition, the integrands in Eqs. (2.40) and (2.41) can be very oscillatory when argument z has a large imaginary part. The following integration scheme is efficient for evaluating these integrals. We use a Gamma function to illustrate the procedure of this scheme.

2.5.1   Integral of Gamma Function

To evaluate the Gamma function given by Eq. (2.43) numerically, the infinite integral must be first changed to a finite integral. For a non-negative η, the integrand

$$f(\alpha) = \alpha^{\eta-1}e^{-\alpha} \tag{2.44}$$

approaches zero very rapidly as shown in Figure 2.2. For a given error criterion δ_g, the upper integral limit T_g can be determined by

$$|f(T_g)| \leq \delta_g \tag{2.45}$$

The infinite integral is now approximately replaced by a finite integral of

$$\Gamma(\eta) \cong \int_0^{T_g} \alpha^{\eta-1}e^{-\alpha}d\alpha \tag{2.46}$$

First, we divide the integral interval $[0, T_g]$ into L_g subregions using points α_j, i.e.,

$$0 = \alpha_1 < \alpha_2 < \cdots < \alpha_j < \cdots < \alpha_{L_g+1} = T_g \tag{2.47}$$

For a sufficiently large L_g, $e^{-\alpha}$ can be approximately expressed by a parabola in each subregion:

$$e_j^{-\alpha} \approx a_j\alpha^2 + b_j\alpha + c_j \quad \text{for } \alpha_j \leq \alpha \leq \alpha_{j+1} \tag{2.48}$$

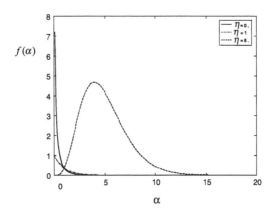

FIGURE 2.2
Integrand of the Gamma function vanishes quickly with increasing argument α. (From Liu, G. R., Han, X., and Lam, K. Y. *Computers and Structures*, 79, 1039, 2001a. With permission.)

where a_j, b_j, c_j are determined by forcing the parabola to pass through the function values at the points α_j, $1/2(\alpha_j + \alpha_{j+1})$ and α_{j+1}.

Next, using Eq. (2.48), Eq. (2.46) can be expressed

$$\Gamma(\eta) = \sum_{j=1}^{L_g} \int_{\alpha_j}^{\alpha_{j+1}} \alpha^{\eta-1}(a_j\alpha^2 + b_j\alpha + c_j)\,d\alpha = \sum_{j=1}^{L_g} I_j \tag{2.49}$$

where I_j can be carried out analytically.

$$I_j = \frac{a_j}{\eta+2}(\alpha_{j+1}^{\eta+2} - \alpha_j^{\eta+2}) + \frac{b_j}{\eta+1}(\alpha_{j+1}^{\eta+1} - \alpha_j^{\eta+1}) + \frac{c_j}{\eta}(\alpha_{j+1}^{\eta} - \alpha_j^{\eta}) \tag{2.50}$$

This technique works for both real and complex valued η. The accuracy of the present integration is dependent only on the approximation of $e^{-\eta}$, and independent of η. For the interval of $[0, T_g]$, a very small number of sampling points can give the function a very accurate approximation using piecewise parabolas, as will be shown later in this chapter. It should be noted that the integrand of Eq. (2.44) is singular when $\mathrm{Re}(\eta) < 1$, but this singular integral can be carried out without any difficulty, as shown in Eq. (2.50), as the integration is done analytically.

2.5.2 Integral of Confluent Hypergeometric Function

The integral in function $\phi(a, b, \eta)$ can be written as (see Eqs. (2.40))

$$II = \int_0^1 e^{\eta\alpha}\alpha^{a-1}(1 - \alpha)^{b-a-1}\,d\alpha \tag{2.51}$$

The integrand is singular when $\alpha = 0$, $\alpha = 1$ for $\mathrm{Re}(a - 1) < 0$ or $\mathrm{Re}(b - a - 1) < 0$. The integral in Eq. (2.51) is divided into two parts

$$II = II_1 + II_2 \tag{2.52}$$

where

$$II_1 = \int_0^{0.5} e^{\alpha\eta}\alpha^{a-1}(1 - \alpha)^{b-a-1}\,d\alpha \tag{2.53}$$

$$II_2 = \int_{0.5}^{1} e^{\alpha\eta}\alpha^{a-1}(1 - \alpha)^{b-a-1}\,d\alpha = \int_0^{0.5} e^{\eta(1-\alpha)}(1 - \alpha)^{a-1}\alpha^{b-a-1}\,d\alpha \tag{2.54}$$

Consider II_1 first. We divide the integral interval $[0, 0.5]$ into L_1 subregions using points α_j.

$$0 = \alpha_1 < \alpha_2 < \cdots < \alpha_j < \cdots < \alpha_{L_1+1} = 0.5 \qquad (2.55)$$

For a sufficiently large L_1, $e^{\alpha\eta}(1 - \alpha)^{b-a-1}$ can be approximately expressed by a parabola within each subregion, i.e.,

$$m(\alpha)_j = [e^{\alpha\eta}(1 - \alpha)^{b-a-1}]_j \approx p_{1j}\alpha^2 + q_{1j}\alpha + r_{1j} \quad \text{for } \alpha_j \le \alpha \le \alpha_{j+1} \qquad (2.56)$$

where p_{1j}, q_{1j} and r_{1j} are determined by forcing the parabola to pass through the function values at the points α_j, $1/2(\alpha_j + \alpha_{j+1})$ and α_{j+1}.

Using Eq. (2.56), Eq. (2.53) can be expressed as

$$II_1 = \sum_{j=1}^{L_1} \int_{\alpha_j}^{\alpha_{j+1}} \alpha^{a-1}(p_{1j}\alpha^2 + q_{1j}\alpha + r_{1j})\, d\alpha = \sum_{j=1}^{L_1} II_{1j} \qquad (2.57)$$

where II_{1j} can be calculated analytically to give

$$II_{1j} = \frac{p_{1j}}{a+2}(\alpha_{j+1}^{a+2} - \alpha_j^{a+2}) + \frac{q_{1j}}{a+1}(\alpha_{j+1}^{a+1} - \alpha_j^{a+1}) + \frac{r_{1j}}{a}(\alpha_{j+1}^{a} - \alpha_j^{a}) \qquad (2.58)$$

Similarly, term $e^{\eta(1-\alpha)}(1 - \alpha)^{a-1}$ in Eq. (2.54) can be approximately expressed by a parabola in subregions:

$$n(t)_j = [e^{\eta(1-\alpha)}(1 - \alpha)^{a-1}]_j \approx p_{2j}\alpha^2 + q_{2j}\alpha + r_{2j} \quad \text{for } \alpha_j \le \alpha \le \alpha_{j+1} \qquad (2.59)$$

Using Eq. (2.59), Eq. (2.54) can be rewritten as

$$II_2 = \sum_{j=1}^{L_2} \int_{\alpha_j}^{\alpha_{j+1}} \alpha^{b-a-1}(p_{2j}\alpha^2 + q_{2j}\alpha + r_{2j})\, d\alpha = \sum_{j=1}^{L_2} II_{2j} \qquad (2.60)$$

where, II_{2j} can also be calculated analytically.

$$II_{2j} = \frac{p_{2j}}{b-a+2}(\alpha_{j+1}^{b-a+2} - \alpha_j^{b-a+2}) + \frac{q_{2j}}{b-a+1}(\alpha_{j+1}^{b-a+1} - \alpha_j^{b-a+1}) + \frac{r_{2j}}{b-a}(\alpha_{j+1}^{b-a} - \alpha_j^{b-a})$$

$$(2.61)$$

It is found in Eqs. (2.58) and (2.61), the singular integral at $\alpha = 0$ and $\alpha = 1$ have been successfully evaluated analytically.

The integral in Eqs. (2.41) can also be treated using the same technique.

2.5.3   Interval Division and Error Control

The simplest way to divide an integral interval, say $[a, b]$, is the even division, where the constant step is $\Delta = \alpha_{j+1} - \alpha_j = \frac{b-a}{L}$ used. However, this might not be efficient and might lead to a large number of divisions. Furthermore, the error is not controllable. An efficient method for dividing the integral interval is proposed by Han and Liu (2001). This method not only can reduce the number of divisions significantly, but also the error can be controlled.

For a given function $y(\alpha)$, its Taylor expansion in a small subregion $[\alpha_j, \alpha_{j+1}]$ can be expressed as

$$y(\alpha) = y(\alpha_{0j}) + y'(\alpha_{0j})(\alpha - \alpha_{0j}) + \frac{1}{2}y''(\alpha_{0j})(\alpha - \alpha_{0j})^2 + \varepsilon_0 \qquad (2.62)$$

where

$$\alpha_{0j} = (\alpha_j + \alpha_{j+1})/2 \qquad (2.63)$$

The leading order error is given by

$$\varepsilon_0 = \frac{1}{6}y'''(\alpha_{0j})(\zeta - \alpha_{0j})^3 \quad \zeta \in [\alpha_j, \alpha_{j+1}] \qquad (2.64)$$

In the present integral technique, the accuracy of the integration depends only on the approximation of the function in the integrand. The error of approximation can be controlled by choosing a proper length of the sub-region $\Delta = \alpha_{j+1} - \alpha_j$ for a given allowable error of approximation ε_0. For example, when $y(\alpha) = e^{-\alpha}$, we can obtain

$$\varepsilon_0 = -\frac{1}{6}(\zeta - \alpha_{0j})^3 e^{-\alpha_{0j}} \qquad (2.65)$$

In each subregion, the integral step is then given by

$$\Delta_j = \alpha_{j+1} - \alpha_j = 2(\alpha_{0j} - \alpha_j) \geq 2|\zeta - \alpha_{0j}| \qquad (2.66)$$

and

$$\varepsilon_0 = \frac{1}{6}(\zeta - \alpha_{0j})^3 e^{-\alpha_{0j}} \leq \frac{1}{48}\Delta_j^3 e^{-\alpha_{0j}} \qquad (2.67)$$

which leads to

$$\Delta_j = \sqrt[3]{48\varepsilon_0/e^{-\alpha_{0j}}} \tag{2.68}$$

in which Δ_j varies in $[a, b]$ for a given ε_0. In some regions, Δ_j can be very large, and the sampling points can be reduced significantly, compared to an even division.

It can be expected that the error of the integration should be even smaller than the error introduced in approximating the integrand because an integral operation generally reduces error. To confirm this, an error analysis study is carried out with the help of MATLAB, by which an integral result for a desired number of accurate digits can be given. The error of integration using the present method is defined as

$$\sigma' = \left|\frac{I_e - I_c}{I_e}\right| \tag{2.69}$$

where I_e is the results obtained from MATLAB with 8 accurate digits, and I_c is the present results. For the Gamma function with $\eta = 0.5$, the error is calculated and the relationship between σ' and ε_0 is shown in Figure 2.3.

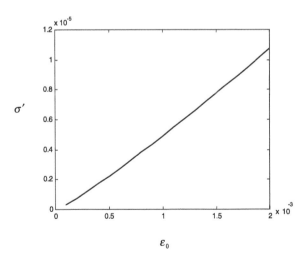

FIGURE 2.3
The integration error is much smaller than the error in the approximation of the integrand. (From Liu, G. R., Han, X., and Lam, K. Y. *Computers and Structures*, 79, 1039, 2001a. With permission.)

It is clearly shown that σ' is about two orders smaller than ε_0. Therefore, it is very conservative to control σ' through the control of ε_0.

2.6　Examples

2.6.1　Confluent Hypergeometric Function

The present integral technique can be easily coded in a computer program. The results obtained by the present method are compared with those by MATLAB for the Gamma function with a real parameter η. Table 2.1 shows the comparison of the results. The Gamma function is calculated with $\eta = 0.5$. Table 2.1 shows that the results are very good. When the even division is used with $L = 100$, the result is accurate up to five digits after the decimal point. With $L = 2000$, the result is exactly equal to the result from MATLAB with seven accurate digits after the decimal point. When the error-controlled division is used, 168 divisions can lead to an accurate result with seven digits. The Gamma function with a complex argument has also been computed. For a Gamma function with $\eta = 1.4 + 3.0i$, Table 2.2 shows the comparison

TABLE 2.1

Comparison of Results Obtained by the Present Method and MATLAB

	Present Method					
	Even Division				Error Controlled Division ($\varepsilon_0 = 1.e^{-6}$) $L_g = 168$	MATLAB Result with 8 Accurate Digits
$\eta = 0.5$	$L = 100$	$L = 200$	$L = 1000$	$L = 2000$		
$\Gamma(\eta)$	1.77245873	1.77245424	1.77245378	1.77245377	1.77245378	1.77245385

Source: Liu, G. R., Han, X., and Lam, K. Y. *Computers and Structures*, 79, 1039, 2001a. With permission.

TABLE 2.2

Comparison of Results Obtained by the Present Method and IMSL

	Present Method					
	Even Division				Error Controlled Division ($\varepsilon_0 = 1.e^{-6}$) $L_g = 168$ CPU: 0.013s	IMSL 8 Accurate Digits CPU: 0.018s
$\eta = 1.4 + 3.0i$	$L = 100$	$L = 200$	$L = 1000$	$L = 2000$		
$Re(\Gamma(\eta))$	−0.00130316	−0.00112999	−0.00113001	−0.00113010	−0.00113010	−0.00113044
$Im(\Gamma(\eta))$	0.06107730	0.06107702	0.06107701	0.06107701	0.06107702	0.06107781

Source: Liu, G. R., Han, X., and Lam, K. Y. *Computers and Structures*, 79, 1039, 2001a. With permission.

TABLE 2.3

Comparison of Results Obtained by the Present Method
and Thompson's Method (1997)

$\psi(a, b, \eta)$	Present Method CPU Time: 0.01s	Thompson's Method CPU Time: 0.01s
$a = 1.0, b = 3.0, \eta = 1.0$	1.99995	1.99999
$a = 2.5, b = 3.5, \eta = 10.0$	3.16232e-3	3.16231e-3

Source: Liu, G. R., Han, X., and Lam, K. Y. *Computers and Structures*, 79, 1039, 2001a. With permission.

TABLE 2.4

Comparison of Results Obtained by the Present Method
and Nardin's Method (1992b)

$\phi(a, b, \eta)$	Present Method	Nardin's Method
$a = 1.0, b = 2.0, \eta = 1.0 + 1.5i$	$1.002916 + 1.207086i$	$1.002920 + 1.207091i$
$a = 6.0 + 1.5i, b = 8.0 + 2.5i, \eta = 1.0 + 1.5i$	$1.014538 + 1.963162i$	$1.014990 + 1.966535i$

Source: Liu, G. R., Han, X., and Lam, K. Y. *Computers and Structures*, 79, 1039, 2001a. With permission.

of the results obtained by the present method against those obtained by the IMSL FORTRAN program. The present method takes less CPU time than that required by the IMSL program.

The evaluations of CHFs for real argument and order are shown in Table 2.3 using the present method together with the algorithm proposed by Thompson (1997). In Thompson's algorithm for $\eta < 60$, the integration is performed using the nine-strip, equal-step, Newton-Cotes formula. The CPU times for evaluating the irregular function $\psi(a, b, \eta)$ using these two methods are nearly the same. The comparisons are performed on an SGI original 2000, and all the CPU times listed are the average results of 15–20 evaluations of integrands with different arguments and orders. CHFs with complex argument and complex order are also evaluated using the present method and the algorithm given by Nardin et al. (1992b); the results are shown in Table 2.4.

The present integration scheme is applied to evaluate confluent hypergeometric functions. Figures 2.4 and 2.5 show functions $\phi(a, b, \eta)$ and $\varphi(a, b, \eta)$ obtained by the present technique with $b = 2.0$, respectively.

2.6.2 Wave Fields in FGM Plates

The present techniques have also been applied to examining waves propagating in FGM plates. In presenting those results, the following dimensionless parameters are used:

$$\left.\begin{array}{l} \bar{w} = w/H, \bar{z} = z/H, \bar{p} = p/c_{33r}, \bar{\sigma} = \sigma/c_{33r}, \bar{c}_{33} = c_{33}/c_{33r} \\ \bar{\omega} = \omega H/c_r, \bar{t} = tc_r/H, \bar{\rho} = \rho/\rho_r, \bar{b}_c = b_c H, \bar{b}_\rho = b_\rho H \end{array}\right\} \tag{2.70}$$

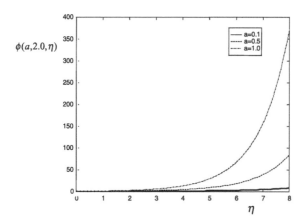

FIGURE 2.4
Confluent hypergeometric function $\phi(a, 2.0, \eta)$. (From Liu, G. R., Han, X., and Lam, K. Y. *Computers and Structures*, 79, 1039, 2001a. With permission.)

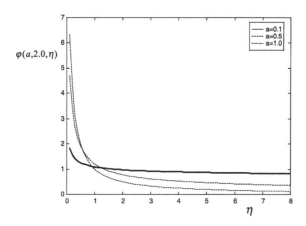

FIGURE 2.5
Confluent hypergeometric function $\varphi(a, 2.0, \eta)$. (From Liu, G. R., Han, X., and Lam, K. Y. *Computers and Structures*, 79, 1039, 2001a. With permission.)

where c_{33r} ($c_{33r} = c_r^2 \rho_r$), c_r and ρ_r are the reference elastic constant, wave velocity, and density, respectively. $\bar{t} = 1$ is the time for the wave of velocity c_r traveling once across the plate of the thickness H. For linear FGM plates, whose material property varies linearly in the thickness direction as defined by Eq. (2.1), the wave velocity at the middle surface is used as the reference wave velocity c_r.

The external force **T** used in Eq. (2.14) should be

$$\mathbf{T} = \{0,\ 0,\ 0, ..., 0,\ 0,\ 0,\ P\} \qquad (2.71)$$

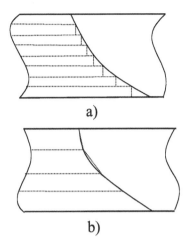

a)

b)

FIGURE 2.6
Two discrete models for FGM plates. (a) Homogeneous element model; (b) model of element
of linear property variation. (From Liu, G. R., Han, X., and Lam, K. Y. *Composites Part B:
Engineering*, 30(4), 383, 1999. With permission.)

where

$$P = 2\sin(\bar{\omega}_f \dot{t}) \qquad (2.72)$$

is the surface traction upon the upper surface. In this study, we set $\bar{\omega}_f = 2\pi$
and $\dot{t}_f = 2\pi/\bar{\omega}_f = 1.0$—namely, the wave is one cycle of a sine function.

For an arbitrary FGM plate, two models, shown in Figure 2.6, can be used.
The first model uses homogeneous elements (see Liu et al., 1991). The second
model uses the present model of elements with linear property variation. In
order to validate the present method, the displacement response of a linear
FGM plate subjected to a single cycle of sinusoidal traction is computed. The
present method should produce the exact results using one element, and the
results are shown in Figure 2.7. The results are compared with those obtained
by a method proposed by Liu et al. (1991), in which the plate is divided into
N homogeneous elements. Two cases of $N = 10$ and $N = 20$ have been computed. From Figure 2.7, it can be seen that more homogeneous elements can
give better results, and many homogeneous elements have to be used in
order to obtain a result with a reasonable accuracy.

Figure 2.8 shows the stress distribution in the thickness direction of a linear
FGM plate at time $\dot{t} = 0.05, 2.05$, and 4.55. Figure 2.9 shows the time history
of the stress at locations $\bar{z} = 0.0, 0.5$, and 1.0 for the linear FGM plate subjected
to the single cycle of sinusoidal traction. Because $\bar{z} = 0.0$ is at the lower free
surface of the plate, the stress is zero as shown in Figure 2.8. On the upper
surface ($\bar{z} = 1.0$), which faces the surface traction, the plate experiences a
compressive stress equating to the traction given in Eq. (2.71), as shown in
Figure 2.9. These results show that the boundary conditions have been satisfied.

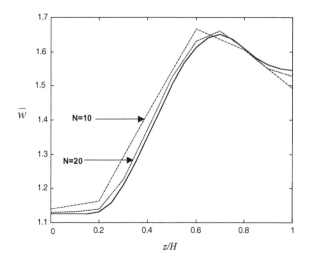

FIGURE 2.7
Comparison of results obtained by the present method (solid line) with those obtained by Liu's (1996) method (dashed lines) for a linear FGM plate subjected to a single cycle of sinusoidal surface traction ($\bar{t} = 0.5$, $\bar{b}_c = 0.5$, $\bar{b}_\rho = 0.25$). (From Liu, G. R., Han, X., and Lam, K. Y. *Composites Part B: Engineering*, 30(4), 383, 1999. With permission.)

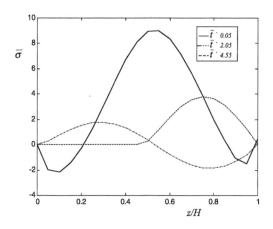

FIGURE 2.8
Stress distributions in the linear FGM plate subjected to a single cycle of sinusoidal surface traction ($\bar{b}_c = 0.5$, $\bar{b}_\rho = 0.25$, $\bar{\omega}_f = 6.28$). (From Liu, G. R., Han, X., and Lam, K. Y. *Composites Part B: Engineering*, 30(4), 383, 1999. With permission.)

At the middle point ($\bar{z} = 0.5$), the stress is zero until the wave propagates onto this point.

To examine how a stress wave is propagating through a FGM plate, the time history of the stress in a linear FGM plate subjected to a single cycle sinusoidal surface traction is computed using the present method. The FGM plate has a linear velocity and constant density. The gradient constant of velocity of the

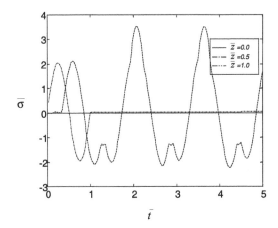

FIGURE 2.9
Time history of the stress for the linear FGM plate subjected to a single cycle sinusoidal surface traction ($\bar{b}_c = 0.5$, $\bar{b}_\rho = 0.25$, $\bar{\omega}_f = 6.28$). (From Liu, G. R., Han, X., and Lam, K. Y. *Composites Part B: Engineering*, 30(4), 383, 1999. With permission.)

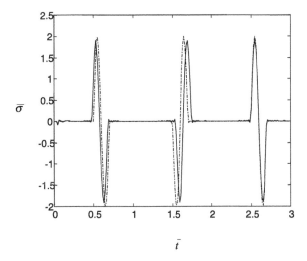

FIGURE 2.10
Comparison of time history of the stress at the middle point of an FGM (solid line) and homogeneous (dashed line) plates subjected to a single cycle of sinusoidal traction on the upper surface ($\bar{\omega}_f = 3.14$, $z = 0.5H$, $\bar{b}_\rho = 0$, $c = c_r(1 + 0.5z/H)$). (From Liu, G. R., Han, X., and Lam, K. Y. *Composites Part B: Engineering*, 30(4), 383, 1999. With permission.)

linear FGM plate is fixed at 0.5 and the wave velocity at the middle point is fixed at $c_r = 5000$ m/s. The results are shown in Figure 2.10 for the stress at the middle point in the FGM plate together with that in a homogeneous plate whose wave velocity is c_r. Because the traction is applied on the upper surface where the wave velocity is the highest, and the wave velocity in the FGM

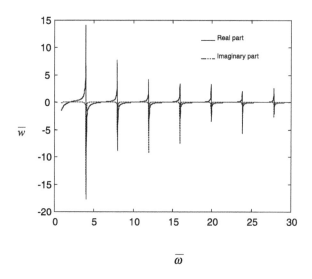

FIGURE 2.11

Displacement on the top surface in the frequency domain for an FGM of linear property variation ($\bar{b}_c = 0.5$, $\bar{b}_\rho = 0.25$). (From Liu, G. R., Han, X., and Lam, K. Y. *Composites Part B: Engineering*, 30(4), 383, 1999. With permission.)

plate is higher than that in the homogeneous plate in the upper half portion of the plate, the stress wave propagates faster. When the stress wave is travelling in the lower portion of the plate, the wave velocity in the homogeneous plate is higher than in the FGM plate. This can be clearly evidenced from Figure 2.10 where the dashed line lags behind during the time $0 < \bar{t} < 1.5$ and then overtakes it during $1.5 < \bar{t} < 3.0$. As time increases, the dashed line will again be overtaken by the solid line, and so on. Figure 2.10 also shows that the amplitude of the stress wave in the FGM plate is slightly smaller than that in the homogenous plate. This is because in the FGM plate the traction is applied on the harder surface that is capable of taking higher stress.

Figure 2.11 shows the frequency spectrum of the displacement on the top surface of an FGM plate whose property varies linearly through the thickness of the plate. The parameters for the FGM plate are $\bar{b}_c = 0.5$, $\bar{b}_\rho = 0.25$. The displacement becomes singular at the natural frequencies. Similar phenomenon is seen in Chapter 1 for the case of fixed-fixed one-dimensional bar.

2.7   Remarks

Wave propagation problems related to FGMs are generally difficult to analyze because the material properties are functions of the coordinates. In this chapter, an analytical method is presented to investigate the response in a FGM plate whose material property (density and elastic constants) varies arbitrarily through the thickness. This method is very efficient and accurate in investigating the dynamic responses of the FGM plate subjected to harmonic and transient excitations as demonstrated by the numerical examples. In addition, a novel, simple, and adaptive integral technique has been discussed for calculating confluent hypergeometric functions arisen from the wave propagation formulation for FGM plates of linear property variation.

3

Free Wave Motion in Anisotropic Laminates

3.1 Introduction

Anisotropic laminates have found wide applications in automobile, aerospace, and many other industries. Impact resistance is one of the major concerns of designers. Hybrid composite materials have been used to improve the energy-absorbing capacity of the laminated plates during wave propagation. The hybrid composite laminate consists of more than one type of composite. Some layers have a higher capacity for resisting static loads, while others have a higher capacity to absorb impact energy. During wave propagation it is desirable for the strain energy to concentrate in layers that have a higher capacity for absorbing impact energy. To achieve this goal, methods for analyzing the distribution of strain energy in the thickness direction of the laminate are a prerequisite.

Several methods are available for analyzing plane waves propagating in unbounded homogeneous layered media and in bounded layered isotropic media, such as the excellent work by Brekhovskikh (1960), Sun, Achenbach, and Herrmann (1968), Murakami and Akiyama (1985), and Thomson (1950), among others.

This chapter formulates an exact method (Liu, Tani, et al., 1990a) for free wave motions in bounded layered anisotropic media that is free of external excitation. It investigates the characteristics of harmonic plane waves in laminates. This class of wave propagation problems is basically three-dimensional (3D) and, therefore, much more complicated compared to the one-dimensional (1D) case discussed in Chapter 1. The procedure is, however, very similar to that of the 1D free wave motion problem. The results obtained for anisotropic laminates are much more interesting. We start with the basic equations for this 3D elasticity. Then we derive the dispersion relations for phase velocities of harmonic plane waves in laminates and formulation for calculating the distribution of strain energy in the thickness direction of the laminates. Last, we present some interesting examples.

3.2   Basic Equations

Consider a laminate that consists of N layers of elastic and anisotropic materials as shown in Figure 3.1. It is assumed that the layers are perfectly bonded, and the material in a layer is homogenous. This assumption holds if the dimension of the inclusions (fibers or particles) is much smaller than the thickness of the layer and the wavelength of the plane waves. At any point in the material, there are three displacement components, u, v, w. Due to the anisotropy of the material, these three displacement components are generally coupled, meaning that all the displacement components are involved in the same set governing system equations. The thickness, the angle of fiber-orientation, and the elastic coefficients of the nth layer are denoted by thickness h_n, ϕ_n, and c_{ijkl}, respectively. The overall thickness of the laminate is H. A rectangular Cartesian coordinate system x_1, x_2, and x_3 is used with x_3 normal to the plane of the laminate defined by x_1 and x_2.

3.2.1   Constitutive Equations

The constitutive relations, which is known as Hook's law in 1D cases, at any point in the nth layer of anisotropic material are given by

$$\sigma_{ij}^{(n)} = c_{ijkl}^{(n)} \varepsilon_{kl}^{(n)} \tag{3.1}$$

where the superscript (n) stands for the nth layer, σ_{ij} and ε_{ij} are the stress and strain tensors, respectively. The indices i and j range over 1, 2, and 3, with repeated indices summed from 1 to 3. This convention of notation has

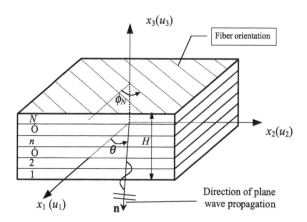

FIGURE 3.1
Plane wave propagating in a laminate and the coordinate system. (From Liu, G. R., et al., *ASME J. Appl. Mech.*, 57, 923, 1990a. With permission.)

been widely used in the formulation in 3D elasticity or solid mechanics for concise equation representation. The fourth order tensor of the elasticity c_{ijkl} satisfies the Green symmetry conditions:

$$c_{ijkl} = c_{klij} = c_{ijlk} = c_{jikl} \tag{3.2}$$

Hence, 21 independent constants are used to describe fully a general anisotropic elastic solid, many more than the two independent constants for isotropic material discussed in Chapter 1.

3.2.2 Equation of Motion

When the solid is free of external body force, the equation of motion or generalized Newton's law is given for any point in the nth layer by

$$\frac{\partial \sigma_{ij}^{(n)}}{\partial x_j} = \rho_n \frac{\partial^2 u_i^{(n)}}{\partial t^2} \tag{3.3}$$

where ρ_n is the mass density of the nth layer, t is the time and $u_i^{(n)}$ is the displacement in the x_i direction. Equation (3.4) is derived in exactly the same way as the 1D dimensional counterpart that is detailed in Chapter 1 by isolating a representative cell, except that the cell in this case is a 3D cube with infinite small dimensions in all x_i directions. Details can be found in any textbook on solid mechanics.

3.2.3 Strain-Displacement Equation

The linear strain-displacement relation for small motion for 3D solids is given by

$$\varepsilon_{ij}^{(n)} = \frac{1}{2}\left(\frac{\partial u_i^{(n)}}{\partial x_j} + \frac{\partial u_j^{(n)}}{\partial x_i}\right) \tag{3.4}$$

The strain-displacement equation is sometimes termed a *kinematics equation* or *geometric equation*, as it gives only the geometric relation between the displacements and the strains at any point in solids.

3.2.4 Continuity Equations Between Layers

Assuming that the layers are perfectly bonded, the continuity conditions of displacements and stresses at the interface between the nth and $(n + 1)$th layers in the stacking sequence shown in Figure 3.1 are

$$\left.\begin{aligned}\mathbf{S}^{(n)}(h_n) &= \mathbf{S}^{(n+1)}(0) \\ \mathbf{D}^{(n)}(h_n) &= \mathbf{D}^{(n+1)}(0)\end{aligned}\right\} \tag{3.5}$$

where

$$
\left.
\begin{aligned}
\mathbf{S}^{(n)}(x_3) &= [\bar{\sigma}_{13} \quad \bar{\sigma}_{23} \quad \bar{\sigma}_{33}]^{(n)} \\
\mathbf{D}^{(n)}(x_3) &= [\bar{u}_1 \quad \bar{u}_2 \quad \bar{u}_3]^{(n)}
\end{aligned}
\right\}
\tag{3.6}
$$

Here the bar designates the amplitudes of the stress and displacement that are functions of x_3. Continuity conditions between layers are also termed *interface conditions*.

3.2.5 Boundary Conditions

Assume the laminate is free of stresses on its surfaces, as we consider only problems of free wave motion. The boundary conditions on the lower surface of the laminate can be expressed as

$$
\mathbf{S}^{(1)}(0) = \mathbf{0}
\tag{3.7}
$$

and that on the upper surfaces are

$$
\mathbf{S}^{(N)}(h_N) = \mathbf{0}
\tag{3.8}
$$

where $\mathbf{0}$ denotes a zero vector.

3.2.6 Wave Equation

When Eqs. (3.1) and (3.4) are substituted into Eq. (3.3), the equation of motion may be expressed in terms of displacements, as follows:

$$
c_{ijkl}^{(n)} \frac{\partial^2 u_k^{(n)}}{\partial x_j \partial x_l} = \rho_n \frac{\partial^2 u_i^{(n)}}{\partial t^2}
\tag{3.9}
$$

which is partial differential equation of 2nd order with constant coefficients. The foregoing equation is the wave equation for 3D solids. All the displacement methods seek solutions to Eq. (3.9) for the displacements, subject to interface and boundary conditions, and then retrieve the strains and stresses after the displacements are obtained.

3.3   Derivation of Dispersion Equation

3.3.1   Bulk Waves in 3D Anisotropic Solids

In our wave propagation problem, we seek for solution to Eq. (3.9) in form of *plane waves* that can be expressed as follows:

$$
u_j^{(n)} = U_j^{(n)} \exp[ik(n_i x_i - ct)]
\tag{3.10}
$$

where U_j ($j = 1, 2, 3$) are the constants related to the amplitudes of displacements, c is the phase velocity, k is the wavenumber of the plane wave, and $n_i = \cos(\bar{x}_i, \bar{n})$ is the direction normal of the plane wave that relates to the direction of wave propagation.

Equation (3.10) can also be rewritten in the form of

$$u_j^{(n)} = U_j^{(n)} \exp[ik(n_i x_i)] \exp(-i\omega t) \tag{3.11}$$

In doing so, relation of $\omega = kc$ has been used. Equation (3.11) clearly indicates that the plane wave is harmonic. Substituting Eq. (3.10) into Eq. (3.9), the time-harmonic terms will drop out, and we obtain

$$\Pi_{ik}^{(n)} U_k^{(n)} = 0 \tag{3.12}$$

where

$$\Pi_{ik}^{(n)} = \Gamma_{ik}^{(n)} - \rho_n c^2 \delta_{ik} \tag{3.13}$$

In Eq. (3.13), δ_{ik} is the Kronecker delta, and Γ_{ik} is the Christoffel stiffness given by

$$\Gamma_{ik}^{(n)} = \Gamma_{ki}^{(n)} = c_{ijkl}^{(n)} n_j n_l \tag{3.14}$$

For Eq. (3.12) to have a nontrivial solution, it is required that

$$\det (\Pi_{ik}^{(n)}) = 0 \tag{3.15}$$

or

$$\begin{vmatrix} \Gamma_{11} - \rho_n c^2 & \Gamma_{12} & \Gamma_{13} \\ \Gamma_{12} & \Gamma_{22} - \rho_n c^2 & \Gamma_{23} \\ \Gamma_{13} & \Gamma_{23} & \Gamma_{33} - \rho_n c^2 \end{vmatrix} = 0 \tag{3.16}$$

The foregoing equation is often called a characteristic equation for waves propagating in 3D anisotropic solids. Equation (3.12) can be rewritten in the matrix form of

$$[\Gamma - \gamma \mathbf{I}]\mathbf{U} = 0 \tag{3.17}$$

where $\mathbf{I}$ is a 3×3 identity matrix, and

$$\gamma = \rho_n c^2 \tag{3.18}$$

$$\mathbf{\Gamma} = \begin{bmatrix} \Gamma_{11} & \Gamma_{12} & \Gamma_{13} \\ \Gamma_{12} & \Gamma_{22} & \Gamma_{23} \\ \Gamma_{13} & \Gamma_{23} & \Gamma_{33} \end{bmatrix} \tag{3.19}$$

$$\mathbf{U} = \begin{Bmatrix} U_1 \\ U_2 \\ U_2 \end{Bmatrix} \tag{3.20}$$

Since matrix $\mathbf{\Gamma}$ is positive definite, we can always obtain three positive eigenvalues γ from Eq. (3.17). From these eigenvalues, we can then obtain three velocities using Eq. (3.18). These three velocities are the velocities of three bulk waves in unbounded 3D anisotropic solids. These bulk waves are not dispersive, similar to the waves in bars discussed in Chapter 1. For an isotropic solid, these three velocities are corresponding to longitudinal wave (P-wave), vertically polarized transverse shear wave (SV-wave), and horizontally polarized transverse shear wave (SH-wave). In addition, the velocities of SV-wave and SH-wave are the same for isotropic materials, and, therefore, there are in fact only two distinct velocities. The velocities of these waves can be easily found from Eq. (3.16). The velocity of the P-wave is

$$c_p = \sqrt{\frac{c_{11}}{\rho}} = \sqrt{\frac{E(1-v)}{\rho(1+v)(1-2v)}} \tag{3.21}$$

The velocity for the SV-wave and SH-wave are given by the same formula of

$$c_{sv} = c_{sh} = c_s = \sqrt{\frac{c_{44}}{\rho}} = \sqrt{\frac{G}{\rho}} \tag{3.22}$$

Details for bulk waves in isotropic solids can be found in Achenbach (1973). For anisotropic materials, there are, in general, three distinct velocities whose values depend on the direction of propagation of the plane wave, n_i. A more detailed discussion on bulk waves in anisotropic solids can be found in Payton (1983).

3.3.2   Lamb Waves in Laminates

We now consider a laminate of finite thickness H with free surfaces, where a harmonic plane wave is propagating within x_1–x_2 plane. Plane waves in plates bounded by two surfaces are termed *Lamb waves*. We first obtain n_3

from Eq. (3.15), for given direction normals of n_1 and n_2. To this end, we write Eq. (3.15) in a polynomial form of n_3 as

$$\det\,(a_{ik} + b_{ik}n_3 + d_{ik}n_3^2) = 0 \tag{3.23}$$

The degree of this polynomial equation of n_3 is six. We can solve it numerically for n_3 and obtain six roots $n_3 = \alpha_l^{(n)}(l = 1,\dots, 6)$. The displacements can now be written as follows:*

$$u_i^{(n)} = \sum_{l=1}^{6} q_{i(l)}^{(n)} A_l^{(n)} \exp(ik\alpha_l^{(n)} x_3)\, \exp[ik(n_1 x_1 + n_2 x_2 - ct)] \tag{3.24}$$

where

$$\left.\begin{aligned}
q_{1(l)}^{(n)} &= 1 \\[4pt]
q_{2(l)}^{(n)} &= \frac{\Gamma_{12}^{(n)}\Gamma_{13}^{(n)} - \Gamma_{23}^{(n)}(\Gamma_{11}^{(n)} - \rho_n c^2)}{\Gamma_{12}^{(n)}\Gamma_{23}^{(n)} - \Gamma_{13}^{(n)}(\Gamma_{22}^{(n)} - \rho_n c^2)} \\[4pt]
q_{3(l)}^{(n)} &= \frac{(\rho_n c^2 - \Gamma_{11}^{(n)})}{\Gamma_{13}^{(n)} - \Gamma_{12}^{(n)}\Gamma_{13}^{(n)} q_{2(l)}^{(n)}} - \frac{\Gamma_{12}^{(n)}}{\Gamma_{13}^{(n)}} q_{2(l)}^{(n)}
\end{aligned}\right\} \tag{3.25}$$

It is easy to verify that Eq. (3.24) satisfies Eq. (3.9). There are six undefined constants in Eq. (3.24). Using Eqs. (3.24) and (3.4), the strain tensors can be expressed as follows:

$$\varepsilon_{jm}^{(n)} = \sum_{l=1}^{6} p_{jm(l)}^{(n)} A_l^{(n)} \exp(ik\alpha_l^{(n)} x_3)\, \exp[ik(n_1 x_1 + n_2 x_2 - ct)] \tag{3.26}$$

where

$$p_{jm(l)}^{(n)} = ik(n_m q_{j(l)}^{(n)} + n_j q_{m(l)}^{(n)})/ 2 \tag{3.27}$$

The stress tensors are

$$\sigma_{jm}^{(n)} = \sum_{l=1}^{6} r_{jm(l)}^{(n)} A_l^{(n)} \exp(ik\alpha_l^{(n)} x_3)\, \exp[ik(n_1 x_1 + n_2 x_2 - ct)] \tag{3.28}$$

---

* A numerically more stable formulation is to divide these six roots into two groups. One relates to waves propagating upward and the other downward. The displacement is then expressed in the form of Eq. (4.27). See Section 4.4 for more details.

where

$$r_{jm(l)}^{(n)} = c_{jmpq}^{(n)} p_{pq}^{(n)} \tag{3.29}$$

Note that in Eqs. (3.27) and (3.29), $n_3 = \alpha_l (l = 1,\ldots,6)$. Equations (3.24) and (3.28) can be written in more concise forms as

$$u_i^{(n)} = \bar{u}_{i(l)}^{(n)} \exp[ik(n_1 x_1 + n_2 x_2 - ct)] \tag{3.30}$$

and

$$\sigma_{i3}^{(n)} = \bar{\sigma}_{i3(l)}^{(n)} \exp[ik(n_1 x_1 + n_2 x_2 - ct)] \tag{3.31}$$

where

$$\bar{u}_i^{(n)} = \sum_{l=1}^{6} q_{i(l)}^{(n)} A_l^{(n)} \exp(ik\alpha_l^{(n)} x_3) \tag{3.32}$$

and

$$\bar{\sigma}_{i3}^{(n)} = \sum_{l=1}^{6} r_{i3(l)}^{(n)} A_l^{(n)} \exp(ik\alpha_l^{(n)} x_3) \tag{3.33}$$

Equations (3.32) and (3.33) can be written in the following matrix form:

$$\left\{ \begin{matrix} \mathbf{S} \\ \mathbf{D} \end{matrix} \right\}^{(n)} = \begin{bmatrix} \mathbf{K}_1 & \mathbf{K}_2 \\ \mathbf{K}_3 & \mathbf{K}_4 \end{bmatrix}^{(n)} \left\{ \begin{matrix} \mathbf{A}_1 \\ \mathbf{A}_2 \end{matrix} \right\}^{(n)} \tag{3.34}$$

where $\mathbf{S}$ and $\mathbf{D}$ are given by Eq. (3.6), $\mathbf{A}_1$ and $\mathbf{A}_2$ are constant vectors defined by

$$\mathbf{A}_1^{(n)} = \left\{ \begin{matrix} A_1 \\ A_2 \\ A_3 \end{matrix} \right\}^{(n)}, \quad \mathbf{A}_2^{(n)} = \left\{ \begin{matrix} A_4 \\ A_5 \\ A_6 \end{matrix} \right\}^{(n)} \tag{3.35}$$

On the upper and lower surfaces of the nth layer, Eq. (3.34) can be expressed as follows.

On the lower surface:

$$\left\{ \begin{matrix} \mathbf{S} \\ \mathbf{D} \end{matrix} \right\}_{x_3=0}^{(n)} = \begin{bmatrix} \mathbf{G}_1 & \mathbf{G}_2 \\ \mathbf{G}_3 & \mathbf{G}_4 \end{bmatrix}^{(n)} \left\{ \begin{matrix} \mathbf{A}_1 \\ \mathbf{A}_2 \end{matrix} \right\}^{(n)} \tag{3.36}$$

On the upper surface:

$$\left\{ \begin{matrix} \mathbf{S} \\ \mathbf{D} \end{matrix} \right\}_{x_3=h_n}^{(n)} = \begin{bmatrix} \mathbf{Q}_1 & \mathbf{Q}_2 \\ \mathbf{Q}_3 & \mathbf{Q}_4 \end{bmatrix}^{(n)} \left\{ \begin{matrix} \mathbf{A}_1 \\ \mathbf{A}_2 \end{matrix} \right\}^{(n)} \tag{3.37}$$

where

$$\left. \begin{aligned} \mathbf{G}_i^{(n)} &= \mathbf{K}_i^{(n)} \big|_{x_3=0} \\ \mathbf{Q}_i^{(n)} &= \mathbf{K}_i^{(n)} \big|_{x_3=h_n} \end{aligned} \right\} \quad (i = 1, 2, 3, 4) \tag{3.38}$$

Note that the above three equations are valid for all the layers ($n = 1, \ldots, N$) in the laminate. We now use the method of matrix transfer to form a system equation for the laminate. For the first layer, $\mathbf{A}_2^{(1)}$ can be expressed in $\mathbf{A}_1^{(1)}$ with the help of Eqs. (3.7) and (3.36)

$$\mathbf{A}_2^{(1)} = -(\mathbf{G}_2^{(1)})^{-1} \mathbf{G}_1^{(1)} \mathbf{A}_1^{(1)} \tag{3.39}$$

Substituting Eq. (3.39) into Eq. (3.37), $\mathbf{S}^{(1)}(h_1)$ and $\mathbf{D}^{(1)}(h_1)$ can be also written in $\mathbf{A}_1^{(1)}$

$$\mathbf{S}^{(1)}(h_1) = \mathbf{Z}_s^{(1)} \mathbf{A}_1^{(1)} \tag{3.40}$$

and

$$\mathbf{D}^{(1)}(h_1) = \mathbf{Z}_d^{(1)} \mathbf{A}_1^{(1)} \tag{3.41}$$

where, in Eq. (3.40)

$$\mathbf{Z}_s^{(1)} = \mathbf{Q}_1^{(1)} - \mathbf{Q}_2^{(1)} (\mathbf{G}_2^{(1)})^{-1} \mathbf{G}_1^{(1)} \tag{3.42}$$

and in Eq. (3.41),

$$\mathbf{Z}_d^{(1)} = \mathbf{Q}_3^{(1)} - \mathbf{Q}_4^{(1)} (\mathbf{G}_2^{(1)})^{-1} \mathbf{G}_1^{(1)} \tag{3.43}$$

Imposing the interface conditions on both displacement and stresses at the interface between the first and second layers, we obtain

$$\mathbf{G}_1^{(2)} \mathbf{A}_1^{(2)} + \mathbf{G}_2^{(2)} \mathbf{A}_2^{(2)} = \mathbf{Z}_s^{(1)} \mathbf{A}_1^{(1)} \tag{3.44}$$

and

$$\mathbf{G}_3^{(2)}\mathbf{A}_1^{(2)} + \mathbf{G}_4^{(2)}\mathbf{A}_2^{(2)} = \mathbf{Z}_d^{(1)}\mathbf{A}_1^{(1)} \tag{3.45}$$

From Eqs. (3.44) and (3.45), the constants of the second layer $\mathbf{A}_1^{(2)}$ and $\mathbf{A}_2^{(2)}$ can be written in terms of $\mathbf{A}_1^{(1)}$, and, with Eq. (3.37), $\mathbf{S}^{(2)}(h_2)$ and $\mathbf{D}^{(2)}(h_2)$ can also be written in $\mathbf{A}_1^{(1)}$ as

$$\mathbf{S}^{(2)}(h_2) = \mathbf{Z}_s^{(2)}\mathbf{A}_1^{(1)} \tag{3.46}$$

and

$$\mathbf{D}^{(2)}(h_2) = \mathbf{Z}_d^{(1)}\mathbf{A}_1^{(1)} \tag{3.47}$$

Applying the continuity condition at the interface between the second and third layers, $\mathbf{S}^{(3)}(h_3)$ and $\mathbf{D}^{(3)}(h_3)$ can also be expressed by $\mathbf{A}_1^{(1)}$. Continuing this procedure, $\mathbf{S}^{(N)}(h_N)$ and $\mathbf{D}^{(N)}(h_N)$ of the Nth layer can be expressed finally by $\mathbf{A}_1^{(1)}$

$$\mathbf{S}^{(N)}(h_N) = \mathbf{Z}_s^{(N)}\mathbf{A}_1^{(1)} \tag{3.48}$$

and

$$\mathbf{D}^{(N)}(h_N) = \mathbf{Z}_d^{(N)}\mathbf{A}_1^{(1)} \tag{3.49}$$

With the boundary condition on the upper surface of the laminate given by Eq. (3.8), we have

$$\mathbf{Z}_s^{(N)}\mathbf{A}_1^{(1)} = 0 \tag{3.50}$$

For a nontrivial solution, the determinant of $\mathbf{Z}_s^{(N)}$ must vanish

$$\det(\mathbf{Z}_s^{(N)}) = 0 \tag{3.51}$$

In general, $\mathbf{Z}_s^{(N)}$ is a complex matrix and Eq. (3.51) means that the absolute value of the determinate of $\mathbf{Z}_s^{(N)}$ must be equal to zero. The determinant of $\mathbf{Z}_s^{(N)}$ is an implicit relation between phase velocity c and wavenumber k. When the wavenumber k is given, the phase velocities can be obtained from Eq. (3.51). This implies that the velocity of the plane wave propagating in the plane of the laminate depends on the wavenumber (or the frequency) of the wave. Consequently, Eq. (3.51) gives the dispersion relations of the phase velocities of plane waves in the laminate. Note that Eq. (3.51) is very complicated; it is not possible to solve it for c analytically in an explicit form, and, hence, numerical root searching tools have to be used to search for c for a given k.

The above procedure from Eq. (3.39) to (3.51) might appear slightly different from the standard matrix transfer method used for vibration analysis, but they are the same in principle.

3.4 Strain Energy Distribution

For any c and k that satisfy Eq. (3.51), the vector $\mathbf{A}_1^{(1)}$ can be determined. By backward substitution, all $\mathbf{A}_1^{(n)}$ and $\mathbf{A}_2^{(n)}(n = 1,\ldots,N)$ can be obtained progressively. The displacement as well as stress and strain tensors can be accordingly determined. These results give the mechanical status in the laminate when plane waves propagate with a wavenumber of k. By selecting the real parts of the results (corresponding cosine form of harmonic excitation), the stresses and strains can be written as follows:

$$[\sigma_{11} \ \sigma_{22} \ \sigma_{33} \ \sigma_{12} \ \varepsilon_{11} \ \varepsilon_{22} \ \varepsilon_{33} \ \varepsilon_{12}]^{T}$$
$$= [\bar{\sigma}_{11} \ \bar{\sigma}_{22} \ \bar{\sigma}_{33} \ \bar{\sigma}_{12} \ \bar{\varepsilon}_{11} \ \bar{\varepsilon}_{22} \ \bar{\varepsilon}_{33} \ \bar{\varepsilon}_{12}]^{T} \cos k\,(n_1 x_1 + n_2 x_2 - ct) \quad (3.52)$$

and

$$[\sigma_{13} \ \sigma_{23} \ \varepsilon_{13} \ \varepsilon_{23}]^{T} = [\bar{\sigma}_{13} \ \bar{\sigma}_{23} \ \bar{\varepsilon}_{13} \ \bar{\varepsilon}_{23}]^{T} \sin k\,(n_1 x_1 + n_2 x_2 - ct) \quad (3.53)$$

Selecting the real part corresponds to a harmonic excitation in cosine form. The strain energy per unit volume in the laminate is defined as

$$E = (\sigma_{11}\varepsilon_{11} + \sigma_{22}\varepsilon_{22} + \sigma_{33}\varepsilon_{33} + 2\sigma_{12}\varepsilon_{12} + 2\sigma_{13}\varepsilon_{13} + 2\sigma_{23}\varepsilon_{23})/2 \quad (3.54)$$

Substituting Eqs. (3.52) and (3.53) into Eq. (3.54), we obtain

$$E = E_c \cos^2 k\,(n_1 x_1 + n_2 x_2 - ct) + E_s \sin^2 k(n_1 x_1 + n_2 x_2 - ct) \quad (3.55)$$

where

$$E_c = (\bar{\sigma}_{11}\bar{\varepsilon}_{11} + \bar{\sigma}_{22}\bar{\varepsilon}_{22} + \bar{\sigma}_{33}\bar{\varepsilon}_{33} + 2\bar{\sigma}_{12}\bar{\varepsilon}_{12})/2 \quad (3.56)$$

and

$$E_s = \sigma_{13}\varepsilon_{13} + \sigma_{23}\varepsilon_{23} \quad (3.57)$$

Using the above three equations, the strain energy distribution in laminates when plane wave is propagating can be computed.

3.5 Examples

In this section, we use the formulae described above to calculate the phase velocity spectra for a unidirectional graphite/epoxy laminate, an angle-ply graphite/epoxy laminate, and a hybrid laminate that consists of carbon/epoxy and glass/epoxy layers. The material properties are shown in Table 3.1 (Murakami and Akiyama, 1985; Takahashi and Chou, 1987). Each layer of these laminates is made of fiber-reinforced composite. In general, it is regarded as transverse isotropic material. Hence, we have

$$E_2 = E_3, \quad v_{21} = v_{31}, \quad v_{12} = v_{13}, \quad v_{23} = v_{32} \tag{3.58}$$

The elastic coefficients c_{ijkl} can be calculated from the moduli shown in Table 3.1 and the fiber-orientation angle ϕ_n using standard formulae given in Vinson and Sierakowski (1987) or Jones (1975).

For an infinite orthotropic body, there are P-wave, SV-wave, and SH-wave. Similarly, the three fundamental modes of Lamb waves are named LP, LSV, and LSH, respectively. Let c_0 be the phase velocity of the P-wave in the strongest layer:

$$c_0 = \sqrt{c_{1111(0)}/\rho_0} \tag{3.59}$$

where $c_{1111(0)}$ is the elastic coefficient of the strongest layer with the fiber-orientation angle equal to zero and ρ_0 is the mass density of this layer.

3.5.1 Dispersion and Anisotropy of Phase Velocities

The phase velocity dispersion curves for Lamb waves in a unidirectional graphite fiber-reinforced composite material are shown in Figure 3.2. When the wave propagates along the principal axis of the material (for $\theta = 0$ and 90 degrees), the LSH wave has no dispersion. The phase velocity of the LSH wave is constant and equals that of the SH-wave. Otherwise ($\theta = 45$ degrees), the LSH wave shows weak dispersion. For the LP wave, the closer the direction of the wave propagation to the fiber-orientation, the stronger the dispersion becomes. When the wave propagates along the direction of fiber-orientation,

TABLE 3.1

Properties of Materials

Material	E_1(GPa)	E_2(GPa)	G_{12}(GPa)	v_{12}	v_{23}	ρ(g/cm^2)
Graphite/epoxy	30.00	0.750	0.375	0.2500	0.2500	1.90
Carbon/epoxy	142.17	9.255	4.795	0.3340	0.4862	1.90
Glass/epoxy	38.49	9.367	3.414	0.2912	0.5071	2.66
Steel	77.4	77.4	29.025	0.3333	0.3333	7.90

Source: Adapted from Liu, G. R. et al., *ASME J. Appl. Mech.,* 57, 923, 1990a. With permission.

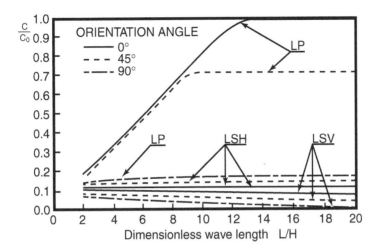

FIGURE 3.2
Phase velocity spectra for unidirectional graphite/epoxy laminates ($\theta = 0$). (From Liu, G. R., et al., *ASME J. Appl. Mech.*, 57, 923, 1990a. With permission.)

the strongest dispersion is observed. For large wavelength ($L > 13H$), the phase velocity of the LP wave approaches that of the P-wave. For the LSV wave, the dispersion becomes weaker when the direction of the wave propagation is closer to the fiber-orientation. When the dimensionless L/H approaches zero, the phase velocity of the LSV wave approaches that of the Rayleigh wave that is observed on the surface of a half space. Because the wavelength is much smaller than the thickness of the laminate, the surface of the laminate behaves like the surface of a half space.

For symmetrically stacked angle-ply laminates [15/−15]s, [30/−30]s, and [45/−45]s which consist of four layers of graphite/epoxy composite, the dispersion occurs for all the fundamental modes of waves, as shown in Figure 3.3. The waves propagate along the x axis. For the LP wave, the dispersion is stronger for a smaller fiber-orientation angle. For the LSH wave, however, the smaller the fiber-orientation angle, the weaker the dispersion becomes.

The dispersion of the hybrid composite laminate that consists of two types of composites of carbon/epoxy and glass/epoxy layers is shown in Figure 3.4. The wave propagates along the x-axis. To achieve strength under static loads, the stronger carbon/epoxy layers are laid on the surfaces of the laminate. To improve the impact strength, glass/epoxy layers with better capability of absorbing the impact energy are laid in the inner part of the laminate. In Figure 3.4, weak dispersion can be observed for the LSH wave and strong dispersion for the LP wave when $L/H < 10$.

Figures 3.2–3.4 show that for the LSV wave, for all of the laminates, the phase velocity decreases with increasing wavelength L/H.

In order to examine the anisotropic feature of the plane waves, phase-velocity of plane waves propagating at various directions ($0 < \theta < 90$ degrees) are computed for the unidirectional laminate. The results are plotted in

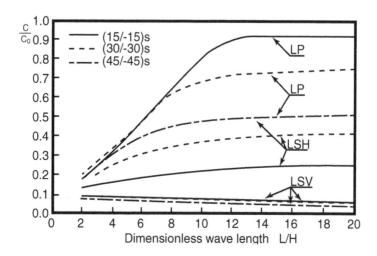

FIGURE 3.3
Phase velocity spectra for symmetric angle-ply graphite/epoxy laminates ($\theta = 0$). (From Liu, G. R., et al., *ASME J. Appl. Mech.*, 57, 923, 1990a. With permission.)

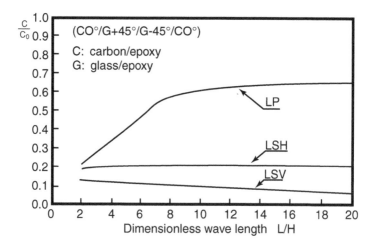

FIGURE 3.4
Phase velocity spectra for hybrid laminate $[C0/G45/G–45/C0]_s$ ($\theta = 0$). (From Liu, G. R., et al., *ASME J. Appl. Mech.*, 57, 923, 1990a. With permission.)

Figure 3.5 for dimensionless wavelengths L/H = 4, 8, and 19. Figure 3.5 shows strong anisotropy for all of the modes of Lamb waves. Since the dispersion of the LSH wave is weak (see Figure 3.2), the curves for L/H = 4, 8, and 19 are too close to be distinguished from each other in this figure. When L/H = 19 (solid line), the shapes of the curves for the LP and LSH waves are close to those given by Murakami and Akiyama (1985), but the values are smaller. Because the laminate discussed by Murakami and Akiyama

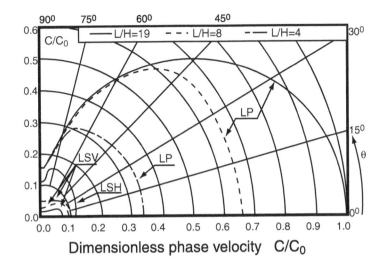

FIGURE 3.5
Phase velocity contours for a unidirectional graphite/epoxy laminate. (From Liu, G. R., et al., *ASME J. Appl. Mech.*, 57, 923, 1990a. With permission.)

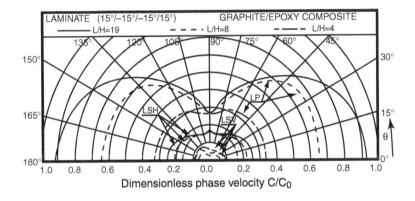

FIGURE 3.6
Phase velocity contours for a ±15° symmetric angle-ply laminate. (From Liu, G. R., et al., *ASME J. Appl. Mech.*, 57, 923, 1990a. With permission.)

was unbounded, the waves were found are nondispersive. When $L/H \to \infty$, the curves for the LP and LSH waves will approach those obtained by Murakami and Akiyama.

Figures 3.6–3.8 show the anisotropy of the phase velocities for the symmetric angle-ply laminates [15/−15]s, [30/−30]s, and [45/−45]s. One angle-ply laminate considered here corresponds to one cell of the unbounded angle-ply laminate discussed by Murakami and Akiyama. Comparing the unbounded laminate with that of finite thickness, it can be found that the curves

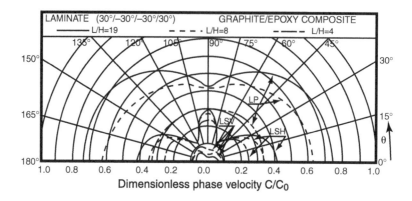

FIGURE 3.7

Phase velocity contours for a ±30° symmetric angle-ply laminate. (From Liu, G. R., et al., *ASME J. Appl. Mech.*, 57, 923, 1990a. With permission.)

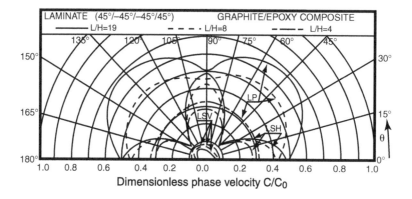

FIGURE 3.8

Phase velocity contours for a ±45° symmetric angle-ply laminate. (From Liu, G. R., et al., *ASME J. Appl. Mech.*, 57, 923, 1990a. With permission.)

of the finite thickness laminate for $L/H = 19$ are close to those of the unbounded laminate for $L/H = \infty$. The phase velocity of the finite thickness laminate is smaller than that of unbounded laminate. The smaller the fiber-orientation angle, the greater is the difference of phase velocity. For the LP wave in laminate [15/−15]s, the difference is about 21% when $L/H = 8$, and about 26% when $L/H = 4$. This implies that the dispersion of the finite laminate is much stronger than that of the unbounded one when the wavelength is short and the number of layers is small.

For the LSV wave in finite thickness laminates, the phase velocities have obviously both anisotropy and dispersion (see Figures 3.5–3.9), but neither anisotropy nor dispersion in unbounded laminate were observed.

Also note that the phase velocities of waves change in the range of $\theta = 0$ to 180 degrees. This means that there is only one symmetric (or antisymmetric)

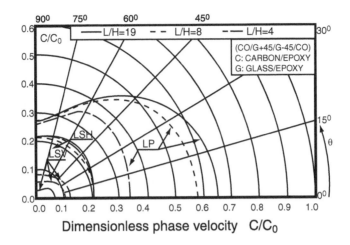

FIGURE 3.9

Phase velocity contour for hybrid laminate [C0/G45/G–45/C0]$_s$. (From Liu, G. R., et al., *ASME J. Appl. Mech.*, 57, 923, 1990a. With permission.)

axis for the phase velocities contours, as shown in Figures 3.6–3.8. However, for the unbounded angle-ply laminate, there are two symmetric or antisymmetric axes, and the phase velocities change only in the range of $\theta = 0$ to 90 degrees.

The anisotropy of the phase velocities of the hybrid composite laminate that consists of carbon/epoxy and glass/epoxy layers are shown in Figure 3.9. Since the stacking sequence of the laminate is antisymmetric, the phase velocities change in the range $\theta = 0$ to 90 degrees. The phase velocities of the LSH wave have both weak dispersion and weak anisotropy.

Figures 3.5–3.9 also show that when the wavelength is large, the phase velocities of the LP and LSH waves depend on the direction of wave propagation, and, when the wavelength is small, the dispersion becomes strong and the anisotropy becomes weak. It seems to suggest that the dispersion weakens the anisotropy of phase velocities of the LP and LSH waves for small wavelength.

In finite thickness laminates, there are an infinite number of wave modes. As an example, the phase velocity spectra for the lower six modes are shown in Figure 3.10 for the graphite/epoxy angle-ply laminate [30/−30]s. The dispersion curves corresponding to three antisymmetric modes plotted by dot lines, and three symmetric modes plotted by solid lines.

3.5.2 Strain Energy Distribution

Using Eqs. (3.56) and (3.57), the distributions of the strain energies E_c and E_s are calculated and shown in Figures 3.11–3.13. For the LP wave, E_c in the laminate [0]$_4$ concentrates within the surface layer. For quasi P-, quasi SV-, and quasi SH-waves in the hybrid laminate [C0/G0/G0/C0] (where letters

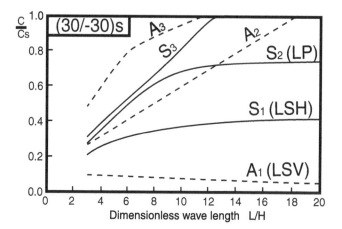

FIGURE 3.10
Phase velocity spectra of higher modes for a ±30° symmetric angle-ply laminate. (From Liu, G. R., et al., *ASME J. Appl. Mech.*, 57, 923, 1990a. With permission.)

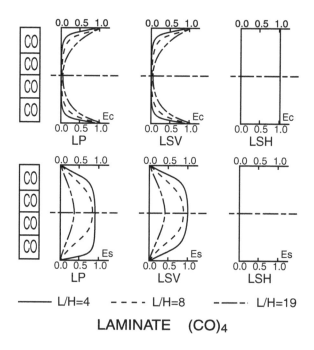

FIGURE 3.11
Distributions of strain energy amplitudes in the thickness direction of a unidirectional carbon/epoxy laminate. (From Liu, G. R., et al., *ASME J. Appl. Mech.*, 57, 923, 1990a. With permission.)

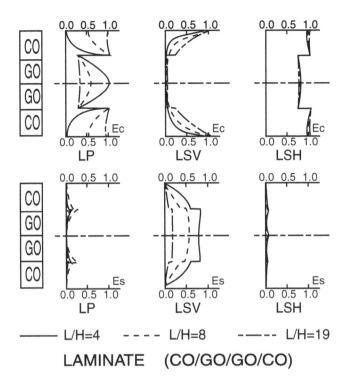

FIGURE 3.12
Distributions of strain energy amplitudes in the thickness direction of a hybrid composite laminate. (From Liu, G. R., et al., *ASME J. Appl. Mech.*, 57, 923, 1990a. With permission.)

C and G denote carbon/epoxy and glass/epoxy layer, respectively, and 0 indicates $\phi = 0$), concentrates within the inner glass/epoxy layers have better capability of absorbing impact energy. The value of E_s in this hybrid laminate is very small compared to E_c. For the LSV wave, E_c concentrates within the surface layers in both laminates $[0]_4$ and [C0/G0/G0/C0], but E_s in the laminate [C0/G0/G0/C0/] concentrates within the two inner layers more than that in laminate $[0]_4$. This is preferred. For the LSH wave, on the other hand, E_s is very small in both laminates $[0]_4$ and [C0/G0/G0/C0], and E_c in the laminate $[0]_4$ is uniform in the thickness direction of the laminate, but, in the laminate [C0/G0/G0/C0], the inner layers contain smaller strain energy than the surface layers. This is not preferable.

Next, the fiber orientation of the inner layers is changed to [C0/G+45/G–45/C0] and the energy distribution in the thickness direction of the laminate is calculated. The results are shown in Figure 3.13. Not only for the LP and LSV waves but also for the LSH wave, the strain energy concentrates within the inner layers. This laminate may possess better capability for absorbing impact energy.

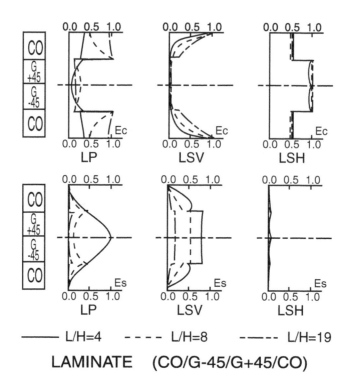

FIGURE 3.13
Distributions of strain energy amplitudes in the thickness direction of a hybrid composite laminate. (From Liu, G. R., et al., *ASME J. Appl. Mech.*, 57, 923, 1990a. With permission.)

It is well known that energy tends to concentrate in the soft layers during wave propagation. However, it is not true for all modes of waves. The strain energies of the LP and LSV waves concentrate in the inner softer layers. On the other hand, the strain energy of the LSH wave concentrates in the inner part where the harder layers are laid. For example, the two inner layers of the laminate $[0]_4$ are harder than those in the laminate $[C0/G0/G0/C0]$, and the energy (E_c) in the two inner layers of the laminate $[0]_4$ is larger than that in laminate $[C0/G0/G0/C0]$. Comparing laminate $[C0/G+45/G-45/C0]$ with laminate $[0]_4$, the same phenomenon can be observed. Since the shear module of the inner layers $[G+45/G-45]$ in the laminate $[C0/G+45/G-45/C0]$ is greater than that of the inner layers $[C0/C0]$ in the laminate $[0]_4$, the strain energy (E_c) that concentrates in the two inner layers of the laminate $[C0/G+45/G-45/C0]$ is more than that of the laminate $[0]_4$. This means that the strain energy of the LSH wave concentrates in the harder layers.

Note that the formulation presented in this chapter has limitations in computing results for very small wavelengths because of numerical truncation errors in the process of searching for the roots of Eq. (3.51). This is because of the use of Eq. (3.24). This can be improved by techniques mentioned in Section 4.4, namely, by writing Eq. (3.24) in the form of Eq. (4.27).

3.6 Remarks

An exact formulation is introduced and used to reveal two very important features of waves in anisotropic laminates: anisotropy and dispersion. It has been found that the phase velocities of Lamb waves in anisotropic laminates are anisotropic, and change in the range of $0 < \theta < 180$ degrees. This anisotropic feature exists for waves of all wavelengths. Moreover, Lamb waves in laminates are dispersive. Stronger dispersion is observed for waves of smaller wavelength.

By analyzing the distribution of strain energy in the thickness direction of laminates, it is possible to determine better stacking sequence of layers and the fiber orientation for each layer in designing a laminate with better capacity to withstand the energy of waves propagating in the laminate.

4

Forced Wave Motion in Composite Laminates

4.1 Introduction

The previous chapter examined Lamb waves in anisotropic composite laminates and revealed two important features of Lamb waves: dispersion and anisotropy. This chapter introduces an exact approach for analyzing harmonic and transient waves in composite laminates that was initially proposed by Liu, Lam, and Tani (1995a).

Methods for analyzing transient wave fields in a plate can be roughly divided into three categories. The first category includes methods based on plate theories (see, for example, Moon, 1973; Sun and Tan, 1984). These methods are effective for calculating responses of anisotropic plates subjected to loads of low frequency, at which the dominant wavelength is much larger than the thickness of the plate. The second category includes finite element type methods, such as semi-numerical methods (SNM), hybrid numerical methods (HNM), and strip element methods (SEM), in which the laminates are divided into layer elements. The SNM was originally proposed by Waas (1972) for layered isotropic structures, extended later by Seale and Kausel (1989) for orthotropic layered half spaces, and by Liu and Achenbach (1994, 1995) for general anisotropic laminated plates. The SNM is very effective for computing forced wave motion in the frequency domain and is covered in Chapter 12 in detail. The HNM suggested by Liu and Tani et al. (1991, 1994) and modified by Liu and Lam (1997) is very efficient for computing transient responses of anisotropic laminates in the time domain. The HNM and its application is covered in Chapters 8–10. The SEM proposed by Liu and Achenbach (1994, 1995) can also be used for analyzing frequency and transient response of anisotropic laminates with or without cracks/flaws. In the SEM, the SNM is employed to obtain the particular solution for the system equation (see Chapter 12). These finite element type methods can be used for much higher frequencies than methods based on plate theories of bending. For excitations of higher frequencies, more elements have to be used. The number of the elements needed is roughly proportional to the frequency of external excitation or wave source.

The third category includes exact methods in which the equation of motion subjected to boundary conditions are solved without any assumptions of mechanics. The difficulties encountered in these methods are in the evaluation of the inverse Fourier integrations to obtain solutions in the spatial domain for a given frequency, because of the complexity of the integrand and the presence of poles on the integration axis. For many cases, searching the poles could be a very computationally extensive task. In overcoming these difficulties, several techniques have been presented. The first technique is to introduce material dissipation so that the poles are shifted off the integration axis. Therefore, the remaining problem is to deal with the integrations of irregular integrands without knowing the location of the poles *a priori*. Xu and Mal (1987) have suggested a scheme to evaluate this kind of irregular integrations and to obtain the Green's function for layered isotropic media. Mal and Lih (1992) have proposed another scheme for evaluating two-dimensional irregular integrations to obtain responses of unidirectional composite plates. The second technique to evaluate the inverse Fourier integration is to remove the poles located on the integration axis as suggested by Kundu and Mal (1985) for layered isotropic media. The third technique is to use complex integral paths proposed by Liu and Tani (1994) for anisotropic gradient piezoelectric plates. The complex-paths approach can be used for both wavenumber-spatial and frequency-time transformation. There are also other exact methods to obtain responses directly in the time domain, such as generalized rays (see Ceranoglu and Pao, 1981; van der Hijden, 1987), normal modes (Weaver and Pao, 1982), and integral transform methods (Green, 1991). One of the advantages of the exact methods over the methods based on plate theories and finite element type methods is the feasibility for computing responses at high frequencies without increasing the dimensions of the system equation. Moreover, the exact solution can be used as a basis for benchmarking the results obtained by approximate methods.

In this chapter, an exact method proposed by Liu, Lam, and Tani (1995a) is introduced to obtain both the frequency and time responses of a general anisotropic laminated plate subject to line loads. First, a matrix formulation for an anisotropic laminate is presented to obtain the response in the wavenumber domain. A complex integral path is then used together with a quadrature scheme for evaluating inverse Fourier integrations to compute responses in the frequency domain. Results in the frequency domain obtained by the present method are compared with those obtained by the SNM for a hybrid-laminate of six anisotropic layers. Responses of the laminate in the time domain are computed by frequency-time transform and compared with those obtained from the HNM.

4.2   Basic Equations

Consider a laminate made of N layers of homogenous elastic materials, as shown in Figure 4.1. It is assumed that the layers are anisotropic and perfectly bonded with each other. The overall thickness of the laminate is denoted by H, and the thickness of the nth layer denoted by h_n. The laminate is subjected

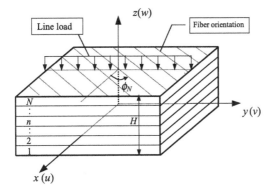

FIGURE 4.1
Coordinate system for a laminate with N layers of anisotropic materials perfectly bonded together. (From Liu et al., *Mechanics of Composite Materials and Structures*, 2, 227, 1995a. With permission.)

to a line load that is independent of the y-axis. Hence, the problem is two-dimensional (2D) and the field variables are independent of the y-axis. Due the anisotropy of the material, the three displacement components, u, v, w are in general coupled. Hence the degree of freedom at any point is still three for our 2D problem. Problems in the frequency domain are considered in this section, but the time harmonic term $\exp(-i\omega t)$, where ω is the angular frequency, has been suppressed for simplicity.

4.2.1 Strain-Displacement Relation

The strain-displacement relations are given in matrix form by

$$\boldsymbol{\varepsilon} = \mathbf{L}\mathbf{U} \tag{4.1}$$

where

$$\boldsymbol{\varepsilon} = \begin{Bmatrix} \varepsilon_{xx} \\ \varepsilon_{yy} \\ \varepsilon_{zz} \\ \gamma_{yz} \\ \gamma_{xz} \\ \gamma_{xy} \end{Bmatrix} = \{\varepsilon_x \quad \varepsilon_y \quad \varepsilon_z \quad \gamma_{yz} \quad \gamma_{xz} \quad \gamma_{xy}\}^T \tag{4.2}$$

is the vector of strains, and

$$\mathbf{u} = \begin{Bmatrix} u \\ v \\ w \end{Bmatrix} = \{u \quad v \quad w\}^T \tag{4.3}$$

is the vector of displacements. Here u, v, and w are the displacement components in the x-, y-, and z- directions, respectively (see Figure 4.1). In Eqs. (4.2) and (4.3), the superscript T stands for the transpose. The differential operator matrix $\mathbf{L}$ is given by

$$\mathbf{L} = \begin{bmatrix} \dfrac{\partial}{\partial x} & 0 & 0 & 0 & \dfrac{\partial}{\partial z} & \dfrac{\partial}{\partial y} \\[2mm] 0 & \dfrac{\partial}{\partial y} & 0 & \dfrac{\partial}{\partial z} & 0 & \dfrac{\partial}{\partial x} \\[2mm] 0 & 0 & \dfrac{\partial}{\partial z} & \dfrac{\partial}{\partial y} & \dfrac{\partial}{\partial x} & 0 \end{bmatrix}^{T} \tag{4.4}$$

Equation (4.1) is a matrix representation of Eq. (3.4). The operator matrix $\mathbf{L}$ can be rewritten in a concise form of

$$\mathbf{L} = \mathbf{L}_x \frac{\partial}{\partial x} + \mathbf{L}_y \frac{\partial}{\partial y} + \mathbf{L}_z \frac{\partial}{\partial z} \tag{4.5}$$

where and the matrices $\mathbf{L}_x$, $\mathbf{L}_y$, and $\mathbf{L}_z$ are constant matrices given by

$$\mathbf{L}_x = \begin{bmatrix} 1 & 0 & 0 & 0 & 0 & 0 \\ 0 & 0 & 0 & 0 & 0 & 1 \\ 0 & 0 & 0 & 0 & 1 & 0 \end{bmatrix}^{T} \tag{4.6}$$

$$\mathbf{L}_y = \begin{bmatrix} 0 & 0 & 0 & 0 & 0 & 1 \\ 0 & 1 & 0 & 0 & 0 & 0 \\ 0 & 0 & 0 & 1 & 0 & 0 \end{bmatrix}^{T} \tag{4.7}$$

$$\mathbf{L}_z = \begin{bmatrix} 0 & 0 & 0 & 0 & 1 & 0 \\ 0 & 0 & 0 & 1 & 0 & 0 \\ 0 & 0 & 1 & 0 & 0 & 0 \end{bmatrix}^{T} \tag{4.8}$$

4.2.2 Stress-Strain Relation

The stress-strain relations are given also in matrix form by

$$\boldsymbol{\sigma} = c\boldsymbol{\varepsilon} \tag{4.9}$$

where

$$\sigma = \begin{Bmatrix} \sigma_{xx} \\ \sigma_{yy} \\ \sigma_{zz} \\ \tau_{yz} \\ \tau_{xz} \\ \tau_{xy} \end{Bmatrix} = \{ \sigma_x \ \sigma_y \ \sigma_z \ \tau_{yz} \ \tau_{xz} \ \tau_{xy} \}^T \qquad (4.10)$$

is the vector of stresses, and

$$c = \begin{bmatrix} c_{11} & c_{12} & \cdots & c_{16} \\ c_{12} & c_{22} & \cdots & c_{26} \\ \vdots & \vdots & \ddots & \vdots \\ c_{16} & c_{26} & \cdots & c_{66} \end{bmatrix} \qquad (4.11)$$

is the matrix of the off-principal-axis stiffness coefficients of the lamina whose expressions in terms of engineering constants can be found in Vinson and Sierakowski (1987).

4.2.3 Equation of Motion

The equation of motion or dynamic equilibrium equation for the laminate in the absence of body force is given in the matrix form of

$$\rho \ddot{U} - L^T \sigma = 0 \qquad (4.12)$$

where the dot represents the differentiation with respect to time and ρ is the mass density of the material. Substituting Eq. (4.9) into Eq. (4.12), we obtain the equation of motion in terms of displacement,

$$\rho \ddot{U} - L^T c L U = 0 \qquad (4.13)$$

which is just a matrix form representation of Eq. (3.9).

The stresses on a plane $z = $ constant can be written as

$$R = \begin{Bmatrix} \sigma_{xz} \\ \sigma_{yz} \\ \sigma_{zz} \end{Bmatrix} = \{ \sigma_{xz} \ \sigma_{yz} \ \sigma_{zz} \}^T = L_z^T c L U \qquad (4.14)$$

4.3   Boundary and Interface Conditions

We assume that the laminate may be loaded on all the $(N - 1)$ interfaces including its two surfaces. Hence, the external force can be written in a general form of

$$\mathbf{T}(x) = \{\mathbf{T}_1(x) \quad \mathbf{T}_2(x) \quad \cdots \quad \mathbf{T}_N(x) \quad \mathbf{T}_{N+1}(x)\}^T \tag{4.15}$$

where $\mathbf{T}_j = \{q_x \; q_y \; q_z\}_j^T$ is the external force vector acting on the jth interface, while $j = 1$ is for the lower surface and $j = N + 1$ is for the upper surface of the laminate. q_x, q_y, and q_z are the external force components in the x, y, and z directions, respectively. The boundary condition for the laminate can be written as follows. On the lower surface of the laminate,

$$-\mathbf{R}_1^L = \mathbf{T}_1 \tag{4.16}$$

At the interfaces,

$$\mathbf{R}_n^U - \mathbf{R}_{n+1}^L = \mathbf{T}_{n+1}, \quad \mathbf{U}_{n-1}^U = \mathbf{U}_n^L, \quad \text{for } 1 \le n \le (N-1) \tag{4.17}$$

On the upper surface of the laminate,

$$\mathbf{R}_N^U = \mathbf{T}_{N+1} \tag{4.18}$$

In Eqs. (4.16) to (4.18), the subscripts in $\mathbf{R}$ and $\mathbf{U}$ indicate the layer numbers, and the superscripts U and L stand, respectively, for the upper and lower surfaces of the layer.

4.4   Displacement in the Wavenumber Domain

We introduce the Fourier transform of spatial-wavenumber with respect to the horizontal spatial coordinate x as follows.

$$\tilde{\mathbf{U}}(z, k) = \int_{-\infty}^{\infty} \mathbf{U}(z, x) e^{ikx} \, dx \tag{4.19}$$

where $i = \sqrt{-1}$, and k is the variable of wavenumber with respect to x. The application of the Fourier transform to Eq. (4.13) leads to the following equation in the wavenumber domain.

$$\rho\tilde{\mathbf{U}} + k^2\mathbf{D}_{xx}\tilde{\mathbf{U}} + i2k\mathbf{D}_{xz}\frac{\partial\tilde{\mathbf{U}}}{\partial z} - \mathbf{D}_{zz}\frac{\partial^2\tilde{\mathbf{U}}}{\partial z^2} = 0 \tag{4.20}$$

in which

$$\mathbf{D}_{xx} = \begin{bmatrix} c_{11} & c_{16} & c_{15} \\ & c_{66} & c_{56} \\ \text{sym.} & & c_{55} \end{bmatrix} \tag{4.21}$$

$$\mathbf{D}_{xz} = \frac{1}{2}\begin{bmatrix} 2c_{15} & c_{14}+c_{56} & c_{13}+c_{55} \\ & 2c_{46} & c_{45}+c_{36} \\ \text{sym.} & & 2c_{35} \end{bmatrix} \tag{4.22}$$

$$\mathbf{D}_{zz} = \begin{bmatrix} c_{55} & c_{45} & c_{35} \\ & c_{44} & c_{34} \\ \text{sym.} & & c_{33} \end{bmatrix} \tag{4.23}$$

are matrices of material constants. We assume that the solution of Eq. (4.20) has the form of

$$\tilde{\mathbf{U}} = \mathbf{d}\,\exp(i\zeta z) \tag{4.24}$$

where $\mathbf{d}$ is the amplitude of the displacement in the wavenumber domain, and ζ is the variable of wavenumber with respect to z. Substituting Eq. (4.24) into (4.20), we obtain

$$[(\rho\omega^2\mathbf{I} - k^2\mathbf{D}_{xx}) + \zeta(2k\mathbf{D}_{xz}) - \zeta^2\mathbf{D}_{zz}]\mathbf{d} = 0 \tag{4.25}$$

where $\mathbf{I}$ is a 3×3 identity matrix. In deriving (4.25), the time dependence $\exp(-i\omega t)$ has been involved. Equation (4.25) can be changed to be the following regular eigenvalue equation (for example, see Dong, 1977).

$$\begin{bmatrix} 0 & \mathbf{I} \\ \mathbf{D}_{zz}^{-1}(\rho\omega^2\mathbf{I} - k^2\mathbf{D}_{xx}) & 2k\mathbf{D}_{zz}^{-1}\mathbf{D}_{xz} \end{bmatrix}\begin{Bmatrix} \mathbf{d} \\ \zeta\mathbf{d} \end{Bmatrix} - \zeta\begin{bmatrix} \mathbf{I} & 0 \\ 0 & \mathbf{I} \end{bmatrix}\begin{Bmatrix} \mathbf{d} \\ \zeta\mathbf{d} \end{Bmatrix} = 0 \tag{4.26}$$

Equation (4.26) can easily be solved for eigenvalue ζ for given k and ω, and the six eigenvalues and the corresponding eigenvectors can be obtained. The first matrix in Eq. (4.26) is non-symmetric and is real-valued for pure elastic materials or complex-valued for materials with dissipation. In general, the six eigenvalues and their eigenvectors are complex-valued. Using the six eigenvalues and their eigenvectors, the displacement in the wavenumber domain can be expressed as

$$\tilde{U} = \sum_{j=1}^{3} C_j^+ d_j^+ \exp(i\zeta_j^+ z) + \sum_{j=1}^{3} C_j^- d_j^- \exp[i\zeta_j^- (z-h)] \qquad (4.27)$$

where C_j are constants to be determined from the boundary conditions on the upper and lower surfaces of the layer, and h (suffix "n" is dropped for convenience) is the thickness of the layer under consideration. In Eq. (4.27), the "+" denotes variables corresponding to the three eigenvalues which satisfy

$$\text{Im}(\zeta_j) > 0$$
$$\text{Re}(\zeta_j) > 0 \quad \text{when } \text{Im}(\zeta_j) = 0 \qquad (4.28)$$

These eigenvalues are related to waves propagating in the positive z direction (upward). The "−" in Eq. (4.27) denotes variables corresponding to the three eigenvalues which satisfy

$$\text{Im}(\zeta_j) < 0$$
$$\text{Re}(\zeta_j) < 0 \quad \text{when } \text{Im}(\zeta_j) = 0 \qquad (4.29)$$

These eigenvalues are related to waves propagating in the negative z-direction (downward).

From Eq. (4.27), it can be seen that the real parts of the arguments for the exponentials are always either negative or zero. Hence the numerical problem of truncation caused by growing exponentials are avoided, and the so-called "numerical truncation problem" can be removed by using Eq. (4.27). Avoiding the numerical truncation problem by this manner has been used by Mal (1988).

Equation (4.27) can be written in the following matrix form:

$$\tilde{U} = V^+ E^+(z) C^+ + V^- E^-(z) C^- = [V^+ E^+ \quad V^- E^-] \begin{Bmatrix} C^+ \\ C^- \end{Bmatrix} \qquad (4.30)$$

where

$$C^+ = [C_1^+ \quad C_2^+ \quad C_3^+]^T, \quad C^- = [C_1^- \quad C_2^- \quad C_3^-]^T \qquad (4.31)$$

$$V^+ = [d_1^+ \quad d_2^+ \quad d_3^+], \quad V^- = [d_1^- \quad d_2^- \quad d_3^-] \qquad (4.32)$$

and

$$E^+(z) = \text{Diag}[e^{i\zeta_1^+ z} \quad e^{i\zeta_2^+ z} \quad e^{i\zeta_3^+ z}]$$

$$E^-(z) = \text{Diag}[e^{i\zeta_1^-(z-h)} \quad e^{i\zeta_2^-(z-h)} \quad e^{i\zeta_3^-(z-h)}]$$

(4.33)

Using Eqs. (4.4) and (4.14), the stress vector on a z = constant plane can be given by

$$R = L_z^T c\left(L_x \frac{\partial}{\partial x} + L_z \frac{\partial}{\partial z}\right)U = D_z \frac{\partial U}{\partial x} + D_{zz} \frac{\partial U}{\partial z}$$

(4.34)

where D_z is given by

$$D_z = \begin{bmatrix} c_{51} & c_{56} & c_{55} \\ c_{41} & c_{46} & c_{45} \\ c_{31} & c_{36} & c_{35} \end{bmatrix}$$

(4.35)

Application of the Fourier transform defined by Eqs. (4.19) to (4.34) leads to

$$\tilde{R} = -ikD_z\tilde{U} + D_{zz}\frac{\partial\tilde{U}}{\partial z}$$

(4.36)

Substituting Eq. (4.30) into (4.36), we obtain

$$\tilde{R} = [P^+E^+ \quad P^-E^-]\begin{Bmatrix} C^+ \\ C^- \end{Bmatrix}$$

(4.37)

where

$$P^+ = -ikD_zV^+ + D_{zz}V_\zeta^+, \quad P^- = -ikD_zV^- + D_{zz}V_\zeta^-$$

(4.38)

in which

$$V_\zeta^+ = [i\zeta_1^+ d_1^+ \quad i\zeta_2^+ d_2^+ \quad i\zeta_3^+ d_3^+], \quad V_\zeta^- = [i\zeta_1^- d_1^- \quad i\zeta_2^- d_2^- \quad i\zeta_3^- d_3^-]$$

(4.39)

Applying the Fourier transform defined by Eq. (4.19) to the boundary conditions Eqs. (4.16) to (4.18), the boundary conditions in the wavenumber domain can be obtained. Using these boundary conditions in the wavenumber domain, we obtain the following global system equation for the entire laminate.

$$AC = \tilde{T}$$

(4.40)

where $\tilde{\mathbf{T}}$ is the Fourier transform of $\mathbf{T}$ defined by Eq. (4.15), and $\mathbf{C}$ consists of the constant vectors for the layers

$$\mathbf{C} = \{\mathbf{C}_1^+ \quad \mathbf{C}_1^- \quad \mathbf{C}_2^+ \quad \mathbf{C}_2^- \quad \cdots \quad \mathbf{C}_N^+ \quad \mathbf{C}_N^-\} \tag{4.41}$$

and the global stiffness matrix $\mathbf{A}$ is given by

$$\mathbf{A} = \begin{bmatrix} -\mathbf{P}_1^+ & -\mathbf{P}_1^-\mathbf{E}_{h1}^- & 0 & 0 & 0 & 0 & 0 & \cdots & 0 \\ \mathbf{V}_1^+\mathbf{E}_{h1}^+ & \mathbf{V}_1^- & -\mathbf{V}_2^+ & -\mathbf{V}_2^-\mathbf{E}_{h2}^- & 0 & 0 & 0 & \cdots & 0 \\ \mathbf{P}_1^+\mathbf{E}_{h1}^+ & \mathbf{P}_1^- & -\mathbf{P}_2^+ & -\mathbf{P}_2^-\mathbf{E}_{h2}^- & 0 & 0 & 0 & \cdots & 0 \\ 0 & 0 & \mathbf{V}_2^+\mathbf{E}_{h2}^+ & \mathbf{V}_2^- & -\mathbf{V}_3^+ & -\mathbf{V}_3^-\mathbf{E}_{h3}^- & 0 & \cdots & 0 \\ 0 & 0 & \mathbf{P}_2^+\mathbf{E}_{h2}^+ & \mathbf{P}_2^- & -\mathbf{P}_3^+ & -\mathbf{P}_3^-\mathbf{E}_{h3}^- & 0 & \cdots & 0 \\ \vdots & \vdots & \cdots & \ddots & \ddots & \ddots & \vdots & \cdots & \vdots \\ 0 & 0 & \cdots & 0 & 0 & 0 & 0 & \mathbf{P}_N^+\mathbf{P}_{hN}^+ & \mathbf{P}_N^- \end{bmatrix}$$

$$\tag{4.42}$$

in which

$$\mathbf{E}_h^+ = \text{Diag}[e^{i\zeta_1^+ h} \quad e^{i\zeta_2^+ h} \quad e^{i\zeta_3^+ h}], \quad \mathbf{E}_h^- = \text{Diag}[e^{-i\zeta_1^- h} \quad e^{-i\zeta_2^- h} \quad e^{-i\zeta_3^- h}] \tag{4.43}$$

The numbers in the subscripts in Eqs. (4.41) and (4.42) stand for the layer numbers.

Solving Eq. (4.40), the constant vector $\mathbf{C}$ can be obtained, and the displacement in the wavenumber domain can be obtained by Eq. (4.30) for all the layers in the laminate. Finally, the displacement can be obtained by performing the following inverse Fourier integration.

$$\mathbf{U}(z, x) = \frac{1}{2\pi} \int_{-\infty}^{\infty} \tilde{\mathbf{U}}(z, k) e^{-ikx} dk \tag{4.44}$$

Note that the integral in the foregoing equation has to be evaluated through numerical means, as the integrand is a very complicated function of wavenumber k.

4.5   A Technique for the Inverse Fourier Integration

For nondissipative materials, numerical evaluation of Eq. (4.44) is not an easy task because there are poles located on the integral axis. We have seen this kind of problem in Chapters 1 and 2. In Chapter 1, the poles can be obtained explicitly and Cauchy's theorem can be applied to evaluate the integration analytically with ease. The integrand in Eq. (4.44) is, however, much more complicated and analytical solution is unlikely. Even if a numerical method is used, a proper treatment has to be done first, as the integrand goes to infinity at the poles on the integral axis. There are a few techniques for evaluating a singular integral like that in Eq. (4.44). The first one is to introduce dissipation in the material, so that the poles are moved off the integral axis. For materials with very small dissipation, the integrand could have very sharp peaks near the poles. Therefore, special schemes for handling rapidly varying integrands at locations have to be used. Details for these techniques can be found in Mal and Lih (1992) and Xu and Mal (1987). For pure elastic materials, the technique proposed by Kundu and Mal (1985) may be used to remove these poles located on the integration axis. In this technique, accurate values of the poles have to be found by numerical means. The use of a complex path in the complex wavenumber plane (Liu and Tani, 1994) can be used for both pure elastic and visco-elastic materials and is introduced in this chapter.

Equation (4.44) can be rewritten as three portions as follows.

$$\mathbf{U} = \frac{1}{2\pi}\int_{-\infty}^{\infty} \tilde{\mathbf{U}} e^{-ikx}\, dk = \frac{1}{2\pi}\left(\int_{-\infty}^{-k_A} \tilde{\mathbf{U}} e^{-ikx}\, dk + \int_{-k_A}^{k_A} \tilde{\mathbf{U}} e^{-ikx}\, dk + \int_{k_A}^{\infty} \tilde{\mathbf{U}} e^{-ikx}\, dk\right) \quad (4.45)$$

where k_A is a positive value which is large enough so that all the real-valued poles are located between $-k_A$ and k_A. The first and last semi-infinite integrals on the right side of Eq. (4.45) can be evaluated by a scheme given by Xu and Mal (1987), or an adaptive scheme suggested by Liu, Lam, and Tani (1995a). Here we deal only with the second finite integral on the right side of Eq. (4.45). This integral can be written as

$$\int_{-k_A}^{k_A} \tilde{\mathbf{U}} e^{-ikx}\, dk = \int_{s(k)} \tilde{\mathbf{U}} e^{-iZx}\, dZ \quad (4.46)$$

where $Z = k + s(k)$, and $s(k)$ is a complex path starting from $-k_A$ and ending at k_A, such as the triangular path suggested by Liu and Tani (1994):

$$s(k) = \begin{cases} \dfrac{-ik_h}{k_A - k_C}(k + k_A), & \text{for } -k_A \le k \le -k_C \\[2mm] \dfrac{ik_h}{k_C}k, & \text{for } -k_C \le k \le k_C \\[2mm] \dfrac{-ik_h}{k_A - k_C}(k - k_A), & \text{for } k_C \le k \le k_A \end{cases} \quad (4.47)$$

where k_C is a positive value smaller than k_A, and k_h is a very small value, as shown in Figure 4.2. Equation. (4.46) is valid only if there is no pole of $\tilde{\mathbf{U}}(k)$ enclosed in the loops enclosed by $s(k)$ and the real k axis (Cauchy's theorem).

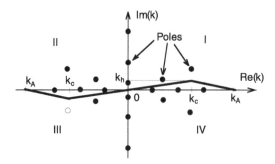

FIGURE 4.2
Poles and a complex triangular integral path. (From Liu et al., *Mechanics of Composite Materials and Structures*, 2, 227, 1995a. With permission.)

Hence, the complex path should be close to the real wavenumber axis to ensure that no poles are included in the loop. Also a precision problem could be present if the complex path is too far from the real wavenumber axis, especially for far-field responses. This is because the exponential terms in the integrand become extremely large for large values of x. The integral on the right hand side of Eq. (4.46) can be written as

$$\int_{s(k)} \tilde{\mathbf{U}} e^{-iZx} \, dZ = \int_{-k_A}^{k_A} \tilde{\mathbf{U}} e^{-i(k+is)x} \left(1 + \frac{ds}{dk}\right) dk = \int_{-k_A}^{k_A} \mathbf{G}(k) e^{-i(k+is)x} \, dk \qquad (4.48)$$

The right side of Eq. (4.48) can be evaluated by ordinary routines using equally spaced sampling points. However, the sampling points could be very large, because $\mathbf{G}(k)$ still varies quite rapidly near the poles, as shown in Figure 4.3. Moreover, it is not easy to control the integral accuracy in using equally spaced routines. The advantage of equally spaced routines is that the value of $\mathbf{G}(k)$ at these sampling points can be used for any x. Hence, it can significantly save computing time when results at many locations are required. To maintain the advantage of equal space routines, and also to reduce the sampling points and control the accuracy, a new adaptive scheme suggested by Liu, Lam, and Tani (1995a) is introduced as follows.

We denote a component of $\mathbf{G}(k)$ by $G(k)$, and discuss the integration of $G(k)$ as an example. Other components of $\mathbf{G}(k)$ can be treated exactly the same way. In the present scheme, $G(k)$ is represented by piece-wise polynomials of the second order, and in each piece the Fourier integration is carried out analytically. The key point in this scheme is to compute the piece-wise polynomials and make the procedure adaptive. The procedure is outlined as follows for integration over a region of $[a, b]$.

1. Calculate $d = (b - a)/(4m)$, where m is any integer, and use d as a primary increment in computing $G(k)$.
2. Calculate $G(a)$, $G(a + d)$, $G(a + 2d)$, $G(a + 3d)$, and $G(a + 4d)$.

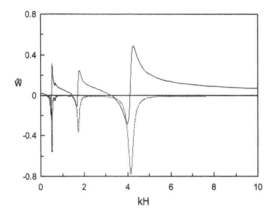

FIGURE 4.3
Dimensionless displacement in wavenumber domain on the rectangular complex path of $k_cH =$ 6.0, $k_AH = 100.0$, and $k_hH = 0.2$. Solid line: Re[$\tilde{w}(k)$]; Dotted line: Im[$\tilde{w}(k)$]. Hybrid laminate [C90/G+45/G–45]$_s$. (From Liu et al., *Mechanics of Composite Materials and Structures*, 2, 227, 1995a. With permission.)

3. Using the three points $G(a)$, $G(a+2d)$, $G(a+4d)$, a second order polynomial, $g(k)$, can be formed, and $g(a+d)$ and $g(a+2d)$ can be obtained.

$$g(a+d) = 0.125[3G(a) + 6G(a+2d) - G(a+4d)] \qquad (4.49)$$

$$g(a+3d) = 0.125[-G(a) + 6G(a+2d) + 3G(a+4d)] \qquad (4.50)$$

Calculate $g(a+d)$ and $g(a+2d)$, and check error in function representation against a predefined allowance ε:

$$\left| \frac{G(a+d) - g(a+d)}{G(a+d)} \right| \le \varepsilon \qquad (4.51)$$

$$\left| \frac{G(a+2d) - g(a+2d)}{G(a+2d)} \right| \le \varepsilon \qquad (4.52)$$

If Eqs. (4.51) and (4.52) are satisfied, we go to the next four steps until reaching b. If one of Eqs. (4.51) and (4.52) is not satisfied, d is halved and go back to the first step. In this case only $G(a+d/2)$, $G(a+3d/2)$, $G(a+5d/2)$, $G(a+7d/2)$ need to be newly computed, and the previously computed Gs and ks can be stored in order for later uses.

4. Finally, the integration region [a, b] is divided into M pieces, and $2M+1$ Gs and ks are obtained.

5. Perform analytically the following integration

$$\int_a^b G(k)e^{-i(k+is)x}\,dk \approx \sum_{j=1}^M \int_{k_j}^{k_{j+2}} g(k)e^{-i(k+is)x}\,dk \qquad (4.53)$$

where

$$g(k) = \left[\left(1-3\frac{k}{d}+2\frac{k^2}{d^2}\right) \quad 4\left(\frac{k}{d}-\frac{k^2}{d^2}\right) \quad \left(-\frac{k}{d}+2\frac{k^2}{d^2}\right)\right] \begin{bmatrix} G_j \\ G_{j+1} \\ G_{j+2} \end{bmatrix} \qquad (4.54)$$

in which $d = k_{j+2} - k_j$, and $G_j = G(k_j)$.

It is convenient to change the integration region in Eq. (4.48) to three subregions, $[-k_A, -k_C]$, $[-k_C, k_C]$, and $[k_C, k_A]$ and, in each region, the integration can be carried out using the above procedure. It should be also noted that if $G(k)$ is symmetric or anti-symmetric with respect to k, only integrations in regions $[0, k_C]$, $[k_C, k_A]$, and $[k_A, +\infty]$ are needed to be evaluated. For a symmetric $G(k)$, we use

$$U = \frac{1}{\pi}\left\{\int_0^{k_A} G(k)\cos[(k+is)x]\,dk + \int_{k_A}^\infty \tilde{U}(k)\cos(kx)\,dk\right\} \qquad (4.55)$$

and for an anti-symmetric $G(k)$, we use

$$U = \frac{-i}{\pi}\left\{\int_0^{k_A} G(k)\sin[(k+is)x]\,dk + \int_{k_A}^\infty \tilde{U}(k)\sin(kx)\,dk\right\} \qquad (4.56)$$

4.6 Response in Time Domain

To obtain responses in the time domain, the exponential window method (EWM) used in signal processing is used here to avoid singularities of the integrand $U(\omega)$ at $\omega = 0$ and at the poles of the cut-off frequencies (at those frequencies the plate stays stationary and there is no propagating wave in the plate). The EWM is basically a numerical approach that avoids the singular integrals in a smart and straightforward manner. It simply replaces

$$\mathbf{U}_t(t) = \frac{1}{2\pi}\int_{-\infty}^\infty \mathbf{U}(\omega)\tilde{F}(\omega)e^{i\omega t}\,d\omega \qquad (4.57)$$

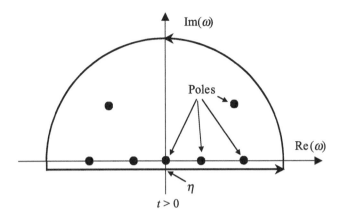

FIGURE 4.4
Contours for integration in Eq. (4.58).

with

$$\mathbf{U}_t(t) = \frac{e^{\eta t}}{2\pi}\int_{-\infty}^{\infty} \mathbf{U}(\omega - i\eta)\tilde{F}(\omega - i\eta)e^{i\omega t}d\omega \qquad (4.58)$$

where $\mathbf{U}_t(t)$ denotes the displacement in the time domain, and $\tilde{F}$ is the Fourier transform of external force, and

$$\tilde{F}(\omega - i\eta) = \int_0^{t_d} e^{-\eta t} f(t)e^{-i\omega t} dt \qquad (4.59)$$

in which t_d is the duration of the external force with the time function $f(t)$, and η is a small positive constant. Equation (4.58) is derived based on the simple fact that the real integral path can be replaced by a complex integral path of $\omega - i\eta$, as shown in Figure 4.4. The reason for choosing the complex path can be justified in the same way described in Section 1.6.

4.7 Poles and Complex Paths

We have encountered two cases involving singular integrals. One is for the inverse Fourier transform from the wavenumber domain back to the spatial domain (Eq. (4.44)). Another is for the inverse Fourier transform from the frequency domain back to the time domain (Eq. (4.57)). These two cases are actually interlinked. In the first case, the integral on the left side of Eq. (4.46) is singular and its solution is not unique and can have many possible values, due to the presence of the poles on the real k axis. Using a complex path or

EWM or introducing material dissipation in dealing with Eq. (4.57), Eq. (4.46) can be actually uniquely defined without the need for any special treatment. The problem lies only with the numerical evaluation of the integral, since the kernel usually behaves very badly. A good understanding of the properties of the integrand and the poles can help to evaluate the integral more efficiently. For a laminate made of dissipative materials, great care has to be taken when using a complex path because the poles may move into the loop enclosed by the triangular complex path and the real k axis, due to the presence of material dissipation. In fact, the triangular complex path does not have to be used because there are no poles on the real k axis, and the integration can be carried out along the real k axis. However, a complex path can help very much in improving the behavior of the integrand and hence to reduce sampling points, especially when the material dissipation is very small. Liu, Lam, and Tani (1995a) investigated the movement of the poles due to the introduction of p (a damping factor introduced by Mal and Lih, 1992). The poles were computed for dissipative and purely elastic materials using the method given by Liu and Achenbach (1994, 1995). The results for the poles near the real k axis are schematically drawn in Table 4.1. It is found that for a laminate of pure elastic materials ($p=0$), there are poles located right on the real k axis, as shown in the first row of Table 4.1. For positive p as given by Mal and Lih (1992), the poles on the positive real k axis move up to quadrant I, and the poles on the negative real k axis move down to

TABLE 4.1

Poles Near the Real Wavenumber Axis and Complex Paths

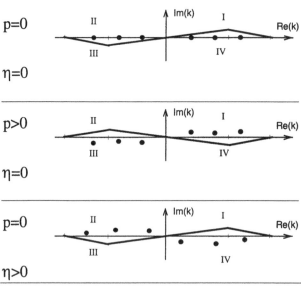

Source: Liu et al., *Mechanics of Composite Materials and Structures*, 2, 227, 1995a. With permission.

quadrant III in the complex k plane. Also, the complex-valued poles are far away from the complex path, when k_h is very small. Hence, a complex path of a positive k_h can be used because no poles will be included in the loop enclosed by the complex path and the real k axis.

When calculating responses in the time domain using Eq. (4.57), $\omega - i\eta$ is used instead of the ω in computing $\tilde{U}(k)$; hence, the poles of $\tilde{U}(k)$ no longer stay on the real k axis. Therefore, the complex path given by Eq. (4.47) does not have to be used for obtaining $U(\omega - i\eta)$. However, a complex path can still be used for the following reasons. The movement of the poles due to the introduction of η is also computed and shown in the bottom row of Table 4.1. It is found that for a small positive η, poles on the positive real k axis move down to quadrant IV, and the poles on the negative real k axis move up to quadrant II in the complex k plane. Therefore, no poles will move into the loop enclosed by the complex path and the real k axis. The double use of the complex path and the EWM can help very much to reduce the sampling points since the kernel is much better behaved than when either complex path or the EWM is used alone. In any case, the quadrature scheme suggested in Section 4.5 should be used for evaluating the inverse integration to obtain U or $U(\omega - i\eta)$.

4.8 Examples

The above approach has been used to compute the responses in both the frequency and time domains for a hybrid laminate denoted by $[C90/G + 45/G-45]_s$. In this notation, the letters C and G stand, respectively, for carbon/epoxy and glass/epoxy layers, while the numbers following the letters indicate the angle of the fiber-orientation with respect to the x-axis. The subscript s indicates that the laminate is symmetrically stacked. Hence the laminate is made of six layers of the same thickness. The material constants are given in Table 3.1. The following dimensionless parameters are used.

$$\bar{x} = x/H, \quad \bar{w} = c(4,4)w/q_0, \quad \bar{u} = c(4,4)u/q_0,$$
$$\bar{\omega} = \omega H/c_s, \quad \bar{t} = tc_s/H, \quad c_s = \sqrt{c(4,4)/\rho}, \tag{4.60}$$

where u and w are, respectively, the displacements in the x and z direction, and c, ρ, and c_s are the material constants matrix, density, and shear wave velocity of a reference material, C90.

In this section, the load is specified to be a line load acting on the upper surface of the laminate at $x = x_0$ in vertical (z) direction. Hence we may write

$$T_1(x) = T_1(x) = \cdots = T_N(x) = 0 \tag{4.61}$$

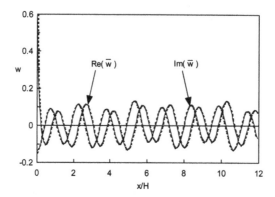

FIGURE 4.5
Dimensionless displacement on the upper surface of the laminate $[C90/G+45/G-45]_s$ subjected to a vertical line harmonic load on the upper surface of the laminate. Solid line: present method; Dotted line: SNM (Liu and Achenbach, 1994). ($x_0 = 0.0$, $q_0 = 1.0$, $\bar{\omega} = 3.14$). (From Liu et al., *Mechanics of Composite Materials and Structures*, 2, 227, 1995a. With permission.)

and

$$\mathbf{T}^T_{N=1}(x) = \{0 \quad 0 \quad q_0\delta(x - x_0)\} \tag{4.62}$$

where δ is the Dirac delta function, and q_0 is a constant representing the amplitude of the external force. In this case, the Fourier transform of the external force can be written as

$$\tilde{\mathbf{T}}^T_{N+1}(x) = \{0 \quad 0 \quad q_0 e^{ikx_0}\} \tag{4.63}$$

In the examples given in this section, q_0 is set to be 1.

Figure 4.5 shows the surface displacement w of the hybrid laminate subjected to a line harmonic load. The dimensionless angular frequency of the load is 3.14. The results are compared with those obtained using SNM (Liu and Achenbach, 1994). In the present method, the complex path given by Eq. (4.45) has been used and k_CH, k_AH, and k_hH are taken to be 6.0, 100.0, and 0.2, respectively. Because $\tilde{w}(k)$ is symmetric with respect to k, w is obtained by using Eq. (4.56). The tolerance ε in Eqs. (4.51) and (4.52) are set to be 1.0×10^{-3}. The primary increment d in the region $[0, k_C]$ is set to be 0.06. This is much smaller than that in the region $[k_C, k_A]$ that is set to be 5.0 because of the presence of sharp peaks in $[0, k_C]$. The final sampling points used are 313 and 37 for $[0, k_C]$ and $[k_C, k_A]$, respectively. Adding 20 samplings used for the semi-infinite integral, a total of 370 sampling points is required for calculating the spectra shown in Figure 4.5. In the SNM, six elements (one element for each layer) are used. A very good agreement can be observed in Figure 4.5.

We have also computed the response of the hybrid laminate with a small damping factor ($p = 0.01$) defined by Mal and Lih (1992). The parameters are

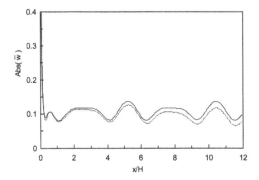

FIGURE 4.6
Dimensionless displacement on the upper surface of the laminate [C90/G+45/G–45]$_s$ subjected to a vertical line harmonic load on the upper surface of the laminate. Results are obtained by the present method. Solid line: pure elastic laminate; Dotted line: visco-elastic laminate with a damping factor of $p=0.01$. ($x_0 = 0.0$, $q_0 = 1.0$, $\bar{\omega} = 3.14$). (From Liu et al., *Mechanics of Composite Materials and Structures*, 2, 227, 1995a. With permission.)

the same as those used for obtaining Figure 4.5, except that $k_h H$ is set to be 0.0, namely, integrating along the real wavenumber axis. Figure 4.6 shows a comparison of the results obtained by the present method without and with a damping factor. For $p = 0.01$, the sampling points are 537 and 37 for regions of $[0, k_C]$ and $[k_C, k_A]$, respectively. A complex path of $k_h H = -0.2$ is also used for the case of $p = 0.01$. Exactly the same results have been obtained. Using the complex path, the sampling points in $[0, k_C]$ have been reduced to 257, which is less than half of integrating along the real k axis. It is found from Figure 4.6 that the effect of the damping on the absolute values of the displacement is larger at observation points farther from the point of excitation.

Figure 4.7 shows responses of the displacement u of the hybrid laminate subject to a transient load with a time history function of

$$f(t) = \begin{cases} \sin(2\pi t/t_d), & 0 < t < t_d \\ 0, & t < 0 \quad \text{or} \quad t_d < t \end{cases} \tag{4.64}$$

The observation point is at $x = 10H$ on the upper surface of the laminate. The results are obtained by the present method and by the HNM (Liu and Tani et al., 1991 or Chapter 8). Equation (4.58) is used in the present method, and the dimensionless shift constant η is set to be 0.05. Because $\tilde{u}(k)$ is anti-symmetric with respect to k, $u(\omega - i\eta)$ is obtained using Eq. (4.56) and the integration is evaluated along the complex path. The sampling points required depend on the angular frequency ω. The higher the ω, the more sampling points are required. For $\bar{\omega} = 3.14$ and 15.0, the total sampling points are 234 and 446. The numbers are nearly doubled when the complex path is not used. In the HNM, six elements (one element for each layer) have been used. An excellent agreement is shown in Figure 4.7.

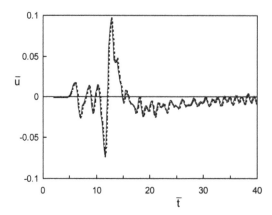

FIGURE 4.7
Dimensionless displacement at $x = 10H$ on the upper surface of the laminate $[C90/G+45/G-45]_s$ subjected to a vertical line load on the upper surface of the laminate. The time function is given by Eq. (4.52). Solid line: present method; Dotted line: HNM. ($x_0 = 0.0$, $q_0 = 1.0$, $\bar{t}_d = 2.0$.) (From Liu et al., *Mechanics of Composite Materials and Structures*, 2, 227, 1995a. With permission.)

4.9   Remarks

In this chapter, an exact formulation for analyzing wave responses of aniso-tropic laminate has been discussed. In this formulation, wave modes for a layer are obtained by solving an eigenvalue equation. This formulation is straightforward and applicable for composite laminates made of arbitrary anisotropic materials. A complex path and a quadrature scheme are used to evaluate the inverse Fourier integrations to obtain responses in the frequency domain to improve the integration efficiency. The present method can also be used for not only pure elastic materials but also materials of dissipation. By investigating the properties of the poles, it is found that the complex path can be used for dissipation materials and can be used together with EWM. This combination has further reduced to half the number of sampling points required. Furthermore, the implementation of complex paths is very easy and only minor changes are needed in the computer program. In summary, the procedure of the present method is as follows.

1. Form matrices for each layer of the laminate.
2. Use the eigenvalues and eigenvectors obtained from Eq. (4.26) to obtain equations for the displacement and stress vector on the upper and lower surfaces for each layer.
3. Assemble these equations using boundary conditions of the lami-nate to obtain Eq. (4.40).

4. Solve Eq. (4.40) to obtain the constant vectors, and then obtain the displacement in the wavenumber domain.

5. To obtain responses in the frequency domain for pure elastic materials, use the complex path and the present quadrature scheme to evaluate the inverse Fourier integration.

6. To obtain responses in the frequency and time domains for dissipative materials, use the complex path and the present quadrature scheme to evaluate the inverse Fourier integration.

7. To obtain responses in the time domain for pure elastic materials, use the complex path and/or EWM and the present quadrature scheme to evaluate the inverse Fourier integration.

The methods described in this and the previous chapters are basically the same in principle, but different in formulation. The method used in the previous chapter was used for analyzing the characteristics of Lamb waves, but it is fully capable for forced wave motion analysis. The method in this chapter is also capable of performing analysis of characteristics of Lamb waves. The method given in the previous chapter is more classic and traditional, and the matrices obtained were much smaller and therefore used more when computers were less powerful and memory was much more expensive. The method in this chapter is basically in matrix form, and the final matrix assembled is much larger than that in the previous chapter. Since fast algorithms have been developed over years for large matrix systems, and computers now are capable of dealing with much larger matrix systems, matrix formulation is more efficient for many problems, including problems of wave propagation. In addition, the global matrix for layered structure is narrowly bonded and hence is efficient to operate. Methods to be covered in the following chapters use concepts of the finite element method, which produces even larger matrices systems. These methods are much more powerful and efficient for complex systems, such as functionally graded material plates, whose mechanical properties vary in space.

The interaction of elastic waves with a fluid-loaded anisotropic laminate is frequently encountered in nondestructive evaluation of materials and underwater explosions. In these practical applications, the acoustic wave generated by a transducer or source of excitation propagates through the upper coupling fluid first and then onto the surface of the layered plate. The method developed in this chapter has been extended to the case of an immersed composite laminate to a Gaussian beam pressure. For more details, readers should consult the reference given by Liu, Wang, and Xi (2000).

5

Characteristics of Waves in Composite Laminates

5.1   Introduction

When waves propagate in an anisotropic laminate, the wave velocity varies with wavelength as demonstrated in Chapter 3. This property of wave-velocity dependence upon wavelength or frequency is known as dispersion of a wave. The study of wave dispersion is of considerable importance because the velocity is no longer constant but changes with frequency or wavelength. Due to the dispersion, waves propagate in the form of groups of waves and at the velocity of the group wave. This means that a wave propagating in a laminate suffers changes in its appearance. Group velocity can often be a better representation, as it is also the velocity of energy propagation, and can actually be detected in experiments because wave energy, which excites the sensor, is propagating in groups. In an anisotropic medium, the phase and group velocities of plane (Lamb) waves also vary with the direction of wave propagation. In isotropic media, wave front or surface are circles if the source of the wave is a point source. In anisotropic media, however, the wave surfaces can be very complicated. This chapter is devoted to address all these issues and interesting phenomena.

Investigation of characteristics of waves in plates has been carried out by many researchers. Thomson (1950) proposed an exact method to determine the characteristics of wave propagation in isotropic laminates. Dong and Nelson (1972) and Nelson and Dong (1973) examined the characteristic solutions to wave propagation in orthotropic laminates using a numerical method. Ohyoshi (1985) investigated the group velocities in sandwich plates using an exact analytical method. For an anisotropic laminate with more than three or four layers, however, it is difficult to obtain exact solutions for group velocities. The difficulty lies not in the formulation but in the cumbersome algebra necessary to generate an explicit dispersion relation. Hence, the method commonly used for estimating the group velocity is by graphical or numerical differentiation of a frequency branch with respect to wave number. The Rayleigh quotient was used to give an approximate explicit dispersion relation from which the group velocity can be obtained directly from the dispersion relation (Nelson and Dong, 1973; Tolstoy, 1965; Lysmer, 1970; Kausel and Roesset, 1981). This method is simple to use and suitable for numerical methods. Moon (1972) calculated the phase velocities and wave surfaces for a thin anisotropic composite plate due to a point impact, using an approximate theory. Sun and co-workers (1973, 1984) investigated the propagation of shock fronts in composite laminated plates.

   This chapter investigates the dispersion of waves propagating in anisotropic laminates and characteristic wave surfaces. The method used in this chapter is different from the exact method in Chapter 3, as the exact method cannot produce an explicit form of relationship between the phase velocity and the frequency. The method of finite element type is used in this chapter, which was originally proposed by Liu et al. (1991b) based on the work by Dong and Nelson (1972), Nelson and Dong (1973), and Kausel (1986).

5.2   Dispersion Equation

Consider a laminate made of layers of anisotropic materials. At any point in the material, there are three displacement components, u, v, w. Due to the anisotropy of the material, these three displacement components are, in general, coupled. The laminate is divided into N layer elements, as shown in Figure 5.1. The thickness, mass density, and matrix of elastic constants of the nth element are denoted, respectively, by h_n, ρ_n, and c_n (defined by Eq. (4.11)). Note that the number of elements is usually more than the number of layers of materials, meaning a layer can be divided into more than one element; this ensures that the material property within one element can be homogeneous. In an element, a governing system equation of free wave motion in matrix form in terms of displacements $U(x, y, z)$ can be given as follows (see Chapter 4):

$$\rho \ddot{U} - L^T cLU = 0 \tag{5.1}$$

Chapters 3 and 4 discussed exact methods to obtain U that satisfies the foregoing equation exactly at any point in the laminate. We now seek alternative ways to obtain an approximate solution to Eq. (5.1). We first assume that the solution for the displacement field within an element has the form of a harmonic plane wave, i.e.,

$$U(x, y, z) = N(z)d \exp(ik_x x + ik_y y - i\omega t) \tag{5.2}$$

where $N(z)$ is the matrix of prescribed shape functions. We use here quadratic interpolation of polynomials, i.e.,

$$N(z) = [(1 - 3\bar{z} + 2\bar{z}^2)I \quad 4(\bar{z} - \bar{z}^2)I \quad (2\bar{z}^2 - \bar{z})I] \tag{5.3}$$

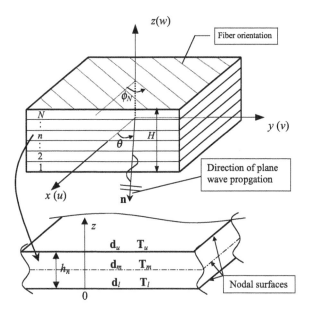

FIGURE 5.1
Composite laminate and the nth isolated element.

in which $\bar{z} = z/h_n$, and $\mathbf{I}$ is a 3×3 identity matrix given by

$$\mathbf{I} = \begin{bmatrix} 1 & 0 & 0 \\ 0 & 1 & 0 \\ 0 & 0 & 1 \end{bmatrix} \tag{5.4}$$

Since the second order shape function is employed, there are three nodal surfaces in a layer element. If N element is used, there are a total of $(N + 1)$ nodal surfaces in the laminate. In Eq. (5.2), k_x and k_y are variables of wavenumber with respect to x and y directions, which can be written as

$$k_x = k \cos \theta, \quad k_y = k \sin \theta \tag{5.5}$$

where k is the wavenumber for the plane wave in direction of propagation θ. The vector $\mathbf{d}$ in Eq. (5.2) is the unknown vector and consists of nodal displacement amplitude vectors at *lower* ($z = 0$), *middle* ($z = 0.5h_n$) and *upper* ($z = 0.5h_n$) nodal surfaces of the element (see Figure 5.1):

$$\mathbf{d} = \begin{Bmatrix} \mathbf{d}_l \\ \mathbf{d}_m \\ \mathbf{d}_u \end{Bmatrix} \tag{5.6}$$

in which

$$\mathbf{d}_i = \begin{Bmatrix} u \\ v \\ w \end{Bmatrix} \quad (i = l, m, u) \tag{5.7}$$

where u, v, and w are the displacement components at, respectively, x, y, and z directions.

The stresses on a plane $z = $ constant can be written in terms of the displacement:

$$\mathbf{R} = \{\sigma_{xz} \quad \sigma_{yz} \quad \sigma_{zz}\}^T = \mathbf{L}_z^T \mathbf{c} \mathbf{L} \mathbf{U} \tag{5.8}$$

The stress vector $\mathbf{S}$ is defined as vectors of $\mathbf{R}$ at lower ($z = 0$), middle ($z = 0.5h_n$) and upper ($z = h_n$) surfaces of the element:

$$\mathbf{S} = \begin{Bmatrix} \mathbf{R}_l = \mathbf{R}|_{z=0} \\ \mathbf{R}_m = \mathbf{R}|_{z=0.5h_n} \\ \mathbf{R}_u = \mathbf{R}|_{z=h_n} \end{Bmatrix} \tag{5.9}$$

Assume an external force of harmonic excitation that can act on the nodal surfaces and has the form

$$\mathbf{T} = \bar{\mathbf{T}} \exp(ik_x x + ik_y y - i\omega t) \tag{5.10}$$

Due to the assumption of the displacement in Eq. (5.2), the equation of motion given in Eq. (5.1) may not be satisfied at all points in the element. Hence, in general,

$$\mathbf{W} = \rho \ddot{\mathbf{U}} - \mathbf{L}^T \mathbf{c} \mathbf{L} \mathbf{U} \neq 0 \tag{5.11}$$

at points in the element. We therefore seek for a vector $\mathbf{d}$ that allows $\mathbf{U}$ to satisfy Eq. (5.1) in an integral (weak) sense following some principle. We choose to use the principle of virtual work

$$\delta \mathbf{d}^T \mathbf{T} = \delta \mathbf{d}^T \mathbf{S} + \int_0^{h_n} \delta \mathbf{U}^T \mathbf{W} \, dz \tag{5.12}$$

Equation (5.12) states that the virtual work performed by the external tractions $\mathbf{T}$ acting at the nodal surfaces is equal to that performed by the stresses $\mathbf{S}$ on the nodal surfaces and the unbalanced internal body forces $\mathbf{W}$ caused by the assumption of the displacements. Equation (5.12) guarantees the balance of the energy over the volume of the element but does not guarantee the balance of the forces and stresses at all the points within the element. Equation (5.12) is, hence, said to be the weak form of equation of motion. In contrast, Eq. (5.1) is termed a strong form of equation of motion. Any **d** that satisfies Eq. (5.12) is said to be satisfying the weak form of the equation of motion.

Substituting Eqs. (5.2) and (5.9)–(5.11) into Eq. (5.12), and after lengthy and tedious but simple algebraic manipulations and a number of integrations over the thickness of the element, we obtain

$$\bar{\mathbf{T}} = [\mathbf{K} - \omega^2 \mathbf{M}]\mathbf{d} \tag{5.13}$$

where **M** is the mass matrix given by

$$\mathbf{M} = \frac{\rho_n h_n}{30} \begin{bmatrix} 4\mathbf{I} & 2\mathbf{I} & -\mathbf{I} \\ 2\mathbf{I} & 16\mathbf{I} & 2\mathbf{I} \\ -\mathbf{I} & 2\mathbf{I} & 4\mathbf{I} \end{bmatrix} \tag{5.14}$$

in which

$$\mathbf{K} = \mathbf{A}_1 k^2 \cos^2\theta + \mathbf{A}_2 k^2 \cos^2\theta \sin\theta + \mathbf{A}_3 k^2 \sin^2\theta$$
$$+ i\mathbf{A}_4 k \cos\theta + i\mathbf{A}_5 k \sin\theta + \mathbf{A}_6 \tag{5.15}$$

and $\mathbf{A}_i$ ($i = 1,...,6$) are found to be

$$\mathbf{A}_1 = \frac{h_n}{30} \begin{bmatrix} 4\mathbf{D}_1 & 2\mathbf{D}_1 & -\mathbf{D}_1 \\ 2\mathbf{D}_1 & 16\mathbf{D}_1 & 2\mathbf{D}_1 \\ -\mathbf{D}_1 & 2\mathbf{D}_1 & 4\mathbf{D}_1 \end{bmatrix} \tag{5.16}$$

$$\mathbf{A}_2 = \frac{h_n}{30} \begin{bmatrix} 4\mathbf{D}_2 & 2\mathbf{D}_2 & -\mathbf{D}_2 \\ 2\mathbf{D}_2 & 16\mathbf{D}_2 & 2\mathbf{D}_2 \\ -\mathbf{D}_2 & 2\mathbf{D}_2 & 4\mathbf{D}_2 \end{bmatrix} \tag{5.17}$$

$$\mathbf{A}_3 = \frac{h_n}{30} \begin{bmatrix} 4\mathbf{D}_3 & 2\mathbf{D}_3 & -\mathbf{D}_3 \\ 2\mathbf{D}_3 & 16\mathbf{D}_3 & 2\mathbf{D}_3 \\ -\mathbf{D}_3 & 2\mathbf{D}_3 & 4\mathbf{D}_3 \end{bmatrix} \tag{5.18}$$

$$\mathbf{A}_4 = \frac{1}{3} \begin{bmatrix} 3(\mathbf{D}_4 - \mathbf{D}_7) & -4\mathbf{D}_4 & \mathbf{D}_4 \\ 4\mathbf{D}_4 & 0 & -4\mathbf{D}_4 \\ -\mathbf{D}_4 & 4\mathbf{D}_4 & -3(\mathbf{D}_4 - \mathbf{D}_7) \end{bmatrix} \tag{5.19}$$

$$\mathbf{A}_5 = \frac{1}{3}\begin{bmatrix} 3(\mathbf{D}_5 - \mathbf{D}_8) & -4\mathbf{D}_5 & \mathbf{D}_5 \\ 4\mathbf{D}_5 & 0 & -4\mathbf{D}_5 \\ -\mathbf{D}_5 & 4\mathbf{D}_5 & -3(\mathbf{D}_5 - \mathbf{D}_8) \end{bmatrix} \tag{5.20}$$

$$\mathbf{A}_6 = \frac{1}{3h_n}\begin{bmatrix} 7\mathbf{D}_6 & -8\mathbf{D}_6 & \mathbf{D}_6 \\ -8\mathbf{D}_6 & 16\mathbf{D}_6 & -8\mathbf{D}_6 \\ \mathbf{D}_6 & -8\mathbf{D}_6 & 7\mathbf{D}_6 \end{bmatrix} \tag{5.21}$$

where the matrices $\mathbf{D}_i$, $(i = 1, 8)$ are given by

$$\mathbf{D}_1 = \begin{bmatrix} c_{11} & c_{16} & c_{15} \\ & c_{66} & c_{56} \\ \text{sym.} & & c_{55} \end{bmatrix}, \quad \mathbf{D}_2 = \frac{1}{2}\begin{bmatrix} 2c_{16} & c_{12} + c_{66} & c_{14} + c_{56} \\ & 2c_{26} & c_{25} + c_{46} \\ \text{sym.} & & 2c_{45} \end{bmatrix} \tag{5.22}$$

$$\mathbf{D}_3 = \begin{bmatrix} c_{66} & c_{26} & c_{46} \\ & c_{22} & c_{24} \\ \text{sym.} & & c_{44} \end{bmatrix}, \quad \mathbf{D}_4 = \frac{1}{2}\begin{bmatrix} 2c_{15} & c_{14} + c_{56} & c_{13} + c_{55} \\ & 2c_{46} & c_{45} + c_{36} \\ \text{sym.} & & 2c_{35} \end{bmatrix} \tag{5.23}$$

$$\mathbf{D}_5 = \frac{1}{2}\begin{bmatrix} 2c_{56} & c_{25} + c_{46} & c_{36} + c_{45} \\ & 2c_{24} & c_{23} + c_{44} \\ \text{sym.} & & 2c_{34} \end{bmatrix}, \quad \mathbf{D}_6 = \begin{bmatrix} c_{55} & c_{45} & c_{35} \\ & c_{44} & c_{34} \\ \text{sym.} & & c_{33} \end{bmatrix} \tag{5.24}$$

$$\mathbf{D}_7 = \begin{bmatrix} c_{15} & c_{56} & c_{55} \\ c_{14} & c_{46} & c_{45} \\ c_{13} & c_{36} & c_{35} \end{bmatrix}, \quad \mathbf{D}_8 = \begin{bmatrix} c_{56} & c_{25} & c_{45} \\ c_{46} & c_{24} & c_{44} \\ c_{36} & c_{23} & c_{34} \end{bmatrix} \tag{5.25}$$

Assembling the matrices of all the elements, the dynamic equilibrium equation for the entire laminate can be assembled to obtain

$$\bar{\mathbf{T}}_t = [\mathbf{K}_t - \omega^2 \mathbf{M}_t]\mathbf{d}_t \tag{5.26}$$

in which $\mathbf{M}_t$ is the global mass matrices, and $\mathbf{K}_t$ is the global stiffness matrix for the laminate given by

$$\mathbf{K}_t = \mathbf{A}_{1t}k^2 \cos^2\theta + \mathbf{A}_{2t}k^2 \cos\theta \sin\theta + \mathbf{A}_{3t}k^2 \sin^2\theta$$
$$+ i\mathbf{A}_{4t}k \cos\theta + i\mathbf{A}_{5t}k \sin\theta + \mathbf{A}_{6t} \tag{5.27}$$

The mass matrix $\mathbf{M}_t$ and matrices $\mathbf{A}_{it}$ ($i = 1,...,6$) are the results of assembling the contributions of $\mathbf{M}$ and $\mathbf{A}_i$ ($i = 1,...,6$) of all the elements. The vectors $\mathbf{V}_t$ and $\overline{\mathbf{T}}_t$ represent, respectively, the nodal displacement amplitudes and the nodal external force applied at the nodal surfaces of the laminate. The dimension of the matrices $\mathbf{M}_t$, $\mathbf{K}_t$, and $\mathbf{A}_{it}$ is $M \times M$, where M is obtained by the following formula

$$\underbrace{M}_{\text{Total DOFs}} = \underbrace{3}_{\substack{\text{DOFs at one nodal surface}}} \times \underbrace{(2N+1)}_{\text{Number of nodal surfaces}} = 6N + 3 \qquad (5.28)$$

where N is the number of the elements used in the discretization of the laminate.

The natural modes of plane waves propagating in the laminate are determined by considering free wave motion and setting the external force vector $\overline{\mathbf{T}}_t$ to zero. We then found the following eigenvalue equation:

$$0 = [\mathbf{K}_t - \omega^2 \mathbf{M}_t]\mathbf{d}_t \qquad (5.29)$$

Matrix $\mathbf{K}_t$ is an explicit function of the wavenumber k, as shown in Eq. (5.27). When the wavenumber k is given, M natural frequency ω_m ($m = 1,...,M$) can be obtained by solving the standard eigenvalue equation, Eq. (5.29). The results in $\omega_m(k)$ give the dispersion relation of the mth wave mode of plane waves in the laminate. The phase velocity of the mth mode, c_{pm}, can be then calculated using

$$c_{pm} = \omega_m / k \qquad (5.30)$$

5.3 Group Velocities

Group velocity c_g is defined as

$$c_g = \frac{d\omega}{dk} \qquad (5.31)$$

It is the slope to a branch of the frequency spectrum as shown in Figure 5.2. The definition of group velocity apparently came from the Stokes's treatment (see, for example, Section 6.5 in Achenbach, 1973). In Stokes' treatment, two plane waves of the same amplitude and slightly different wavenumber and hence different frequency are considered matching together in the same direction θ. The first plane wave with k_1 and ω_1 can be expressed in the following form using Eq. (5.2).

$$\mathbf{U}_1 = \mathbf{A}\exp(ik_1 r - i\omega_1 t) \qquad (5.32)$$

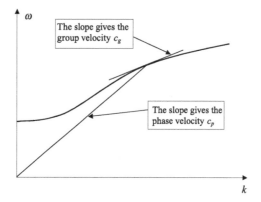

FIGURE 5.2
Group velocity is the gradient of the ω-k curve.

where **A** is the amplitude vector of the wave, and r is obtained as follows with the help of Eq. (5.5):

$$r = x \cos\theta + y \sin\theta \tag{5.33}$$

Similarly, the second plane wave with k_2 and ω_2 can be expressed as

$$\mathbf{U}_2 = \mathbf{A} \exp(ik_2 r - i\omega_2 t) \tag{5.34}$$

The superimposition of these two plane waves leads to

$$\mathbf{U}_{12} = \mathbf{U}_1 + \mathbf{U}_2 = \mathbf{A}[\exp(ik_1 r - i\omega_1 t) + \exp(ik_2 r - i\omega_2 t)]$$

$$= 2\mathbf{A} \cos\left[\frac{k_1 - k_2}{2}r + \frac{\omega_1 - \omega_2}{2}t\right] \exp\left(i\frac{k_1 + k_2}{2}r - i\frac{\omega_1 + \omega_2}{2}t\right) \tag{5.35}$$

Let

$$\bar{k} = \frac{k_1 + k_2}{2}, \quad \bar{\omega} = \frac{\omega_1 + \omega_2}{2}$$

$$\Delta k = \frac{k_1 - k_2}{2}, \quad \Delta\omega = \frac{\omega_1 - \omega_2}{2} \tag{5.36}$$

We have

$$\mathbf{U}_{12} = 2\mathbf{A} \cos[\Delta k r + \Delta\omega t] \exp(i\bar{k} r - i\bar{\omega} t) \tag{5.37}$$

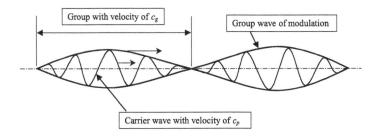

FIGURE 5.3
Formation of group by two plane waves.

This result is schematically illustrated in Figure 5.3. The amplitude of the superimposed waves is modulated by a term of $\cos[\Delta kr + \Delta \omega t]$. This modulation forms groups that are propagating with a velocity of

$$\frac{dr}{dr} = \frac{\Delta \omega}{\Delta k} \tag{5.38}$$

which is found by simply setting

$$\Delta kr + \Delta \omega t = \text{constant} \tag{5.39}$$

The group velocity is then found by taking the limit of $\Delta k \to 0$

$$c_g = \lim_{\Delta k \to 0} \frac{\Delta \omega}{\Delta k} = \frac{d\omega}{dk} \tag{5.40}$$

which is seen in the beginning of this section.

Figure 5.3 and Eq. (5.37) show that the carrier wave has the same wave-number and frequency of the original plane waves, but its amplitude has to follow the movement of the group. The energy cannot travel through the nodes that are moving with the group. Therefore, the group velocity is also the velocity of the energy propagation.

For nondispersion waves, we have $\omega = ck$, which, substituting into Eq. (5.31), leads to $c_g = c$. Therefore, for nondispersion waves, the group velocity is the same as the phase velocity.

From Eq. (5.29), we can find that $\mathbf{M}_t$ is a symmetric matrix and $\mathbf{K}_t$ is a Hermitian matrix for a given real wave number. Hence, the frequency obtained from Eq. (5.29) must be real, but the corresponding eigenvector may be complex, and the left and right eigenvectors are complex conjugate. For the mth mode, the Rayleigh quotient can be written as

$$\omega_m^2 = \frac{\mathbf{\psi}_m^L \mathbf{K}_t \mathbf{\psi}_m^R}{\mathbf{\psi}_m^L \mathbf{M}_t \mathbf{\psi}_m^R} \tag{5.41}$$

where ψ_m^L and ψ_m^R are the mth transposed left and right eigenvectors obtained by solving the eigenvalue equation, Eq. (5.29). Taking the derivative of Eq. (5.41) with respect to k and noting that $\mathbf{M}_t$ is independent of k, we obtain

$$\frac{\partial \psi_m^L}{\partial k}[\mathbf{K}_t - \omega_m^2 \mathbf{M}_t]\psi_m^R + \psi_m^L[\mathbf{K}_t - \omega_m^2 \mathbf{M}_t]\frac{\partial \psi_m^R}{\partial k} + \psi_m^L\left[\frac{\partial \mathbf{K}_t}{\partial k} - 2\omega_m\frac{\partial \omega_m}{\partial k}\mathbf{M}_t\right]\psi_m^R = 0$$

(5.42)

The vectors $\partial \psi_m^L / \partial k$ and $\partial \psi_m^R / \partial k$ have the same size as ψ_m^L or ψ_m^R; hence, they can be expressed, respectively, in terms of all the transposed left eigenvectors and the right eigenvectors:

$$\frac{\partial \psi_m^L}{\partial k} = \sum_{n=1}^{M}\chi_n^k\psi_m^L, \quad \frac{\partial \psi_m^R}{\partial k} = \sum_{n=1}^{M}\lambda_n^k\psi_m^R$$

(5.43)

where χ_n^k and λ_n^k are arbitrary scaling factors. Substituting Eq. (5.43) into Eq. (5.42), using the orthogonality condition of the left and right eigenvectors, and with the help of Eq. (5.41), the first two terms in Eq. (5.42) become zero. Hence the group velocity of the mth mode can be given by

$$c_{gm} = \frac{\psi_m^L \mathbf{K}_{t,k} \psi_m^R}{2\omega_m \psi_m^L \mathbf{M}_t \psi_m^R}$$

(5.44)

where $\mathbf{K}_{t,k} = \dfrac{\partial \mathbf{K}_t}{\partial k}$ and is found using Eq. (5.27) as follows.

$$\mathbf{K}_{t,k} = 2k\mathbf{A}_{1t}\cos^2\theta + 2k\mathbf{A}_{2t}\cos\theta\sin\theta + 2k\mathbf{A}_{3t}\sin^2\theta + i\mathbf{A}_{4t}\cos\theta + i\mathbf{A}_{5t}\sin\theta$$

(5.45)

Different ways to obtain equations similar to Eq. (5.44) were given by Lysmer (1970) and Kausel and Roesset (1981).

5.4 Phase Velocity Surface

The phase velocity surface (PVS) shows the dependence of the phase velocity of a plane wave on the direction of wave propagation. The phase velocity of the mth mode of the plane wave is defined as

$$c_m(\theta) = \omega_m(\theta)/k$$

(5.46)

Considering Eq. (5.41), we obtain the function in polar coordinates for phase velocity surface for the mth mode of a plane wave propagating in direction θ with wavenumber k as

$$f_c(c_m,\ \theta)\ =\ \varphi_m^L \mathbf{K}_t \varphi_m^R - c_m^2 k^2 \varphi_m^L \mathbf{M}_t \varphi_m^R\ =\ 0 \qquad (5.47)$$

Plotting $f_c(c_m,\ \theta) = 0$ on polar coordinates gives the PVS.

5.5 Phase Slowness Surface

The phase slowness surface (PSS) shows the dependence of the relative arrival time of a plane wave on the direction of wave propagation θ; a normal vector on this surface is proportional to a position vector of the wave front from a point wave source (see proof in Section 5.6). The phase slowness for the mth mode is defined as the reciprocal of the phase velocity

$$s_m\ =\ 1/c_m \qquad (5.48)$$

With the help of Eq. (5.47), we have the phase slowness surface for the mth mode as

$$f_s(s_m,\ \theta)\ =\ s_m^2 \varphi_m^L \mathbf{K}_t \varphi_m^R - k^2 \varphi_{mt}^L \mathbf{M}_t \varphi_m^R\ =\ 0 \qquad (5.49)$$

Plotting $f_c(c_m,\ \theta) = 0$ on polar coordinates gives the PSS.

5.6 Phase Wave Surface

Consider now a point source (excitation) of waves in an anisotropic laminate. The excitation will generate waves propagating in all directions in the x-y plane that is the plane of the laminate. The waves propagating in all directions will form a *wave front* or *wave surface*. The wave surface for the mth wave mode can be constructed by superimposing all the plane fronts of plane waves in all the directions with phase velocity $c_m(\theta)$ as shown in Figure 5.4. It is, in fact, the envelope of all straight lines representing the front of plane waves in all the directions. The phase wave surface (PWS) used here represents the phase wave front emanating from a point wave source. The PWS for the mth mode is found by calculating the envelope formed by straight

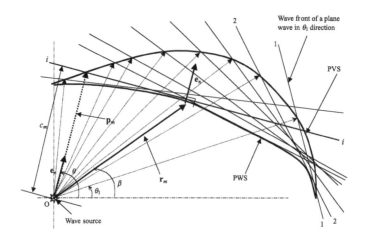

FIGURE 5.4
Relationship between PVS and PWS: $c_m = (\mathbf{r}_m \bullet \mathbf{e}_n)$.

lines of plane wave fronts originating from the same point source. From Figure 5.4, we can easily determine the relationship of the following vectors:

$$c_m = (\mathbf{r}_m \bullet \mathbf{e}_n) \tag{5.50}$$

where $\mathbf{e}_n$ denotes the unit vector of $\mathbf{p}_m$ whose tip draws the PVS, and $\mathbf{r}_m$ is a vector whose tip draws the PWS (see Figure 5.4). Let $\mathbf{s}_m$ be the vector whose tip draws the PSS. We then have

$$\mathbf{s}_m = \mathbf{e}_n / c_m \tag{5.51}$$

because the PSS is drawn exactly in the same way as PVS, except the length of the vector is $1/c_m$ instead of c_m. The direction of $\mathbf{s}_m$ is the same as that of $\mathbf{p}_m$, that is, $\mathbf{e}_n$. Therefore, we have

$$\mathbf{r}_m \bullet \mathbf{s}_m = \mathbf{r}_m \bullet \left[\frac{1}{c_m} \mathbf{e}_n \right] = \frac{1}{c_m} (\mathbf{r}_m \bullet \mathbf{e}_n) = \frac{1}{c_m} c_m = 1 \tag{5.52}$$

This means that $\mathbf{r}_m$ is parallel to $\mathbf{s}_m$. Hence $\mathbf{r}_m$ should be propotional to the gradient of a point on the PSS:

$$\mathbf{r}_m = \alpha \nabla \mathbf{f}_s \tag{5.53}$$

where α is a constant which can be determined as follows by substituting Eqs. (5.53) and (5.51) into (5.52).

$$\alpha = c_m/(\mathbf{e}_n \bullet \nabla \mathbf{f}_s) \tag{5.54}$$

Substituting the foregoing equation into Eq. (5.53) leads to

$$\mathbf{r}_m = c_m \nabla \mathbf{f}_s/(\mathbf{e}_n \bullet \nabla \mathbf{f}_s) \tag{5.55}$$

where

$$\nabla \mathbf{f}_s = f_{s,s_m} \mathbf{e}_n + f_{s,\theta}/s_m \mathbf{e}_t \tag{5.56}$$

$\mathbf{e}_n = [\cos\theta \ \sin\theta]$ and $\mathbf{e}_t$ are respectively the normal and tangential unit vectors to the phase wave surface formed by the trace of the tip of the vector $\mathbf{r}_m$. Substituting Eq. (5.56) into (5.55), we find

$$\mathbf{r}_m = c_m \mathbf{e}_n + c_m^2 f_{s,\theta}/f_{s,s_m} \mathbf{e}_t \tag{5.57}$$

Hence

$$|\mathbf{r}_m| = c_m \sqrt{1 + \alpha_s^2}, \quad \psi_s = \tan^{-1}\alpha_s \tag{5.58}$$

where

$$\alpha_s = c_m f_{s,\theta}/f_{s,s_m} \tag{5.59}$$

Differentiating Eq. (5.49) with respect to s_m and θ respectively, we obtain

$$f_{s,s_m} = 2\varphi_m^L \mathbf{K}_t \varphi_m^R/c_m \tag{5.60}$$

$$f_{s,\theta} = \varphi_m^L \mathbf{K}_{t,\theta} \varphi_m^R/c_m^2 \tag{5.61}$$

where

$$\mathbf{K}_{t,\theta} = -\mathbf{A}_1 k^2 \sin 2\theta + \mathbf{A}_2 k^2 \cos 2\theta + \mathbf{A}_3 k^2 \sin 2\theta - \mathbf{A}_4 ik \sin\theta + \mathbf{A}_5 ik \cos\theta \tag{5.62}$$

The phase wave surface can be drawn in the polar coordinate $(|\mathbf{r}_m|, \beta)$, where $\beta = \psi_s + \theta$ in reference to the x-axis (see Figure 5.4).

5.7   Group Velocity Surface

The group velocity surface (GVS) shows the dependence of the energy propagation velocity of a plane wave on the direction of propagation. Using
definition Eq. (5.31) and Eq. (5.41), we obtain the group velocity for the mth
mode of the plane wave as

$$c_{gm} = \frac{\varphi_m^L \mathbf{K}_{t,k} \varphi_m^R}{2\omega_m \varphi_m^L \mathbf{M}_t \varphi_m^R} \tag{5.63}$$

where

$$\mathbf{K}_{t,k} = 2\mathbf{A}_1 k \cos^2\theta + \mathbf{A}_2 \sin 2\theta + 2\mathbf{A}_3 k \sin^2\theta + i\mathbf{A}_4 \cos\theta + \mathbf{A}_5 i \sin\theta \tag{5.64}$$

From Eq. (5.63), we have the group velocity surface for the mth mode which
can be expressed as

$$f_g(c_{gm}, \theta) = \varphi_m^L \mathbf{K}_{t,k} \varphi_m^R - 2c_{gm}\omega_m \varphi_m^L \mathbf{M}_t \varphi_m^R \tag{5.65}$$

Plotting $f_g(c_{gm}, \theta) = 0$ on polar coordinates gives the GVS.

5.8   Group Slowness Surface

The group slowness surface (GSS) shows the dependence of the relative
arrival time of groups of a plane wave on the direction of wave propagation;
a normal vector on this surface is proportional to a vector that represents
the direction of energy flow from a point wave source. The group slowness
for the mth mode is defined as the reciprocal of the group velocity

$$q_m = 1/c_{gm} \tag{5.66}$$

From Eq. (5.63), we obtain the group slowness surface for the mth mode

$$f_q(q_m, \theta) = q_m \varphi_m^L \mathbf{K}_{t,k} \varphi_R^m - 2\omega_m \varphi_m^L \mathbf{M}_t \varphi_m^R \tag{5.67}$$

Plotting $f_q(q_m, \theta) = 0$ on polar coordinates gives the GSS.

5.9 Group Wave Surface

The group wave surface (GWS) shows the group propagation front emanating from a point wave source. The group wave surface for the mth mode can be obtained in the similar manner, except using GVS and GSS instead of PVS and PSS.

$$\mathbf{R}_m = c_{gm}\nabla f_q/(\mathbf{e}_q \bullet \nabla f_q) \tag{5.68}$$

where

$$\nabla f_q = f_{q,q_m}\mathbf{e}_n + f_{q,\theta}/q_m\mathbf{e}_t \tag{5.69}$$

$\mathbf{e}_n = [\cos\theta \ \sin\theta]$ and $\mathbf{e}_t$ are respectively the normal and tangential unit vectors to the group wave surface. Substituting Eq. (5.69) into Eq. (5.68), we find

$$\mathbf{R}_m = c_{gm}\mathbf{e}_n + c_{gm}^2 f_{q,\theta}/f_{q,q_m}\mathbf{e}_t \tag{5.70}$$

Hence

$$|\mathbf{R}_m| = c_{gm}\sqrt{1 + \alpha_q^2}, \quad \psi_q = \tan^{-1}\alpha_q \tag{5.71}$$

where

$$\alpha_q = c_{gm}f_{q,\theta}/f_{q,q_m} \tag{5.72}$$

Differentiating Eq. (5.67) with respect to q_m and θ, respectively, we have

$$f_{q,\theta} = q_m\varphi_m^L\mathbf{K}_{t,k\theta}\varphi_R^m - \varphi_m^L\mathbf{K}_{t,\theta}\varphi_m^R/\omega_m \tag{5.73}$$

where

$$\mathbf{K}_{t,k\theta} = -2\mathbf{A}_1k \sin 2\theta + 2\mathbf{A}_2 \cos 2\theta + 2\mathbf{A}_3k \sin 2\theta - i\mathbf{A}_4 \sin\theta + i\mathbf{A}_5 \cos\theta \tag{5.74}$$

The phase wave surface can be drawn in the polar coordinate ($|\mathbf{R}_m|$, β), where $\beta = \psi_q + \theta$ in reference to the x-axis.

Up to now, six characteristic wave surfaces have been formulated. These six wave surfaces reveal very important characteristics of the Lamb waves in laminates.

5.10 Examples

In this section, the following dimensionless parameters are used:

$$\xi = kH, \quad \lambda = L/H, \quad \Omega = \omega H/c_0, \quad c_0 = \sqrt{c_{11(0)}/\rho_0} \qquad (5.75)$$

where L stands for wave length, and ξ, λ, and Ω are the dimensionless wavenumber, dimensionless wavelength, and dimensionless frequency, respectively. c_0, $c_{11(0)}$, and ρ_0 are the longitudinal wave velocity in the direction of fiber orientation, the elastic coefficient in the fiber direction, and the density of a reference layer in a laminate, respectively. For the hybrid composite laminate, the carbon/epoxy layer is chosen to be the reference layer. The material constants are given in Table 3.1. The matrix of the off-principal-axis stiffness coefficients **c** of the lamina can be expressed in terms of engineering constants. The standard formulae are given by Vinson and Sierakowski (1987). The elastic coefficient matrix **c** for the graphite/epoxy laminates can be found in the reference given by Moon (1972).

5.10.1 Dispersion Curves

Figures 5.5 and 5.6 show, respectively, the frequency spectra of the lowest 19 modes and the group velocity spectra of the lowest 12 modes for laminate $[C0/G+45/G–45]_s$ where the wave propagates in the x-direction. From these

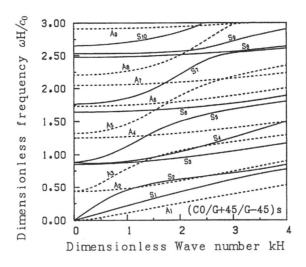

FIGURE 5.5

Frequency spectra for a hybrid laminate ($\theta = 0$; S: symmetric mode; A: anti-symmetric mode). (From Liu et al., *ASME J. Vibration and Acoustics*, 113, 279, 1991b. With permission.)

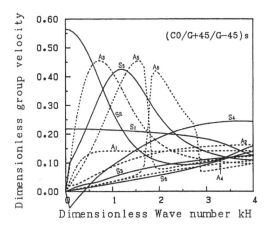

FIGURE 5.6
Group velocity spectra for a hybrid laminate ($\theta = 0$; S: symmetric mode; A: anti-symmetric mode). (From Liu, G. R., et al., *ASME J. Vibration and Acoustics*, 113, 279, 1991b. With permission.)

figures we can clearly see the dispersive behaviors of waves propagating in anisotropic laminates. In Figure 5.6, negative group velocities (S_4) are found for the small wave numbers. That is to say, the energy propagates in the direction that is opposite to that of the phase propagation. A similar phenomenon was observed by Lysmer (1970) in an isotropic plate.

5.10.2 Results for Graphite/Epoxy Laminates

The six characteristic surfaces of the first four modes in a graphite/epoxy $\pm30°$ laminate are shown in Figure 5.7 for the large wavelength ($\lambda = 20$). In order to illustrate the shape of the curves as completely as possible, each of these is plotted using a different scale factor. From Figure 5.7, the following remarks can be noted.

- The first two symmetric modes S_1 and S_2 for large wavelength agree with the results obtained by the approximate theory of Moon (1972).

- There are small differences between PVS and GVS, PSS and GSS, PWS and GWS for the S_1 and S_2 modes. This implies that S_1 and S_2 have weak dispersion for the large wavelength.

- There are considerable differences between PVS and GVS, PSS and GSS, PWS and GWS for the other modes (including the higher symmetric and antisymmetric modes, which are not shown in these figures) because of strong dispersion.

In Moon's approximate method, the laminate is treated as an anisotropic homogeneous plate using laminated plate theory. For this reason, the staking

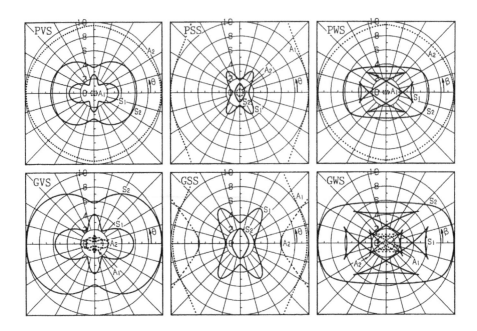

FIGURE 5.7
Characteristic surfaces for graphite/epoxy ±30° laminate ($\xi = \pi/10$; $\lambda = 20$; S: symmetric mode; A: anti-symmetric mode) Scale factors: PVS, $0.123c_0$; PSS, $1.746/c_0$; PWS, $0.123c_0$; GVS, $0.0776c_0$; GSS, $1.057/c_0$; GWS, $0.0776c_0$. (From Liu et al., *ASME J. Vibration and Acoustics*, 113, 279, 1991b. With permission.)

sequence of the layers is not effecting the results as shown in Figure 5.7. Assume that the laminate consists of 4 layers and has the sequence denoted by [30/−30/−30/30] and that the other parameters are the same as those in Figure 5.7. We calculate the six characteristic surfaces which are shown in Figure 5.8. Comparing Figure 5.7 with Figure 5.8 shows that for modes S_1 and S_2, the six characteristic surfaces shown in Figures 5.7 and 5.8 are almost the same. For the other modes, however, they are significantly different. This means that the sequence of layers has little influence on modes S_1 and S_2 when the wavelength is large, but has obvious influence on the other modes even if the wavelength is large. The influences of the sequence of layers can be observed also from the asymmetry of the curves. In Figure 5.7, the curves have two symmetric axes, but the curves in Figure 5.8 have no symmetric axis because the material properties of the laminate [30/−30/−30/30] have no symmetric axis in the plane of the laminate.

The characteristic surfaces in a laminate [30/−30/−30/30] for the small wavelength ($\lambda = 4$) are shown in Figure 5.9. PVS, PSS, and PWS are different from GVS, GSS, and GWS, respectively, for all the modes. Furthermore, all the curves are significantly different from those given by the approximate theory because all the modes have dispersion.

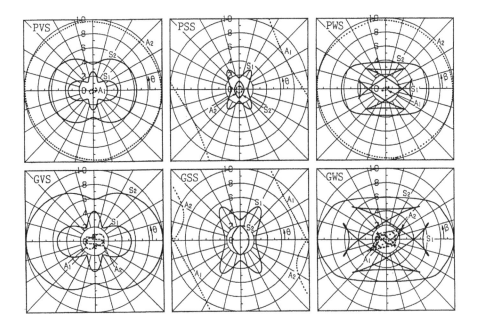

FIGURE 5.8
Characteristic surfaces for graphite/epoxy [30/−30/−30/30] laminate ($\xi = \pi/10$; $\lambda = 20$; S: symmetric mode; A: anti-symmetric mode) Scale factors: PVS, $0.119c_0$; PSS, $1.903/c_0$; PWS, $0.119c_0$; GVS, $0.0738c_0$; GSS, $1.11/c_0$; GWS, $0.0738c_0$. (From Liu et al., *ASME J. Vibration and Acoustics*, 113, 279, 1991b. With permission.)

5.10.3   Results for a Hybrid Composite Laminate

The characteristic surfaces of the lowest four modes in the laminate [C0/G+ 45/G−45]$_s$ are shown in Figure 5.10. From this figure, it can be seen that the shapes of the PVS, PSS, and PWS are, respectively, different from those of the GVS, GSS, and GWS of the corresponding mode. For instance, in the mode S_2, the largest phase velocity of the plane wave is observed at about $\theta = 146°$ or $\theta = −34°$ (see curve PVS), while the largest group velocity is observed at about $\theta = 116°$ or $\theta = −64°$ (see curve GVS). This means that the direction in which the plane wave propagates fastest is different from the direction with the fastest energy propagation. This is one of the important phenomena in the propagation of waves in anisotropic laminates. Such discrepancy in direction can be observed also for the point source wave (cf. curves PWS and GWS). Another interesting phenomenon can be observed from curve GVS. The three group velocities of modes A_1, A_2, and S_2 at about $\theta = 19°$ or $\theta = 161°$ are approximately equal. The energy of the three wave modes propagates with the same velocity in this direction. Hence, it is possible that the energy of the plane wave propagating in this direction may be more concentrated than that in other directions. For the point wave source, the four group wave fronts are closer to each other in the direction of $\beta = 0°$ or $\beta = 180°$.

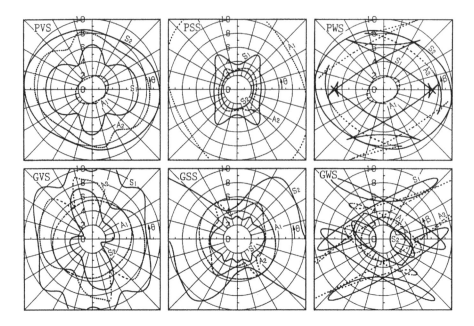

FIGURE 5.9

Characteristic surfaces for graphite/epoxy [30/−30/−30/30] laminate ($\xi = \pi/2$; $\lambda = 4$; S: symmetric mode; A: anti-symmetric mode) Scale factors: PVS, $0.0416c_0$; PSS, $1.057/c_0$; PWS, $0.0416c_0$; GVS, $0.0151c_0$; GSS, $1.692/c_0$; GWS, 0.01610. (From Liu et al., *ASME J. Vibration and Acoustics*, 113, 279, 1991b. With permission.)

5.11 Remarks

This chapter explained techniques for dealing with the dispersion of waves in composite laminates. Using numerical examples, it demonstrated that anisotropic laminates are dispersive medium. The velocity of wave is dependent on the wavelength or frequency. Waves propagating in the laminate form groups of waves. The group wave shape and velocity considerably differ from those of the phase waves. These findings are very helpful in understanding wave and energy propagation in composite laminates.

In practice, the present method also has other applications. In winter, ice often forms on the wings of aircraft and aerospace structures. We can analyze the thickness of the ice through studying the pertinent dispersion behaviors (Rose, 1999). The fluid presence has significant influence on composite structures such as submarines. Xi, Liu, Lam, and Shang (2000) have proposed a method for analyzing waves in fluid-loaded composite laminates.

Effects of the dispersion and anisotropy on the characteristics of waves propagating in composite laminates are examined. The dispersion caused by the

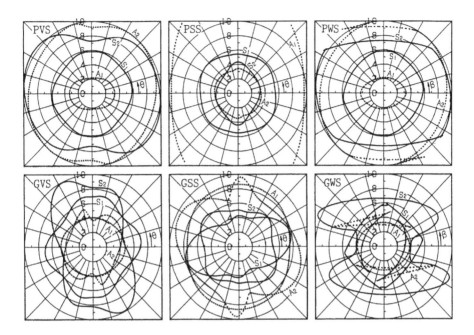

FIGURE 5.10
Characteristic surfaces for hybrid laminate $[C0/G{+}45/G{-}45]_s$ ($\xi = \pi/2$; $\lambda = 4$; S: symmetric mode; A: anti-symmetric mode) Scale factors: PVS, $0.037c_0$; PSS, $0.85/c_0$; PWS, $0.037c_0$; GVS, $0.03c_0$; GSS, $0.95/c_0$; GWS, $0.037c_0$. (From Liu et al., *ASME J. Vibration and Acoustics*, 113, 279, 1991b. With permission.)

presence of boundaries and interfaces of the laminate leads to a difference between the phase and group velocities. The dispersion also results in a difference between phase velocity surface and group velocity surface, between phase slowness surface and group slowness surface, and between phase wave surface and group wave surface. Furthermore, the anisotropy of the laminate leads to a dependence of the phase and group velocities on the direction of propagation and causes a difference between phase velocity surface and phase wave surface and between group velocity surface and group wave surface.

The characteristic surfaces have been successfully obtained by the use of layer elements and the application of the Rayleigh quotient, by which the phase and group velocities can be expressed explicitly as a function of wave number and the direction of plane wave propagation. The numerical results allow the following remarks to be made.

- The approximate method of Moon (1972) can give phase wave surfaces of the lowest two symmetric modes for a large wavelength.
- For an anisotropic laminate, the direction in which the plane wave propagates fastest is not the direction in which the energy propagates fastest.

- It is possible that the energy propagation in the hybrid laminate concentrates in the same direction because many modes propagate with almost the same group velocity in those directions.
- There are six characteristic wave surfaces in anisotropic laminates. With these six surfaces, the characteristics of the dispersive and anisotropic waves in laminates can be clearly revealed.

Finally, with regard to the methods presented in Chapter 3 and this chapter, the authors' experience is that the method given in this chapter is much more efficient. With the method in this chapter, there is no concern about missed roots (wave modes) in computation, and many modes can be found without difficulty. It is also very stable numerically. However, one has to pay special attention to the number of the elements used. If the number of elements used is too small, the results will not be accurate. The general guideline is that thickness of the element should be at least smaller than one quarter of the wavelength, if the quadratic shape function is used. The number should be at least doubled if linear shape functions are used. Finally, the lowest 20% of the all the wave modes is much more accurate than the reset modes.

6

Free Wave Motion in Anisotropic Laminated Bars: Finite Strip Element Method

6.1 Introduction

In engineering practice, laminated composite bars are common structural components. Many of them have a rectangular cross-sectional area. This chapter addresses problems of free wave motion in such anisotropic laminated bars. The difference between this topic and that from Chapter 1 is the recognition of the fact that the cross-section of the bar is not homogeneous and the field variables could vary in the cross-section of the bar. Therefore, the problem now is basically three dimensional. The difference between this topic and the topics in Chapters 3–5 is that the laminate is now bounded not only by the upper and lower surfaces, but also by the left and right edge surfaces, as shown in Figure 6.1. The channel for waves to propagate is an infinite long rectangular tunnel filled with layers of anisotropic materials.

Many researchers have investigated similar problems. Mindlin and Fox (1960) derived a special solution for bars of rectangular cross-section with certain ratios of width to depth that yields a discrete set of points in the frequency-versus-wavenumber branches. Aalami (1973) and Nigro (1966) obtained solutions for isotropic and orthotropic bars of rectangular cross-section using Ritz's method. Finite element solutions for orthotropic bars have also been given by Koshiba, Tanifuji, and Suzuki (1974), and Koshiba and Suzuki (1976).

This chapter presents an application of finite strip element method (Cheung, 1976) for the analysis of harmonic waves propagating along anisotropic laminated bars. This approach was originally proposed by Liu and co-workers (1990b). In this approach, the laminated bar is divided into N layer elements in the thickness direction, and the displacement field within each element is approximated by (linear) polynomials in the thickness direction and by a series in the width direction. The principle of virtual work is applied to establish the eigenvalue equations, whose solution gives the dispersion relation of harmonic waves in the laminated bar. As an example, the dispersion relations are computed using the present method for unidirectional

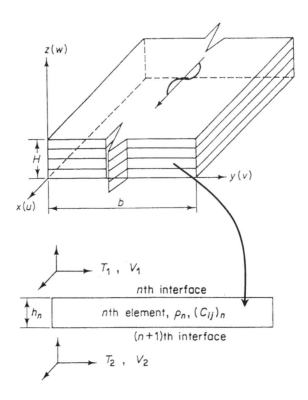

FIGURE 6.1
A laminated bar and the coordinate system. (From Liu, G. R., et al., *J. Sound and Vibration*, 139(2), 313, 1990b. With permission.)

carbon/epoxy laminated bars and hybrid composite laminated ones which consist of carbon/epoxy and glass/epoxy composite layers. Results for bars with different thickness-to-width ratios are also investigated.

6.2   System Equation

Consider plane harmonic waves propagating in a laminated bar that consists of layers of arbitrary anisotropic materials, as shown in Figure 6.1. H and b denote the thickness and width of the bar, respectively. The bar is divided into N elements in the thickness direction with $N - 1$ nodal surfaces and two surfaces of the bar. The number of the elements is usually more than the number of layers of the laminated bar. A layer in the bar should be divided into at least one element to ensure that the material property in an element is homogenous. More elements are required for waves of high frequencies. A general rule of thumb is that the thickness of any element should be less than one eighth of the wavelength of waves in the bar, if linear shape function is used. The thickness, elastic coefficient matrix, and density of the nth element are denoted by h_n, $c_n = (c_{ij})_n$ $(i, j = 1,...,6)$ and ρ_n, respectively, as shown in Figure 6.1. The wave is propagating in the x direction. The governing equation and the basic equations are exactly the same as that in Chapter 5.

Following the same argument in Chapter 5, we approximate the displacement field within the element in the form of

$$\mathbf{U} = \sum_{m=1}^{M} \mathbf{N}_m \mathbf{d}_m(z, y) \exp(ikx - i\omega t) = \mathbf{N}\mathbf{d}(z, y)\exp(ikx - i\omega t) \quad (6.1)$$

where $i = \sqrt{-1}, t, k,$ and ω are time, wavenumber, and angular frequency, respectively, and M is the terms of series. In Eq. (6.1), the matrix of shape functions is given by

$$\mathbf{N} = [\mathbf{N}_1 \quad \mathbf{N}_2 \quad \cdots \quad \mathbf{N}_m \quad \cdots \quad \mathbf{N}_M] \quad (6.2)$$

where

$$\mathbf{N}_m = \mathbf{N}_z \mathbf{Y}_{dm} \quad (m = 1,...,M) \quad (6.3)$$

in which $\mathbf{N}_z$ is a linear shape function of polynomials.

$$\mathbf{N}_z = [z/h_n \mathbf{I} \quad (1 - z/h_n)\mathbf{I}] \quad (6.4)$$

Here $\mathbf{I}$ denotes the 3×3 identity matrix. It can be easily found that the dimension of $\mathbf{N}$ is $(3 \times 6M)$. In Eq. (6.1), $\mathbf{d}$ is the unknown vector of displacements arranged in the following form.

$$\mathbf{d} = [\mathbf{d}_1 \quad \mathbf{d}_2 \quad \cdots \quad \mathbf{d}_m \quad \cdots \quad \mathbf{d}_M]^T \quad (6.5)$$

where

$$\mathbf{d}_m = [\mathbf{d}_{1m} \quad \mathbf{d}_{2m}]^T \quad (m = 1,...,M) \quad (6.6)$$

Therefore, the dimension of $\mathbf{d}$ is $(6M \times 1)$. In Eq. (6.3),

$$\mathbf{Y}_{dm} = \begin{bmatrix} \mathbf{Y}_m & 0 \\ 0 & \mathbf{Y}_m \end{bmatrix} \quad m = 1,..., M \quad (6.7)$$

where

$$
\mathbf{Y}_m = \begin{bmatrix} \mathbf{Y}_{um} & 0 & 0 \\ 0 & \mathbf{Y}_{vm} & 0 \\ 0 & 0 & \mathbf{Y}_{wm} \end{bmatrix} \tag{6.8}
$$

in which $\mathbf{Y}_{um}$, $\mathbf{Y}_{vm}$, and $\mathbf{Y}_{wm}$ are functions of y predetermined according to the boundary conditions at two edges of the bar.

Using Eq. (6.1), the displacement at the upper and lower surfaces of the element can be written as follows (see Figure 6.1):

$$
\mathbf{V}^T = \{\mathbf{V}_1^T \quad \mathbf{V}_2^T\} = \mathbf{Yd}\exp(ikx - i\omega t) \tag{6.9}
$$

where $\mathbf{Y} = [\mathbf{Y}_{d1} \quad \mathbf{Y}_{d2} \quad \cdots \quad \mathbf{Y}_{dM}]$.

The stresses on a plane z = constant can be written in terms of the displacement

$$
\mathbf{R} = \{\sigma_{xz} \quad \sigma_{yz} \quad \sigma_{zz}\}^T = \mathbf{L}_z^T\mathbf{cLU} \tag{6.10}
$$

The stress vector $\mathbf{S}$ consists of vectors of $\mathbf{R}$ at *lower* ($z = 0$), and *upper* ($z = h_n$) surfaces of the element:

$$
\mathbf{S} = \begin{Bmatrix} \mathbf{R}_l = \mathbf{R}|_{z=0} \\ \mathbf{R}_u = \mathbf{R}|_{z=h_n} \end{Bmatrix} \tag{6.11}
$$

With the help of Eq. (6.1), the nodal surface stresses $\mathbf{S}$ can be expressed as

$$
\mathbf{S} = \sum_{m=1}^{M}\mathbf{G}_m\mathbf{d}_m\exp(ikx - i\omega t) = \mathbf{Gd}\exp(ikx - i\omega t) \tag{6.12}
$$

where

$$
\mathbf{G} = [\mathbf{G}_1 \quad \mathbf{G}_2 \quad \cdots \quad \mathbf{G}_m \quad \cdots \quad \mathbf{G}_M] \tag{6.13}
$$

in which

$$
\mathbf{G}_m = \left[(ik\mathbf{Q}_1 + \mathbf{Q}_2/h)\mathbf{Y}_{dm} + \mathbf{Q}_3\frac{\partial\mathbf{Y}_{dm}}{\partial y}\right] \tag{6.14}
$$

In Eq. (6.14), matrices $\mathbf{Q}_i$ (i = 1, 2, 3) are given by

$$\mathbf{Q}_1 = \begin{bmatrix} \mathbf{Q}'_1 & \mathbf{Q}'_1 \\ \mathbf{Q}'_1 & -\mathbf{Q}'_1 \end{bmatrix}, \quad \mathbf{Q}_2 = \begin{bmatrix} \mathbf{Q}'_2 & \mathbf{Q}'_2 \\ -\mathbf{Q}'_2 & \mathbf{Q}'_2 \end{bmatrix}, \quad \mathbf{Q}_3 = \begin{bmatrix} \mathbf{Q}'_3 & \mathbf{Q}'_3 \\ \mathbf{Q}'_3 & -\mathbf{Q}'_3 \end{bmatrix} \quad (6.15)$$

$$\mathbf{Q}'_1 = \begin{bmatrix} c_{15} & c_{56} & c_{55} \\ c_{14} & c_{46} & c_{45} \\ c_{13} & c_{36} & c_{35} \end{bmatrix}, \quad \mathbf{Q}'_2 = \begin{bmatrix} c_{55} & c_{45} & c_{35} \\ c_{45} & c_{44} & c_{43} \\ c_{35} & c_{43} & c_{33} \end{bmatrix}, \quad \mathbf{Q}'_3 = \begin{bmatrix} c_{56} & c_{25} & c_{45} \\ c_{46} & c_{24} & c_{44} \\ c_{36} & c_{23} & c_{34} \end{bmatrix} \quad (6.16)$$

The displacement field assumed in Eq. (6.1) cannot satisfy the strong form of the equation of motion. We, therefore, seek for $\mathbf{d}$ that satisfies the weak form of the equation of motion by applying the principle of virtual work:

$$\int_0^b \delta \mathbf{V}^T \mathbf{T} \, dy = \int_0^b \delta \mathbf{V}^T \mathbf{S} \, dy + \int_0^b \int_0^{h_n} \delta \mathbf{U}^T \mathbf{W} \, dz \, dy \quad (6.17)$$

This implies that the virtual work performed by the external tractions $\mathbf{T}$ is equal to that performed by the interface stresses $\mathbf{S}$ and the unbalanced internal unbalanced distributed forces $\mathbf{W}$ caused by the assumption of the displacement field.

The external force $\mathbf{T}$ can be given in the following form of harmonic excitation:

$$\mathbf{T} = \bar{\mathbf{T}} \exp(ikx - i\omega t) \quad (6.18)$$

Substituting Eqs. (6.1), (6.9), and (6.18) into Eq. (6.17), after a lengthy but simple algebraic, differential operation followed by integrals over the element (in the thickness direction), we have

$$\mathbf{F} = [\mathbf{K}^* - \omega^2 \mathbf{M}] \mathbf{d} \quad (6.19)$$

where

$$\mathbf{F} = \int_0^b \mathbf{Y}^T \bar{\mathbf{T}} \, dy \quad (6.20)$$

The stiffness matrix in Eq. (6.19) consists of two terms as

$$\mathbf{K}^* = \mathbf{P} - \mathbf{K} \quad (6.21)$$

where matrix $\mathbf{P}$ is given by

$$\mathbf{P} = \int_0^b \mathbf{Y}^T \overline{\mathbf{G}} \, dy = \begin{bmatrix} \mathbf{P}_{11} & \mathbf{P}_{12} & \cdots & \mathbf{P}_{1M} \\ \mathbf{P}_{21} & \mathbf{P}_{22} & \cdots & \mathbf{P}_{2M} \\ \vdots & \vdots & \ddots & \vdots \\ \mathbf{P}_{M1} & \mathbf{P}_{M2} & \cdots & \mathbf{P}_{MM} \end{bmatrix} \tag{6.22}$$

in which

$$\mathbf{P}_{mn} = \int_0^b \mathbf{Y}_m^T \mathbf{G}_n \, dy = ik\mathbf{P}_{1mn} + \mathbf{P}_{2mn}/h_n + \mathbf{P}_{3mn}$$

$$\mathbf{P}_{jmn} = \int_0^b \mathbf{Y}_m^T \mathbf{Q}_j \mathbf{Y}_n \, dy \quad (j = 1, 2, 3) \tag{6.23}$$

To simplify the notation, we define an operation. If the elements of matrices $\mathbf{z}$, $\mathbf{a}$, and $\mathbf{b}$ are denoted by z_{ij}, a_{ij}, and b_{ij}, respectively, the multiplication between the corresponding elements of these two matrices, $z_{ij} = a_{ij} \times b_{ij}$, is denoted in the matrix level as $\mathbf{z} = \mathbf{a}*\mathbf{b}$. Using this definition, $\mathbf{P}_{jmn}$ can be expressed as follows:

$$\mathbf{P}_{1mn} = \mathbf{Q}_1 * \mathbf{I}_{1mn} \tag{6.24}$$

$$\mathbf{P}_{2mn} = \mathbf{Q}_2 * \mathbf{I}_{1mn} \tag{6.25}$$

$$\mathbf{P}_{3mn} = \mathbf{Q}_3 * \mathbf{I}_{3mn} \tag{6.26}$$

where

$$\mathbf{I}_{1mn} = \begin{bmatrix} \mathbf{I}_{0mn} & 0 \\ 0 & \mathbf{I}_{0mn} \end{bmatrix} \tag{6.27}$$

in which

$$\mathbf{I}_{0mn} = \int_0^b \begin{bmatrix} \mathbf{Y}_{um}\mathbf{Y}_{un} & \mathbf{Y}_{um}\mathbf{Y}_{vn} & \mathbf{Y}_{um}\mathbf{Y}_{wn} \\ \mathbf{Y}_{vm}\mathbf{Y}_{un} & \mathbf{Y}_{vm}\mathbf{Y}_{vn} & \mathbf{Y}_{vm}\mathbf{Y}_{wn} \\ \mathbf{Y}_{wm}\mathbf{Y}_{un} & \mathbf{Y}_{wm}\mathbf{Y}_{vn} & \mathbf{Y}_{wm}\mathbf{Y}_{wn} \end{bmatrix} dy \tag{6.28}$$

The matrix $\mathbf{I}_{3mn}$ can be obtained using Eq. (6.28) by replacing $\mathbf{Y}_{im}\mathbf{Y}_{in}$ ($i = u, v, w$) by $\mathbf{Y}_{im}\mathbf{Y}'_{in}$.

The matrix $\mathbf{K}$ in Eq. (6.21) is defined by

$$\mathbf{K} = \begin{bmatrix} \mathbf{K}_{11} & \mathbf{K}_{12} & \cdots & \mathbf{K}_{1M} \\ \mathbf{K}_{21} & \mathbf{K}_{22} & \cdots & \mathbf{K}_{2M} \\ \vdots & \vdots & \ddots & \vdots \\ \mathbf{K}_{M1} & \mathbf{K}_{M2} & \cdots & \mathbf{K}_{MM} \end{bmatrix} \tag{6.29}$$

where

$$\mathbf{K}_{mn} = \mathbf{K}_{1mn} + \mathbf{K}_{2mn} + \mathbf{K}_{3mn} + \mathbf{K}_{4mn} + \mathbf{K}_{5mn} \tag{6.30}$$

The matrices $\mathbf{K}_{imn}$ ($i = 1,\ldots,5$) are given by

$$\mathbf{K}_{1mn} = \mathbf{B}_1 * \mathbf{I}_{1mn} \tag{6.31}$$

$$\mathbf{K}_{2mn} = \mathbf{B}_2 * \mathbf{I}_{1mn} \tag{6.32}$$

$$\mathbf{K}_{3mn} = \mathbf{B}_3 * \mathbf{I}_{2mn} \tag{6.33}$$

$$\mathbf{K}_{4mn} = \mathbf{B}_4 * \mathbf{I}_{3mn} \tag{6.34}$$

$$\mathbf{K}_{5mn} = \mathbf{B}_5 * \mathbf{I}_{3mn} \tag{6.35}$$

where $\mathbf{I}_{2mn}$ can be obtained using Eq. (6.28) by replacing $\mathbf{Y}_{im}\mathbf{Y}_{in}$ ($i = u, v, w$) by $\mathbf{Y}_{im}\mathbf{Y}'_{in}$, and $\mathbf{B}_i$ ($i = 1,\ldots,5$) are given by

$$\mathbf{B}_1 = -\frac{k^2 h_n}{6}\begin{bmatrix} 2\mathbf{D}_{xx} & \mathbf{D}_{xx} \\ \mathbf{D}_{xx} & 2\mathbf{D}_{xx} \end{bmatrix} \tag{6.36}$$

$$\mathbf{B}_2 = ik\begin{bmatrix} \mathbf{D}_{xz} & -\mathbf{D}_{xz} \\ \mathbf{D}_{xz} & -\mathbf{D}_{xz} \end{bmatrix} \tag{6.37}$$

$$\mathbf{B}_3 = \frac{h_n}{6}\begin{bmatrix} 2\mathbf{D}_{yy} & \mathbf{D}_{yy} \\ \mathbf{D}_{yy} & 2\mathbf{D}_{yy} \end{bmatrix} \tag{6.38}$$

$$\mathbf{B}_4 = \frac{ikh_n}{3}\begin{bmatrix} 2\mathbf{D}_{xy} & \mathbf{D}_{xy} \\ \mathbf{D}_{xy} & 2\mathbf{D}_{xy} \end{bmatrix} \tag{6.39}$$

$$\mathbf{B}_5 = \begin{bmatrix} \mathbf{D}_{yz} & -\mathbf{D}_{yz} \\ \mathbf{D}_{yz} & -\mathbf{D}_{yz} \end{bmatrix} \tag{6.40}$$

In Eq. (6.19), the mass matrix $\mathbf{M}$ can be written as follows:

$$\mathbf{M} = \rho \int_0^b \int_0^h \mathbf{N}^T \mathbf{N} \, dz \, dy = \begin{bmatrix} \mathbf{M}_{11} & \mathbf{M}_{12} & \cdots & \mathbf{M}_{1M} \\ \mathbf{M}_{21} & \mathbf{M}_{22} & \cdots & \mathbf{M}_{2M} \\ \vdots & \vdots & \ddots & \vdots \\ \mathbf{M}_{M1} & \mathbf{M}_{M2} & \cdots & \mathbf{M}_{MM} \end{bmatrix} \tag{6.41}$$

where

$$\mathbf{M}_{mn} = \rho \int_0^b \int_0^h \mathbf{N}_m^T \mathbf{N}_n \, dz \, dy = \int_0^b \mathbf{Y}_{dm}^T \mathbf{R} \mathbf{Y}_{dn} \, dy \tag{6.42}$$

and

$$\mathbf{R} = \rho \int_0^h \mathbf{N}_z^T \mathbf{N}_z \, dz \tag{6.43}$$

By assembling the matrices in Eq. (6.19) of all the elements in the similar manner of the finite element method, the dynamic system equation for the entire bar becomes

$$\mathbf{F}_t = [\mathbf{K}_t^* - \omega^2 \mathbf{M}_t] \mathbf{d}_t \tag{6.44}$$

where the subscript t indicates that the matrix is for the whole bar. The dimension of the matrices $\mathbf{K}_t^*$ and $\mathbf{M}_t$ is $L \times L$ ($L = 3M(N + 1)$), where M is the number of terms of the series, and N is the number of the elements used. If the bar is free from traction at the interfaces, the characteristic equation can be expressed by

$$0 = [\mathbf{K}_t^* - \omega^2 \mathbf{M}_t] \mathbf{d}_t \tag{6.45}$$

For a given wavenumber k, L natural frequencies can be obtained from Eq. (6.45), which is the dispersion relationship of interest.

6.3 Examples

Two kinds of boundary conditions at both edges ($y = 0$ and b) of the bar are considered below.

A. Simply supported boundary condition defined by

$$u = 0, \quad w = 0, \quad \sigma_{yy} = 0, \tag{6.46}$$

which means that the displacement in the x and z direction is zero, and normal stress in the y direction vanishes at the two edges.

B. Clamped boundary conditions defined by

$$u = 0, \quad v = 0, \quad w = 0, \tag{6.47}$$

which means that all the displacement components must be zero at the two edges.

For an orthotropic laminate bar, the simply supported boundary condition (A) can be satisfied, if we assume

$$Y_{um} = \sin(m\pi y/b), \quad Y_{vm} = \cos(m\pi y/b), \quad Y_{wm} = \sin(m\pi y/b) \tag{6.48}$$

For an arbitrary anisotropic laminated bar, the clamped boundary condition (B) can be satisfied, if we assume

$$Y_{um} = \sin(m\pi y/b), \quad Y_{vm} = \sin(m\pi y/b), \quad Y_{wm} = \sin(m\pi y/b) \tag{6.49}$$

The above mentioned method is employed to compute the dispersion relation of bars. Figure 6.2 shows the calculated frequency spectra for a unidirectional carbon/epoxy bar with very large width of $b = 100H$. Figure 6.3 shows the exact results for plane waves propagating in the x-direction obtained using the method for an unbounded laminate of same stacking configuration. When the wavenumber is large enough, it can be seen from Figures 6.2 and 6.3 that the curves for the bar with a large width approach those for the unbounded laminate. This comparison validates the present numerical results.

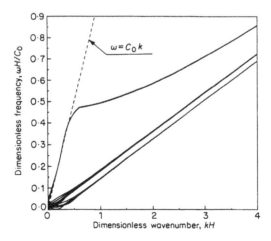

FIGURE 6.2

Frequency spectra for unidirectional bar [C0] with a clamped boundary condition ($b = 100H$). (From Liu, G. R., et al., *J. Sound and Vibration*, 139(2), 313, 1990b. With permission.)

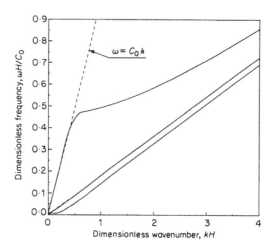

FIGURE 6.3
As Figure 6.2, but $b \rightarrow \infty$. (From Liu, G. R., et al., *J. Sound and Vibration*, 139(2), 313, 1990b. With permission.)

Frequency spectra for orthotropic bars (unidirectional carbon/epoxy denoted by [C0]), orthotropic hybrid laminated bars [C0/G90/G90/C0], and general anisotropic hybrid laminated bars [C0/G+45/G−45/C0] have been calculated and are plotted in Figures 6.2–6.11 for real wavenumbers. The bars are two or ten times as wide as they are thick. Material properties of the composites are given in Table 3.1. In Figures 6.2–6.11, $\omega H / c_0$ is the dimensionless frequency, and kH is a dimensionless wavenumber. The dotted lines $\omega = c_0 k$ and $\omega = c_s k$ express the frequency spectra for the infinite body of carbon/epoxy material whose fiber orientation is in the x direction. Further, c_0 and c_s are defined by

$$c_0 = \sqrt{c_{11(0)}/\rho_c}, \quad c_s = \sqrt{c_{66(0)}/\rho_c}$$

where $c_{11(0)}$ and $c_{66(0)}$ are the elastic coefficients of carbon/epoxy whose fiber orientation coincides with the x-axis, and ρ_c is the density of carbon/epoxy.

The number of elements N and the number of terms in the series M need to be chosen according to the desired accuracy. If only the lowest few eigenvalues and wave modes are required, the calculation can converge using a few terms of the series ($M = 3$ to 6). It is recommended, however, that the element number N be large enough to give sufficiently accurate results. The results shown in Figures 6.3–6.11 are obtained using $M = 6$ and $N = 16$. The lowest 10 eigenvalues have been plotted for laminated bars of $b = 2H$ and $b = 10H$.

The frequency spectra of bars [C0] are shown in Figures 6.4 and 6.5 for simply supported boundary conditions and in Figures 6.8 and 6.9 for perfectly clamped boundary conditions. The spectral features are significantly different on each side of the straight line $\omega = c_0 k$. At a specified value of kH,

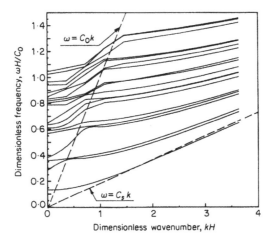

FIGURE 6.4
Frequency spectra for unidirectional bar [C0] with a simply supported boundary condition ($b = 2H$). (From Liu, G. R., et al., *J. Sound and Vibration*, 139(2), 313, 1990b. With permission.)

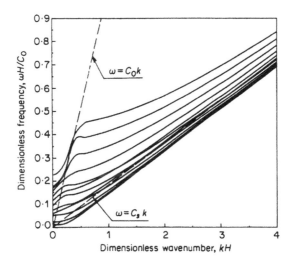

FIGURE 6.5
As Figure 6.4, but $b = 10H$. (From Liu, G. R., et al., *J. Sound and Vibration*, 139(2), 313, 1990b. With permission.)

the natural frequencies of the bars with $b = 2H$ are higher than those of the bars with $b = 10H$.

Figures 6.6 and 6.7 show the frequency spectra of orthotropic hybrid laminated bars [C0/G90/G90/C0] with simply supported boundary conditions at two edges of bars of $b = 2H$ and $b = 10H$, respectively. Figures 6.10 and 6.11 show the same for general anisotropic bars [C0/G+45/G−45/C0] with clamped boundary conditions.

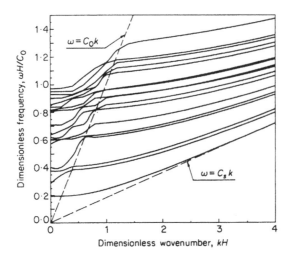

FIGURE 6.6
Frequency spectra for orthotropic laminated bar [C0/G90/G90/C0] with a simply supported boundary condition ($b = 2H$). (From Liu, G. R., et al., *J. Sound and Vibration*, 139(2), 313, 1990b. With permission.)

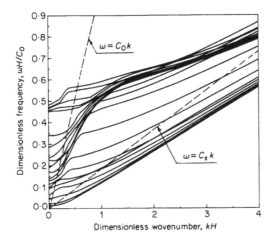

FIGURE 6.7
As Figure 6.6, but $b = 10H$. (From Liu, G. R., et al., *J. Sound and Vibration*, 139(2), 313, 1990b. With permission.)

In general, if the functions of series Y_{im} ($i = u, v, w$) can be found to satisfy the boundary conditions, the frequency spectra can be obtained accurately and effectively. Instead of the trigonometrical function, the following beam function also can be used for the clamped boundary condition:

$$Y_{im} = \sin(m\pi y/b) + \cos(m\pi y/b) + \alpha_m \sinh(m\pi y/b) - \cosh(m\pi y/b) \quad (6.50)$$

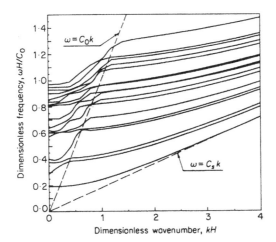

FIGURE 6.8
Frequency spectra for unidirectional bar [C0] with a clamped boundary condition ($b = 2H$). (From Liu, G. R., et al., *J. Sound and Vibration*, 139(2), 313, 1990b. With permission.)

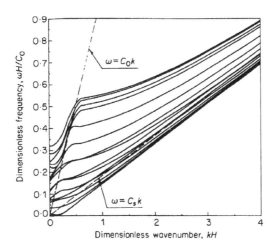

FIGURE 6.9
As Figure 6.8, but $b = 10H$. (From Liu, G. R., et al., *J. Sound and Vibration*, 139(2), 313, 1990b. With permission.)

where

$$\alpha_m = [\cosh(m\pi) - \cos(m\pi)]/\sinh(m\pi) \qquad (6.51)$$

Table 6.1 lists the lowest seven natural frequencies obtained for unidirectional carbon/epoxy by the two kinds of functions of series for comparison. A close agreement between the corresponding frequencies can be observed

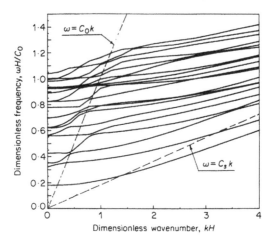

FIGURE 6.10
Frequency spectra for hybrid laminated bar [C0/G45/G-45/C0] with a clamped boundary condition ($b = 2H$). (From Liu, G. R., et al., *J. Sound and Vibration*, 139(2), 313, 1990b. With permission.)

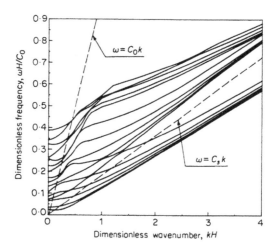

FIGURE 6.11
As Figure 6.10, but $b = 10H$. (From Liu, G. R., et al., *J. Sound and Vibration*, 139(2), 313, 1990b. With permission.)

from Table 6.1. The results obtained by the beam function are slightly larger than those using the trigonometrical function. Other new types of functions developed recently in the area of the finite strip element method can also be used for more complex boundary condition of bars at edges.

TABLE 6.1

Dimensionless Natural Frequencies Obtained for [C0] ($kH = 2.0$, $b = 2H$)

Mode	1	2	3	4	5	6	7
Trigonometric function	0.3822	0.5096	0.5342	0.6942	0.7104	0.7790	0.8738
Beam function	0.3857	0.5162	0.5407	0.7064	0.7149	0.7864	0.8775

Source: Adapted from Liu, G. R., et al., *J. Sound and Vibration*, 139(2), 313, 1990b. With permission.

6.4   Remarks

The solution for wave propagation in an unbounded laminated plate consists of infinite harmonic waves in the thickness direction only. These solutions can be obtained using the exact methods given in Chapter 1. For a bar with rectangular section, however, the exact solution must be formulated using infinite harmonic waves in the two directions of the cross-section. It is very difficult to determine the large number of harmonic waves that satisfy the boundary conditions at all surfaces of the bar. Hence, approximate methods have been developed, such as the Ritz method and the finite element method. In these methods, the cross-section of the bar is divided into elements in two directions. In the alternative method presented here, the bar is divided into elements only in the thickness direction, and the variation of the displacement in the width is expressed using a series of functions. This method makes it possible to obtain accurate results using only a few terms of the series, and, therefore, only comparatively small matrices need to be solved to obtain good results. When the width of the bar is much larger than its thickness, many elements would be needed in the finite element method to get good results, whereas with the present method a small number of plate elements can be used and the calculation is convergent within a few terms of the series. Hence, the present method considerably reduces the computation work for a solution with comparable accuracy. Furthermore, the present method is particularly suitable for the analysis of a laminated bars made of many thin layers with a large width-to-thickness ratio.

Finally, we mention that linear shape function is used for the thickness direction in the present formulation. A higher order of shape functions can, of course, be applicable for the present problem.

7

Free Wave Motion in Composite Laminated Bars: Semi-Exact Method

7.1 Introduction

The previous chapter discusses a finite strip element method to analyze free wave motion in anisotropic laminated bars with simple and clamped boundary conditions. In the strip element method, the laminated bars are divided into a number of layer elements, and the displacement field within each element is approximated using a linear shape function in the thickness direction and by a series in the width direction. The difficulty with this method is to choose the functions of series, which must satisfy the boundary conditions at the edges. This chapter introduces the semi-exact method, which does not need to choose any function satisfying any condition *a priori*. This method was proposed originally by Liu and co-workers (1990c, 1991c).

In the semi-exact method, the bar is divided into layer elements in the thickness direction. The displacement is approximated using polynomial shape functions only in the thickness direction to reduce the system equation to a set of dimension-reduced equations, which can be solved analytically. Then the dispersion relations of wave modes are obtained using the boundary conditions on the left and right sides of the bar. Numerical results are obtained for isotropic bars, unidirectional carbon/epoxy bars, and a hybrid laminated bar. This method is also applied to analyze waves on the edges of anisotropic plates. The results are compared with those obtained by other methods. It is found that the semi-exact method can give more accurate results.

7.2 System Equation

Consider a composite laminated bar that consists of layers of arbitrary anisotropic materials as shown in Figure 6.1. H and b denote the thickness and width of the bar, respectively. Following the same procedure in Chapter 6, the bar is divided into N elements in the thickness direction with $N-1$ nodal surfaces. The number of the element is usually more than the number of the layers of the laminated bar. A layer in the bar should be divided into at least one element to ensure that the material property in an element is homogenous. The thickness, elastic coefficient matrix, and density of the nth element are denoted by h_n, $c_n = (c_{ij})_n$ $(i, j = 1,...,6)$ and ρ_n, respectively. Harmonic plane waves are propagating in the x direction. The governing equation of motion and the basic equations are the same as those in Chapter 5. We approximate the displacement field within the element in the form of

$$\mathbf{U} = \mathbf{N}(z)\mathbf{d}\, \exp(i\eta y)\, \exp(ikx - i\omega t) \qquad (7.1)$$

where $\mathbf{d}$ is the unknown vector of displacement amplitudes at the nodal surfaces, and $\mathbf{N}$ is the matrix of linear shape functions given by

$$\mathbf{N} = [z/h_n \mathbf{I} \quad (1 - z/h_n)\mathbf{I}] \qquad (7.2)$$

in which $\mathbf{I}$ is a 3×3 identity matrix. In Eq. (7.1) ω is the angular frequency, k is wavenumber, and η is the characteristic value in the y direction. Using Eq. (7.1), the displacement on the upper and lower surfaces of the element can be written in terms of $\mathbf{d}$ as follows.

$$\mathbf{V}^T = \{\mathbf{V}_1^T \quad \mathbf{V}_2^T\} = \mathbf{d}^T \exp(i\eta y)\, \exp(ikx - i\omega t) \qquad (7.3)$$

The interface stresses $\mathbf{S}$ can be expressed as

$$\mathbf{S} = \mathbf{P}\mathbf{d}\, \exp(i\eta y)\, \exp(ikx - i\omega t) \qquad (7.4)$$

where $\mathbf{P}$ is obtained using Eqs. (6.10) and (6.11).

$$\mathbf{P} = [ik\mathbf{Q}_1 + \mathbf{Q}_2/h_n + i\eta\mathbf{Q}_3] \qquad (7.5)$$

These matrices $\mathbf{Q}_i$ $(i = 1, 2, 3)$ are the same as those given in Eq. (6.13) in the previous chapter.

The external tractions $\mathbf{T}$ can accordingly be expressed as follows:

$$\mathbf{T} = \bar{\mathbf{T}}\, \exp(i\eta y)\, \exp(ikx - i\omega t) \qquad (7.6)$$

The displacement field assumed by Eq. (7.1) cannot exactly satisfy the strong form of the equation of motion. It must, therefore, satisfy a weak form of the equation of motion based on the principle of virtual work:

$$\delta \mathbf{d}^T \mathbf{T} = \delta \mathbf{d}^T \mathbf{S} + \int_0^{h_n} \delta \mathbf{U}^T \mathbf{W}\, dz \qquad (7.7)$$

Equation (7.7) states that the virtual work performed by the external tractions **T** is equal to that performed by the interfacial stresses **S** and the unbalanced internal body forces **W** caused by the assumption to the displacements.

Substituting Eqs. (7.1), (7.3), (7.4), and (7.6) into (7.7), performing lengthy but simple algebraic and differential operations and integrations over the element, we obtain the following approximated system equation for an element:

$$\bar{\mathbf{T}} = [\mathbf{A}_1 + \eta \mathbf{A}_2 + \eta^2 \mathbf{A}_3]\mathbf{d} \tag{7.8}$$

where

$$\mathbf{A}_1 = ik\mathbf{Q}_1 + \mathbf{Q}_2/h_n - \mathbf{B}_1 - \mathbf{B}_2 - \omega^2 \mathbf{M} \tag{7.9}$$

and

$$\mathbf{A}_2 = i\mathbf{Q}_3 - \mathbf{B}_4 - \mathbf{B}_5 \tag{7.10}$$

$$\mathbf{A}_3 = \mathbf{B}_3 \tag{7.11}$$

The matrices **M** and **B**$_i$ ($i = 1,...,5$) are given by

$$\mathbf{M} = \frac{\rho h_n}{6}\begin{bmatrix} 2\mathbf{I} & \mathbf{I} \\ \mathbf{I} & 2\mathbf{I} \end{bmatrix} \tag{7.12}$$

$$\mathbf{B}_1 = -\frac{k^2 h_n}{6}\begin{bmatrix} 2\mathbf{D}_{xx} & \mathbf{D}_{xx} \\ \mathbf{D}_{xx} & 2\mathbf{D}_{xx} \end{bmatrix} \tag{7.13}$$

$$\mathbf{B}_2 = ik\begin{bmatrix} \mathbf{D}_{xz} & -\mathbf{D}_{xz} \\ \mathbf{D}_{xz} & -\mathbf{D}_{xz} \end{bmatrix} \tag{7.14}$$

$$\mathbf{B}_3 = \frac{h_n}{6}\begin{bmatrix} 2\mathbf{D}_{yy} & \mathbf{D}_{yy} \\ \mathbf{D}_{yy} & 2\mathbf{D}_{yy} \end{bmatrix} \tag{7.15}$$

$$\mathbf{B}_4 = \frac{ikh_n}{3}\begin{bmatrix} 2\mathbf{D}_{xy} & \mathbf{D}_{xy} \\ \mathbf{D}_{xy} & 2\mathbf{D}_{xy} \end{bmatrix} \tag{7.16}$$

Assembling the matrices of adjacent elements in a manner similar to the finite element method, the dynamic system equation for the entire bar can be obtained as

$$\bar{\mathbf{T}}_t = [\mathbf{A}_{1t} + \eta\mathbf{A}_{2t} + \eta^2\mathbf{A}_{3t}]\mathbf{d}_t \qquad (7.17)$$

where the subscript t means that the matrix is for the entire bar. The dimension of the matrices $\mathbf{A}_{1t}$, $\mathbf{A}_{2t}$, and $\mathbf{A}_{2t}$ is $M \times M$ ($M = 3N + 3$), where N is the number of the elements used.

Considering free wave motion, the bar is free from external traction on all the nodal surfaces, we have the following characteristic equation in terms of η as

$$0 = [\mathbf{A}_{1t} + \eta\mathbf{A}_{2t} + \eta^2\mathbf{A}_{3t}]\mathbf{d}_t \qquad (7.18)$$

Equation (7.18) can be changed into following general eigenvalue equation:

$$\begin{bmatrix} \mathbf{0} & \mathbf{I} \\ -\mathbf{A}_{1t} & -\mathbf{A}_{2t} \end{bmatrix} \begin{Bmatrix} \mathbf{d}_t \\ \eta\mathbf{d}_t \end{Bmatrix} = \eta \begin{bmatrix} \mathbf{I} & \mathbf{0} \\ \mathbf{0} & \mathbf{A}_{3t} \end{bmatrix} \begin{Bmatrix} \mathbf{d}_t \\ \eta\mathbf{d}_t \end{Bmatrix} \qquad (7.19)$$

Solving Eq. (7.19), we have $2M$ characteristic values η_i ($i = 1,\dots,2M$) and the corresponding $2M$ characteristic vectors. We note the upper half portion of the characteristic vector corresponding to η_i as

$$\mathbf{G}_i^T = [g_{1i} \quad g_{2i} \quad \cdots \quad g_{Mi}] \qquad (7.20)$$

The vector of displacements at the interface can now be expressed as

$$\mathbf{V}_t = \mathbf{G}(y)\mathbf{C}\exp(ikx - i\omega t) \qquad (7.21)$$

where

$$\mathbf{V}_t = \{u_1 \quad v_1 \quad w_1 \quad u_2 \quad v_2 \quad w_2 \quad \cdots \quad u_{N+1} \quad v_{N+1} \quad w_{N+1}\}^T \qquad (7.22)$$

$$\mathbf{G}(y) = \begin{bmatrix} g_{11}Y_1(y) & g_{12}Y_2(y) & \cdots & g_{1(2M)}Y_{2M}(y) \\ g_{21}Y_1(y) & g_{22}Y_2(y) & \cdots & g_{2(2M)}Y_{2M}(y) \\ \vdots & \vdots & \ddots & \vdots \\ g_{M1}Y_1(y) & g_{M2}Y_1(y) & \cdots & g_{M(2M)}Y_{2M}(y) \end{bmatrix} \qquad (7.23)$$

where

$$Y_m(y) = \exp(i\eta_m y), \quad m = 1,\dots,2M \qquad (7.24)$$

and $\mathbf{C} = [C_1 \; C_2 \; \cdots \; C_{2M}]^T$ are constants determined by the boundary condition at $y = 0$ and $y = b$. The stresses on the x-z plane on the upper and lower surfaces of the element can be given by

$$\mathbf{R} = \mathbf{R}_v \mathbf{V} + \mathbf{R}_d \frac{\partial \mathbf{V}}{\partial y} \tag{7.25}$$

where

$$\mathbf{R}_v = i k \mathbf{R}_1 + \mathbf{R}_2 / h_n \tag{7.26}$$

$$\mathbf{R}_d = \mathbf{R}_3 \tag{7.27}$$

The matrices $\mathbf{R}_i$ ($i = 1, 2, 3$) are given by

$$\mathbf{R}_1 = \begin{bmatrix} \mathbf{R}_1' & 0 \\ 0 & -\mathbf{R}_1' \end{bmatrix}, \quad \mathbf{R}_2 = \begin{bmatrix} \mathbf{R}_2' & -\mathbf{R}_2' \\ -\mathbf{R}_2' & \mathbf{R}_2' \end{bmatrix}, \quad \mathbf{R}_3 = \begin{bmatrix} \mathbf{R}_3' & 0 \\ 0 & \mathbf{R}_3' \end{bmatrix} \tag{7.28}$$

with

$$\mathbf{R}_1' = \begin{bmatrix} c_{41} & c_{46} & c_{45} \\ c_{16} & c_{66} & c_{65} \\ c_{21} & c_{26} & c_{25} \end{bmatrix}, \quad \mathbf{R}_2' = \begin{bmatrix} c_{45} & c_{44} & c_{43} \\ c_{65} & c_{64} & c_{63} \\ c_{25} & c_{24} & c_{23} \end{bmatrix}, \quad \mathbf{R}_3' = \begin{bmatrix} c_{46} & c_{42} & c_{44} \\ c_{66} & c_{62} & c_{64} \\ c_{26} & c_{22} & c_{24} \end{bmatrix} \tag{7.29}$$

Assembling these matrices in Eq. (7.25) of all the elements, the stresses on any surface of $y =$ constant of the bar can be written as follows:

$$\mathbf{R}_t = \mathbf{R}_{vt} \mathbf{V}_t + \mathbf{R}_{dt} \frac{\partial \mathbf{V}_t}{\partial y} \tag{7.30}$$

where

$$\frac{\partial \mathbf{V}_t}{\partial y} = \frac{\partial \mathbf{G}(y)}{\partial y} \mathbf{C} \exp(ikx - i\omega t) \tag{7.31}$$

in which

$$\frac{\partial \mathbf{G}(y)}{\partial y} = \begin{bmatrix} g_{11} \dfrac{\partial Y_1(y)}{\partial y} & g_{12} \dfrac{\partial Y_2(y)}{\partial y} & \cdots & g_{1(2M)} \dfrac{\partial Y_{2M}(y)}{\partial y} \\[2mm] g_{21} \dfrac{\partial Y_1(y)}{\partial y} & g_{22} \dfrac{\partial Y_2(y)}{\partial y} & \cdots & g_{2(2M)} \dfrac{\partial Y_{2M}(y)}{\partial y} \\[2mm] \vdots & \vdots & \ddots & \vdots \\[2mm] g_{M1} \dfrac{\partial Y_1(y)}{\partial y} & g_{M2} \dfrac{\partial Y_2(y)}{\partial y} & \cdots & g_{M(2M)} \dfrac{\partial Y_{2M}(y)}{\partial y} \end{bmatrix} \tag{7.32}$$

and

$$\frac{\partial Y_m(y)}{\partial y} = i\eta_m \exp(i\eta_m y), \quad (m = 1,\ldots,2M) \tag{7.33}$$

Substituting Eqs. (7.21) and (7.31) into Eq. (7.30), we have

$$\mathbf{R}_t = \mathbf{G}_s(y)\mathbf{C}\exp(ikx - i\omega t) \tag{7.34}$$

where

$$\mathbf{G}_s(y) = \mathbf{R}_{vt}\mathbf{G}(y) + \mathbf{R}_{dt}\frac{\partial \mathbf{G}(y)}{\partial y} \tag{7.35}$$

Using the simply supported boundary condition

$$\sigma_{xy} = 0, \quad \sigma_y = 0, \quad \sigma_{yz} = 0; \tag{7.36}$$

we obtain

$$\mathbf{G}_R\mathbf{C} = \mathbf{0} \tag{7.37}$$

where

$$\mathbf{G}_R = \begin{bmatrix} \mathbf{G}_s(0) \\ \mathbf{G}_s(b) \end{bmatrix} \tag{7.38}$$

For a nontrivial solution, the determinant of $\mathbf{G}_R$ must vanish, i.e.,

$$\det(\mathbf{G}_R) = \mathbf{0} \tag{7.39}$$

Using the clamped boundary condition on the edges, we have

$$u = 0, \quad v = 0, \quad w = 0; \tag{7.40}$$

which leads to

$$\mathbf{G}_V\mathbf{C} = \mathbf{0} \tag{7.41}$$

where

$$\mathbf{G}_R = \begin{bmatrix} \mathbf{G}(0) \\ \mathbf{G}(b) \end{bmatrix} \tag{7.42}$$

Also, for a nontrivial solution, the determinant of $\mathbf{G}_V$ must vanish, i.e.,

$$\det(\mathbf{G}_V) = \mathbf{0} \tag{7.43}$$

The elements of $\mathbf{G}_R$ and $\mathbf{G}_V$ are functions of the wavenumber k and the frequency ω. When the wavenumber k is given, the natural frequency can be obtained from Eq. (7.39) or Eq. (7.43). Consequently Eqs. (7.39) and (7.43) are the dispersion relations of harmonic plane waves propagating in the x direction in a bar for the traction-free and clamped boundary conditions, respectively. In practice, for any given wavenumber, the natural frequencies can be found numerically using root searching procedures. For a conservation system, the natural frequency will be real because the materials are assumed to be without damping. However, the characteristic value and the vector obtained from Eq. (7.19) are generally complex. Hence the numerical computations from Eqs. (7.19) to (7.43) should be performed in complex variables.

7.3   Examples of Harmonic Waves in Bars

First, we calculate natural frequencies and the corresponding wave modes for an orthotropic bar of rectangular cross section, with the clamped boundary condition given by Eq. (7.30). The material of the bar is a unidirectional fiber reinforced composite noted as carbon/epoxy. The natural frequencies are compared with those obtained using the finite strip element method. The natural frequencies and the corresponding modes for an isotropic bar of square cross section with Poisson's ratio of 0.3 have also been calculated. The results are compared with those given by Aalami (1973). Finally, we calculate the natural frequencies and the corresponding modes for an aniso-tropic laminated bar with stacking sequence of [C0/G+45/G–45/C0] (where C and G stand for the carbon/epoxy and glass/epoxy layer, respectively, and the numbers following the letters express the fiber-orientation angle with respect to the x axis). The material constants for the composites are given in Table 3.1. The following dimensionless parameters are used in the computation.

$$\xi = kH, \quad \Omega_k = \omega H/c_s, \quad \Omega_c = \omega H/c_0$$
$$c_0 = \sqrt{c_{11(0)}/\rho_c}, \quad c_s = \sqrt{G_k/\rho_k}$$

where ξ, Ω_k, c_s, G_k, and ρ_k are the dimensionless wavenumber, the dimensionless frequency, the transverse wave velocity in the infinite body, the shear modulus, and the density of the isotropic material, respectively. Ω_e and c_0 are the dimensionless frequency of the composite material and the longitudinal

TABLE 7.1

Convergence of Natural Frequencies of Nonpropagating
Modes in a Unidirectional Carbon/Epoxy Bar Clamped
at $y = 0$, b ($\xi = 0.0$, $b = 2H$)

Methods	N	Wave Modes			
		1	2	3	4
Finite strip element method (Chapter 6)	16	0.1917	0.2852	0.3770	0.4033
Semi-exact method (Chapter 7)	2	0.2020	0.2844	0.3923	0.4057
	5	0.1942	0.2844	0.3806	0.4031
	8	0.1925	0.2844	0.3776	0.4024
	9	0.1922	0.2844	0.3771	0.4023
	10	0.1920	0.2844	0.3768	0.4021

Source: Liu, G. R., et al. *Wave Motion*, 12, 361, 1990c. With permission.

wave velocity in the fiber direction of the unbounded carbon/epoxy composite body, respectively. $c_{11(0)}$ is the elastic coefficient c_{11} of the carbon/epoxy composite when the fiber orientation is parallel to the x-axis, and ρ_c is the mass density of the carbon/epoxy composite.

7.3.1 Results for a Clamped Bar

To verify the convergence of the method, the natural frequencies for $\xi = 0.0$, 2.0, 4.0 are calculated for an orthotropic rectangular bar clamped on the two sides, using different number of elements N. The bar is made of carbon/epoxy composite with $b = 2H$. Table 7.1 shows the variations of the natural frequencies of the first four modes at $\xi = 0$ computed using different numbers of elements. Very fast convergence has been observed. Since $\xi = 0$, the corresponding modes are not propagating, and hence it is called *non-propagating modes*. At a non-propagating mode, the disturbance in the bar stays stationary and does not vary with the x coordinate. The corresponding natural frequencies are called cut-off frequencies at which the modes of waves stop propagating.

Table 7.2 and Table 7.3 show the variations of the natural frequencies of the first six modes for $\xi = 2.0$ and $\xi = 4.0$ obtained using different numbers of elements, respectively. The results are compared with those calculated using the finite strip element method discussed in the previous chapter, where the bar was divided into 16 finite strip elements. From Table 7.1 to Table 7.3, it can be seen that the results converge very quickly. When N is larger than 5, good results can be obtained for the natural frequencies of the lowest few wave modes. The results obtained by these two methods are in very good agreement.

The shapes of the first nine wave modes for $\xi = 2.0$ are shown in Figure 7.1. For a mode, the mode shape is presented in a perspective figure for the displacement u and in a plane figure for the displacements v and w.

TABLE 7.2

Convergence of Natural Frequencies of Wave Modes in a Unidirectional
Carbon/Epoxy Bar Clamped at $y = 0$, b ($\xi = 2.0$, $b = 2H$)

Methods	N	Wave Modes					
		1	2	3	4	5	6
Finite strip element method (Chapter 6)	16	0.3822	0.5096	0.5342	0.6942	0.7104	0.7790
Semi-exact method (Chapter 7)	5	0.3822	0.5155	0.5342	0.7026	0.7205	0.7896
	6	0.3865	0.5136	0.5339	0.7000	0.7142	0.7859
	8	0.3846	0.5116	0.5334	0.6967	0.7113	0.7819
	9	0.3840	0.5107	0.5332	0.6957	0.7105	0.7807
	10	0.3836	0.5101	0.5331	0.6949	0.7099	0.7798
	12	0.3830	0.5094	0.5330	0.6938	0.7091	0.7786

Source: Liu, G. R., et al. *Wave Motion*, 12, 361, 1990c. With permission.

TABLE 7.3

Convergence of Natural Frequencies of Wave Modes in a Unidirectional
Carbon/Epoxy Bar Clamped at $y = 0$, b ($\xi = 4.0$, $b = 2H$)

Methods	N	Wave Modes					
		1	2	3	4	5	6
Finite strip element method (Chapter 6)	16	0.7201	0.7976	0.8217	0.9272	0.9448	0.9890
Semi-exact method (Chapter 7)	5	0.7312	0.8077	0.8227	0.9387	0.9511	1.0037
	8	0.7252	0.8019	0.8216	0.9317	0.9461	0.9945
	9	0.7240	0.8007	0.8214	0.9301	0.9453	0.9928
	10	0.7230	0.7998	0.8212	0.9291	0.9447	0.9915

Source: Liu, G. R., et al. *Wave Motion*, 12, 361, 1990c. With permission.

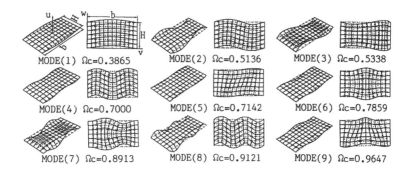

MODE(1) $\Omega c = 0.3865$ MODE(2) $\Omega c = 0.5136$ MODE(3) $\Omega c = 0.5338$

MODE(4) $\Omega c = 0.7000$ MODE(5) $\Omega c = 0.7142$ MODE(6) $\Omega c = 0.7859$

MODE(7) $\Omega c = 0.8913$ MODE(8) $\Omega c = 0.9121$ MODE(9) $\Omega c = 0.9647$

FIGURE 7.1
Mode shapes of a unidirectional carbon/epoxy bar ($\xi = 2.0$, $b = 2H$, clamped at $y = 0$, b). (From
Liu, G. R., et al. *Wave Motion*, 12, 361, 1990c. With permission.)

7.3.2 Results for a Free Bar

Harmonic waves in a bar of squared cross-section ($b = H$) with free lateral surfaces are investigated. The bar is made of isotropic material with a Poisson's ratio of 0.3. The natural frequencies and mode shapes are obtained for $\xi = \pi$ using six elements. The natural frequencies for the lowest 13 modes are listed in Table 7.4, together with these obtained by Aalami (1973) for comparison. Aalami's results are obtained using 100 finite elements. The natural frequencies obtained by the present semi-exact method agree well with those given by Aalami (1973) except for mode 9. The shapes of 12 modes are plotted in Figure 7.2. Comparing the shape of the modes, it can be found that the ninth mode does not appear within the lowest 10 modes in Aalami's

TABLE 7.4

Comparison of Dimensionless Natural Frequencies of Wave Modes in an Isotropic Square Bar with Free Lateral Surface ($\xi = \pi$, $v = 0.3$)

Modes	1	2	3	4	5	6	7	8	9	10	11	12	13
Aalami (1973)	2.393	2.393	2.922	4.456	4.464	4.491	4.999	4.999	—	6.088	6.088	—	—
Chapter 7	2.381	2.396	2.908	4.428	4.443	4.511	4.900	4.968	5.938	6.037	6.113	6.600	6.690

Source: Liu, G. R., et al. *Wave Motion*, 12, 361, 1990c. With permission.

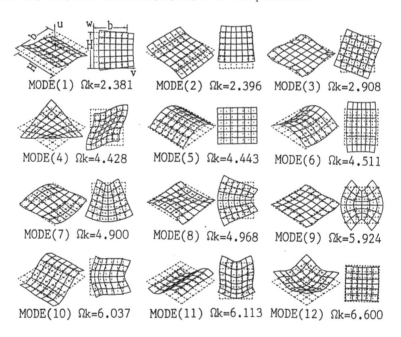

MODE(1) Ωk=2.381   MODE(2) Ωk=2.396   MODE(3) Ωk=2.908

MODE(4) Ωk=4.428   MODE(5) Ωk=4.443   MODE(6) Ωk=4.511

MODE(7) Ωk=4.900   MODE(8) Ωk=4.968   MODE(9) Ωk=5.924

MODE(10) Ωk=6.037   MODE(11) Ωk=6.113   MODE(12) Ωk=6.600

FIGURE 7.2

Mode shapes of an isotropic square bar ($\xi = \pi$, $v = 0.3$, free from any lateral restraint). (From Liu, G. R., et al. *Wave Motion*, 12, 361, 1990c. With permission.)

TABLE 7.5

Dimensionless Natural Frequencies of Wave Modes in a Hybrid Laminated
Square [C0/G45/G–45/C0] Bar with Free Lateral Surfaces

Mode	1	2	3	4	5	6	7	8	9	10	11
$\xi = 0.0$	0.4660	0.5521	0.5899	0.6428	0.6662	0.7911	0.8260	0.9430	0.9567	0.9691	1.0405
$\xi = 2.0$	0.2989	0.3062	0.6224	0.6787	0.6871	0.7258	0.7353	0.8737	0.8999	0.9963	1.0637

Source: Liu, G. R., et al. *Wave Motion*, 12, 361, 1990c. With permission.

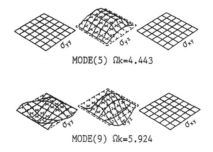

MODE(5) Ωk=4.443

MODE(9) Ωk=5.924

FIGURE 7.3
Distribution of stresses on plane *x-z*. (From Liu, G. R., et al. *Wave Motion*, 12, 361, 1990c. With permission.)

results, and the shape of the fifth mode obtained here is different from that obtained by Aalami (1973). To varify the fifth and the ninth modes, we calculate the distribution of the stresses $\mathbf{R}_t$ along the *y* direction and plot the results in Figure 7.3. The free boundary condition is indeed satisfied, which confirms the correctness of these modes obtained using the present semi-exact method.

7.3.3 Anisotropic Laminated Bar

The results for a hybrid laminated square bar of [C0/G+45/G–45/C0] are computed and discussed in this subsection. The same problem was investigated in the previous chapter, but for a clamped boundary condition using the finite strip element method. In this section, the natural frequencies for the free boundary condition are obtained using 8 elements. Eleven natural frequencies for $\xi = 0.0$ and $\xi = 2.0$ are listed in Table 7.5. Shapes of the lowest twelve modes for $\xi = 0.0$ are shown in Figure 7.5. On comparing these mode shapes with those for the square isotropic bar shown in Figure 7.2, it can be seen that the mode shapes of waves in the bar made of hybrid composite material are more complex than those made of isotropic material.

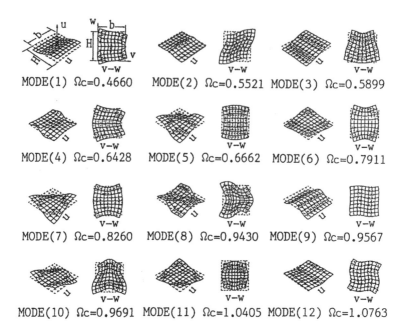

MODE(1) $\Omega c=0.4660$ MODE(2) $\Omega c=0.5521$ MODE(3) $\Omega c=0.5899$

MODE(4) $\Omega c=0.6428$ MODE(5) $\Omega c=0.6662$ MODE(6) $\Omega c=0.7911$

MODE(7) $\Omega c=0.8260$ MODE(8) $\Omega c=0.9430$ MODE(9) $\Omega c=0.9567$

MODE(10) $\Omega c=0.9691$ MODE(11) $\Omega c=1.0405$ MODE(12) $\Omega c=1.0763$

FIGURE 7.4

Mode shapes of a hybrid laminated square bar [C0/G45/G–45/C0] ($\xi = 0.0$, free from any lateral restraint). (From Liu, G. R., et al. *Wave Motion*, 12, 361, 1990c. With permission.)

7.4 Edge Waves in Semi-Infinite Laminates

Rayleigh waves are surface waves on the surface of half space. These waves are very important in seismology. They are widely used in nondestructive evaluation (NDE) (Achenbach and Rajapakse, 1987), and they have also been widely employed in surface acoustic wave (SAW) devices (Ash and Paige, 1985). Waves confined to the edges of plates are important in understanding problems of edge impact of plate-like structures, such as jet engine fan blades. Ohyoshi and Miura (1986) detected a very strong surface wave at the edge of an isotropic plate when a small, hard plastic ball impacts the edge of the plate. For composite laminates, delaminations can easily originate at the edges during wave propagation in the plate. Hence, the analysis of edge waves propagating in an anisotropic laminates is of considerable importance.

It is very difficult to find the exact solution for edge wave motion, especially for anisotropic laminates. A finite element solution was given by Lagasse (1973). However, the finite element solution does not converge for small wavenumbers. An approximate solution for the first anti-symmetric mode was obtained based on a plate theory (Sinha, 1974). It is noted that the solution of the plate theory is not suitable for large wavenumbers. Therefore, an effective method is required. In the following, the semi-exact method

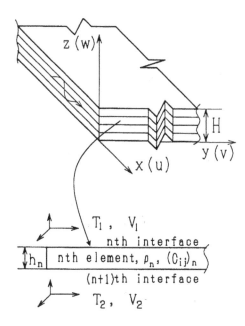

FIGURE 7.5
Semi-infinite laminate, isolated element and the coordinate system. (From Liu, G. R., et al. *Wave Motion*, 13, 243, 1991. With permission.)

is used to study the surface waves propagating along the edges of semi-infinite isotropic plates and anisotropic laminates.

Consider now an edge of a composite laminate as shown in Figure 7.5. We use the semi-exact method to analyze harmonic plane waves propagating on the edge of the laminate in the x-direction. In the formulation, equations up to Eq. (7.19) given in this chapter can be adopted exactly as they are. After solving the eigenvalue equation, Eq. (7.19), and obtaining $2M$ ($M = 3N + 3$) characteristic values η_i ($i = 1,...,2M$) and the corresponding $2M$ characteristic vectors, we need to apply nonreflecting boundary conditions to ensure that there are no waves coming (or reflected) from $y = \infty$. In other words, we should have only leftward-going wave modes. The nonreflecting boundary condition in this case should be

$$\mathrm{Im}(\eta_i) > 0 \tag{7.44}$$

The reasoning for the imposition of the above equation is very similar with that in Chapters 1, 4, and 12 when determining the indentation to include or exclude the poles. Note from Eq. (7.1) that the displacement has the term of $\exp[i(\eta y - \omega t)]$. In order to have only waves of leftward going, Eq. (7.44) needs to be satisfied. As all the eigenvalues η_i obtained from Eq. (7.19) corresponding to waves propagating in both negative and positive directions of y, Eq. (7.44) will naturally exclude exactly half of the eigenvalues and their corresponding wave modes. The reset of the equations from Eqs. (7.20) to

(7.43) is valid, except that only half of the wave modes that satisfy Eq. (7.44) should be used. Accordingly, the number of the boundary conditions is also halved because there is only one edge for a semi-infinite laminate.

7.4.1 Verification of Results for Edge Waves

Edge waves in isotropic bars are investigated first. To verify the convergence of the natural frequencies with an increasing number of elements N, the natural frequencies of the first anti-symmetric mode versus the number of elements have been computed for dimensionless wavenumbers $\xi = 2.00$ and π. The results are listed in Table 7.6. The Poisson's ratio of the isotropic material of the bar is 0.3. From Table 7.6, it can be seen that the results converge very fast, and when $N = 4$ satisfactory results can be obtained. For the first symmetric mode, the frequency converges even faster. Hence the following results are calculated using $N = 6$ for isotropic and unidirectional laminates and $N = 4$ for hybrid composite laminates. Figure 7.6 shows the

TABLE 7.6

Convergence of Natural Frequencies for Edge
Waves in a Semi-Infinite Isotropic Bar
($v = 0.3$, Wave Mode A1)

Number of Elements N	2	4	6	8
$\eta = 2.0$	1.348	1.312	1.303	1.300
$\eta = \pi$	2.536	2.463	2.442	2.433

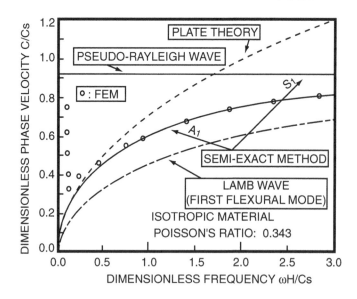

FIGURE 7.6
Comparison of phase-velocity spectra. (From Liu, G. R., et al. *Wave Motion*, 13, 243, 1991. With permission.)

TABLE 7.7

Convergence of Natural Frequencies for Edge Waves in a Semi-Infinite Isotropic Plate ($v = 0.3$, Wave Mode A1)

Wavenumber ξ	0.25	0.5	0.75	1.0	1.5	2.0	2.5	3.0	3.5	4.0
Pseudo-Rayleigh Wave (Ohyoshi et al., 1986)	0.9203	0.9203	0.9203	0.9203	0.9203	0.9203	0.9203	0.9203	0.9203	0.9203
Present S_1 mode	0.9203	0.9202	0.9201	0.9200	0.9196	0.9191	0.9186	0.9180	0.9174	0.9168

dispersion curves obtained by the semi-exact method for the isotropic plate with $v = 0.343$, together with those obtained based on a plate theory (Sinha, 1974). The results are also listed for the first symmetric mode of edge waves obtained using the finite element method (Lagasse, 1973) for the first symmetric mode obtained by the so-called pseudo-Rayleigh theory.

For the first anti-symmetric mode (A1), the agreement between the two sets of results obtained by the semi-exact method and the FEM is very good at high frequencies. They show, however, striking differences at lower frequencies. When the frequency approaches zero, the phase-velocity obtained by the FEM tends to become unbounded. The results obtained by the plate theory are in good agreement with those obtained by the semi-exact method at low frequencies. However, significant difference is found at higher frequencies between the results obtained by the semi-exact method and the method based on plate theory. The dispersion curve for the first anti-symmetric mode (A1) of Lamb waves (in an infinite plate) is also shown in Figure 7.6, for comparison. The shape of the curve of the Lamb wave is somewhat similar to that of the first anti-symmetric mode obtained by the semi-exact method. However, the phase-velocities of edge waves are larger than that of the Lamb wave.

For the first symmetric mode, the phase velocities obtained by the semi-exact method are close to those of the pseudo-Rayleigh wave that is nondispersive. It is difficult to make a distinction between them in Figure 7.6. Hence the data are also listed in Table 7.7. The phase velocity of the first symmetric mode has very weak dispersion and is a little smaller than that of the pseudo-Rayleigh wave. The larger the wavenumber, the greater the difference between the phase velocity of the first symmetric mode and that of the pseudo-Rayleigh. However, even if $\xi = 4$, the difference is only about 0.35%.

7.4.2 Effect of Poisson's Ratio on Edge Waves

The frequency spectra of the first anti-symmetric mode (A1) have been calculated for Poisson's ratios of $v = 0.05, 0.20$, and 0.45 and plotted in Figure 7.7. For a given wavenumber, the frequencies increase with the increasing Poisson's ratio. The first symmetric wave mode (S1) exhibits very weak dispersion, regardless of Poisson's ratio. The phase-velocity of wave mode S1 is close to that of the pseudo-Rayleigh wave. This implies that the pseudo-Rayleigh wave can give a very good approximation of the first symmetric mode of the edge wave.

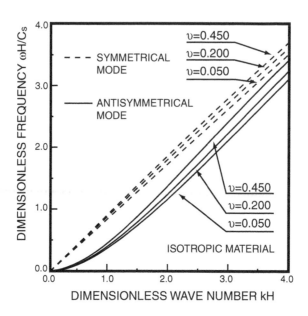

FIGURE 7.7
Influence of Poisson's ratio on frequency spectra for modes S1 and A1. (From Liu, G. R., et al. *Wave Motion*, 13, 243, 1991. With permission.)

For the first anti-symmetric mode, the greater the Poisson's ratio, the weaker the dispersion becomes. It should be emphasized that a much weaker dispersion is observed when $\xi > 3.0$, and the frequency spectra can be approximated by a straight line. This means that when $\xi > 3.0$, the group velocity of the first anti-symmetric mode of the edge wave c_{ga} is approximately a constant. Due to this important finding, we can derive an approximate formula for estimating the group velocity for isotropic materials of different Poisson's ratios.

First, we approximate the frequency spectra for each of three Poisson's ratios in the wavenumber range of $3.75 < \xi < 4.00$ by straight lines and calculate the three group velocities. Then, expressing the group velocity in linear function of Poisson's ratio using polynomial regression, the following equation has been obtained:

$$c_{ga} = 0.9801 + 0.1595v - 0.1826v^2 \qquad (7.45)$$

The mode shapes of the first anti-symmetric and symmetric modes are, respectively, shown in Figure 7.8 for $v = 0.343$ and in Figure 7.9 for $v = 0.050$. The displacement u is plotted by perspective drawings and the displacements v and w are plotted by plane drawings. The displacement fields in the range of $0 < y < 5H$ are plotted. The anti-symmetric and symmetric modes are shown in the left and right sides of Figures 7.8 and 7.9, respectively.

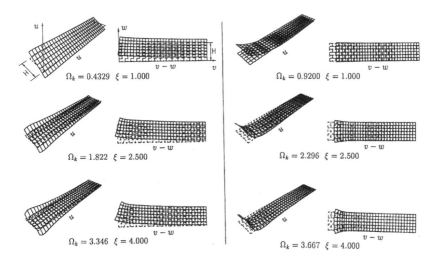

FIGURE 7.8
Mode shapes of a semi-infinite isotropic plate (left side: anti-symmetric mode; right side: symmetric mode; Poisson's ratio $v = 0.343$). (From Liu, G. R., et al. *Wave Motion*, 13, 243, 1991. With permission.)

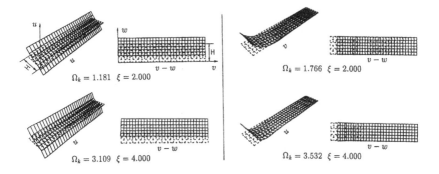

FIGURE 7.9
Mode shapes of a semi-infinite isotropic plate (left side: anti-symmetric mode; right side: symmetric mode; Poisson's ratio $v = 0.50$). (From Liu, G. R., et al. *Wave Motion*, 13, 243, 1991. With permission.)

From Figures 7.8 and 7.9, it can be found that, as the wave number increases, the wave becomes more strongly concentrated on the neighborhood of the edge for both anti-symmetric and symmetric modes. Comparing Figures 7.8 and 7.9, it can be seen that the larger the Poisson's ratio, the more the displacements concentrate near the edge of the plate for the first anti-symmetric mode. The concentration of the displacements for the first symmetric mode is, however, almost independent of Poisson's ratio of the material.

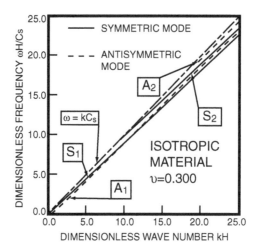

FIGURE 7.10
Frequency spectra of higher order modes of edge waves. (From Liu, G. R., et al. *Wave Motion*, 13, 243, 1991. With permission.)

7.4.3 Higher Modes of Edge Waves

McCoy and Mindlin (1963) obtained the second symmetric mode based on a second-order equation. The frequency spectra for the four lowest modes (A1, S1, A2, S2) of the edge waves have been calculated using the semi-exact method and are plotted in Figure 7.10, where the following remarks for A2 and S2 can be noted.

- The frequency spectra for A2 and S2 are above a certain frequency (the cut-off frequency).
- Both A2 and S2 exibits only slight dispersion.

The mode shapes of A2 and S2 are shown in Figure 7.11. The displacements of these modes of edge waves are strongly confined to the edge of the plate.

7.4.4 Results for Anisotropic Semi-Infinite Laminates

The frequency spectra for the lowest two modes of edge waves in composite semi-infinite laminates have been computed and are plotted in Figure 7.12. For the unidirectional laminate [C0], the modes can be divided into the symmetric mode S1 and the anti-symmetric mode A1 because the laminate has a symmetric middle plane with respect to material properties. From Figure 7.12 it can be observed that S1 has almost no dispersion. For the anti-symmetric mode, A1, a strong dispersion is observed at small wavenumbers

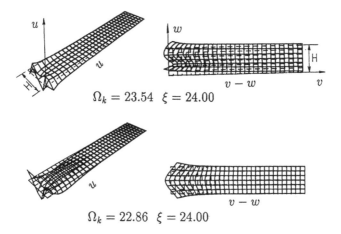

FIGURE 7.11
Mode shapes of higher order modes (A2 and S2) of edge waves in a semi-infinite isotropic plate ($v = 0.3$). (From Liu, G. R., et al. *Wave Motion*, 13, 243, 1991. With permission.)

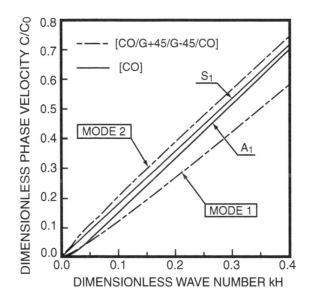

FIGURE 7.12
Frequency spectra of edge waves in composite laminates. (From Liu, G. R., et al. *Wave Motion*, 13, 243, 1991. With permission.)

and a slight dispersion is observed at large wavenumbers. For the hybrid laminate [C0/G+45/G–45/C0], the displacements become complicated and the mode shape is neither symmetric nor anti-symmetric. The two lowest modes demonstrate a slight dispersion at large wavenumbers and a stronger dispersion at small wavenumbers.

7.5   Remarks

The exact solution for a bar with a rectangular cross section may be composed of solutions for infinite harmonic waves that satisfy the boundary conditions on all the lateral surfaces. However, the composition cannot be easily done in practice. Hence, approximate methods have been developed, such as Ritz's method and the finite element method. In these methods, the displacement field is approximated in the cross section of the bar and the approximate solution will consist of element-wise solutions in two directions. In this chapter, the bar is divided into elements only in the thickness direction in which the material properties of the bar vary. The displacement field is approximated in the thickness direction in elements. In the width direction, however, analytical formulation is employed. Because of this nature of the method, it is called a semi-exact method. Using this method, an accurate solution can be obtained using only a few elements because the displacement and the stress boundary conditions are satisfied accurately on the boundaries. In particular, it is a good method for an inhomogeneous laminated bar. The laminated bar is made of many layers and an exact analysis becomes very difficult due to the presence of the interfaces. By using the semi-exact method, these difficulties can be overcome by dividing the laminated bar into layer elements. The numerical results allow us to make the following remarks.

- The results obtained by the present method for an isotropic bar with a square cross section agree fairly well with those obtained by Aalami (1973). However, the ninth mode obtained here does not appear within the 10 modes of Aalami's results and the mode shape of the fifth mode is different.

- The results for an orthotropic bar are in good agreement with those obtained in the previous chapter using the finite strip element method, which confirms the accuracy of the semi-exact method.

- The mode shapes of anisotropic laminated bars are much more complex than those of isotropic bars.

When the FEM is used to analyze edge waves, the semi-infinite plate should be considered as a wide strip. One side of the strip is free and the other side is completely clamped for the simulation of the nonreflecting boundary condition at $y \to \infty$, where the displacements vanish. However, at lower frequencies, the displacements approach zero very slowly as y increases (see Figure 7.8) even when $y = 5H$, the displacements are still quite large. Hence, the completely clamped boundary condition cannot be expected to lead to satisfactory results. When the frequency closes to zero, the phase-velocity becomes infinite. Further, when the Poisson's ratio is small, the displacements also approach zero very slowly as y increases (see

Figure 7.9). Consequently, a very wide strip must be modeled if the FEM is used to obtain good results.

Because the plate is not divided into elements in the y direction, the semi-exact method is very suitable for the problem of wave propagation in a semi-infinite laminates. For example, if we let $N = 2$ and use the symmetric condition, the phase velocity of the Rayleigh waves can be obtained using a large value of thickness H. The phase velocity of pseudo-Rayleigh wave can also be obtained using the same model but with a small value of thickness H. This is because the pseudo-Rayleigh wave is a wave at the edge of a very thin semi-infinite plate, and it is assumed that the stresses on the z-plane vanish (plane stress). The stresses on the z-plane for the first symmetric mode obtained by the semi-exact method are not zero, but they are very small. Hence, the semi-exact method confirms that the pseudo-Rayleigh wave is a very good approximation of the first symmetric mode of the edge waves. As shown in this chapter, this method can be used effectively for the analysis of characteristics of edges waves in anisotropic laminates. Finally, the results for the edge wave give us the following findings.

- The larger the wave number, the more the wave is confined to the neighborhood of the edge of the semi-infinite plate.
- A very good approximation of the first symmetric mode of the edge wave can be given by the pseudo-Rayleigh theory.
- For the first anti-symmetric mode there is a slight dispersion at high frequency and the group velocity can be obtained using Eq. (7.42) for plates with different Poisson's ratios.

Compared to the finite strip element method given in the previous chapter, the semi-exact method is much more adaptive to the boundary conditions on the edges of bars. The semi-exact method can be used for any types of conditions on the boundary of the edges of the bar because there is no need to predefine functions of series that must satisfy the boundary conditions at the edges. Its application to the semi-infinite plates is an excellent example. The disadvantage of the semi-exact method compared to the finite strip element or finite element method is that an efficient algorithm is needed to search the roots of Eq. (7.43), and there is no guarantee that one will not miss any roots in the search process.

8

Transient Waves in Composite Laminates

8.1 Introduction

For aircraft and automotive structures made of laminated composites, the analysis of transient response of anisotropic laminated plates is very important for structural design, monitoring, and nondestructive evaluation. The studies of transient waves excited by an impulsive force have been summarized by Achenbach (1973), Miklowitz (1978), and Pao (1977). For the problem of transient waves excited by a line load, the method of generalized rays, i.e., an exact method, has been used for isotropic plates (Miklowitz, 1978) and for isotropic laminated plates (Pao, 1977; Wiggins and Helmberger, 1974). The reflectivity method (Booth and Crampin, 1983; Fryer and Frazer, 1984) has been used for stratified media. Lu and Felsen (1986) proposed a hybrid method that combined the method of generalized rays and the method of normal modes (Weaver and Pao, 1982) for an isotropic plate. Furthermore, the approximate method, such as the method of the head of the pulse approximation (Scott and Miklowitz, 1967), has been used for isotropic and anisotropic laminated plates. Moreover, the approximate method based on plate theory has been used for anisotropic plates (Moon, 1973).

For the problem of transient waves excited by a point load, the method of generalized rays (Pao, 1977; Miklowitz, 1978; Ceranoglu and Pao, 1981) and the method of normal modes (Weaver and Pao, 1982; Santosa and Pao, 1989) have been used for isotropic plates. Moreover, the approximate methods based on plate theories have been used for isotropic plates (Ohyoshi, 1989) and for anisotropic laminated plates (Sun and Tan, 1984). Indeed, the method of generalized rays can give a highly accurate short-time solution of transient waves in an isotropic plate. However, when the observation is at a large distance, or when the plate consists of many anisotropic layers, a very large number of generalized ray integrals must be evaluated and this method becomes impractical. To overcome this shortcoming, van der Hijden (1987) proposed an improved method of generalized rays using an "energy-based criterion" in selecting rays for integration. The method of normal modes has been shown to be effective for isotropic plates; however, this method is seldom applied to an anisotropic laminated plate due to the difficulty of finding the eigenfunctions (eigenvalues and eigenvectors). The difficulty does not lie in the formulation, but in the cumbersome calculation necessary to search for eigenvalues. Chapter 3 discusses an exact method to obtain the eigenvalues of anisotropic laminated plates. The eigenvalues have been successfully obtained from transcendental equations with the aid of a search procedure. The limitations of the procedure are

- The search process is very time consuming, and
- There is no guarantee against missing any roots (eigenfunctions).

These difficulties must be overcome before the method of normal modes can be applied to anisotropic laminated plates.

This chapter introduces a hybrid numerical method (HNM) that combines the layer element method with the method of Fourier transforms. The HNM was initially proposed by Liu et al. (1991a). Chapter 5 demonstrates that the layer element method is very efficient and reliable in computing the wave modes. The HNM uses first the layer element method to obtain the modes in the wavenumber domain and then uses modal analysis to express the displacement in a summation of wave modes. Inverse Fourier transform is finally carried out using FFT or a quadrature technique presented in this chapter to obtain the displacement in the spatial domain. The procedure of the HNM is summarized below.

1. Apply the principle of virtual work to develop an approximate dynamic system equation that is a set of partial differential equations (PDEs).
2. Perform a (spatial-wavenumber) Fourier transform to this set of PDEs to obtain a set of differential equations with respect only to time in the wavenumber domain.
3. Formulate the displacement in the wavenumber domain using the modal analysis procedure for given transient loads.
4. Numerically perform the inverse (wavenumber-spatial) Fourier transform using the FFT or a method presented herein to obtain the displacement response field in the space-time domain.

This chapter details the above procedure and develops the formulation for the HNM. Examples of the displacement responses of isotropic plates and a hybrid laminate excited by step-impact loads are computed using the code developed of the hybrid numerical method. The results are validated and investigated in the following steps.

First, to validate the hybrid numerical method, we make a comparison among the numerical results obtained by the present method and the other methods. For a line step-impact load, the displacement responses of an isotropic plate with Poisson's ratio of $\frac{1}{3}$ are calculated and the results are compared with those obtained by the method of the head of the pulse approximation (Scott and Miklowitz, 1967). For a point step-impact load, the displacement responses of an isotropic plate with Poisson's ratio of 0.21 are calculated and the results are compared with those obtained by the method of generalized rays (Ceranoglu and Pao, 1981).

Next, the displacement responses of a hybrid laminated composite plate excited by line and point step-impact loads are computed. This hybrid laminate consists of two carbon/epoxy layers and four glass/epoxy layers with a symmetric stacking sequence. The stacking sequence of the laminated layers is denoted by $[C0/G+45G-45]_s$, where 0, +45, and −45 indicate the angle of fiber-orientation with respect to the x axis. The material constants of the two kinds of composite materials are listed in Table 3.1. Numerical results for the displacement responses of laminates excited by line and point step-impact loads are presented using different integration schemes.

8.2 HNM Formulation

Consider a composite laminate with any number of anisotropic layers and the overall thickness of H. We assume that the layers are homogenous in material property and perfectly bonded together. The laminate is excited by a point dynamic load, as shown in Figure 8.1. The composite laminate is first

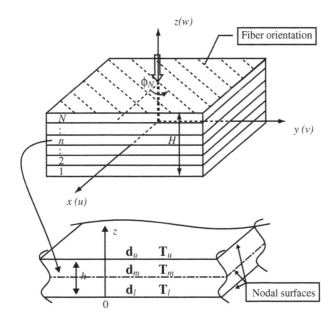

FIGURE 8.1
An anisotropic laminate and the nth isolated layer element.

divided into N strip elements in the thickness direction. The thickness, mass density, and matrix of elastic constants of an element are denoted, respectively, by h, ρ, and $\mathbf{c}$ (defined by Eq. (4.8)), which could be different from element to element. Note that the number of elements is usually more than the number of layers of materials, meaning a layer can be divided into more than one element. This is to ensure that the material property in one element can be homogeneous. In an element, the governing system equation of motion in matrix form is given as

$$\mathbf{W} = \rho\ddot{\mathbf{U}} - \mathbf{L}^{T}\mathbf{c}\mathbf{L}\mathbf{U} - \mathbf{f} \tag{8.1}$$

where $\mathbf{f}$ is the distributed external force (or body force) applied within the element, and the matrix of material constant $\mathbf{c}$, operator matrix $\mathbf{L}$ are defined in Section 4.4. The exact solution of $\mathbf{U}$ should lead to $\mathbf{W} = 0$. Chapter 4 demonstrates that the exact solution can be sought. Here we formulate a numerical method that can produce approximate results more efficiently. To this end, we first approximate the displacement field within the element in the form of

$$\mathbf{U} = \mathbf{N}(z)\mathbf{d}(x, y, t) \tag{8.2}$$

where $\mathbf{N}(z)$ is the matrix of shape functions. We use here the quadratic shape function of polynomials:

$$\mathbf{N}(z) = [(1 - 3\bar{z} + 2\bar{z}^{2})\mathbf{I} \quad 4(\bar{z} - \bar{z}^{2})\mathbf{I} \quad (2\bar{z}^{2} - \bar{z})\mathbf{I}] \tag{8.3}$$

in which $\bar{z} = z/h$, and $\mathbf{I}$ is a 3×3 identity matrix. In Eq. (8.2), $\mathbf{d}$ is a vector consisting of nodal displacement vectors, which are functions of the axes x, y and time t, at $z = 0$, $z = 0.5h$, and $z = h$ as follows:

$$\mathbf{d} = \begin{Bmatrix} \mathbf{V}_{l} \\ \mathbf{V}_{m} \\ \mathbf{V}_{u} \end{Bmatrix} \tag{8.4}$$

with

$$\mathbf{V}_{i}^{T} = \{u \quad v \quad w\}_{i} \quad (i = l, m, u) \tag{8.5}$$

where u, v, and w are the displacement components at, respectively, x, y, and z directions.

The stresses on a plane z = constant in the element can be written as

$$\mathbf{R} = \{\sigma_{xz} \quad \sigma_{yz} \quad \sigma_{zz}\}^T = \mathbf{L}_z^T \mathbf{cLU} \tag{8.6}$$

where and $\mathbf{L}_z^T$ is defined in Section 4.2. The stress vector $\mathbf{S}$ is defined as vectors of $\mathbf{R}$ evaluated at *lower* ($z = 0$), *middle* ($z = 0.5h_n$), and *upper* ($z = 0.5h_n$) nodal surfaces of the element:

$$\mathbf{S} = \begin{Bmatrix} \mathbf{R}_l \\ \mathbf{R}_m \\ \mathbf{R}_u \end{Bmatrix} \tag{8.7}$$

Assuming external force excitation at the nodal surface of the element can be written as

$$\mathbf{T} = \begin{Bmatrix} \mathbf{T}_l \\ \mathbf{T}_m \\ \mathbf{T}_u \end{Bmatrix} \tag{8.8}$$

The assumption of (8.2) will, in general, result in $\mathbf{W} \neq 0$ at points in the element. We therefore need to have $\mathbf{W} = 0$ to be satisfied in a weak sense by introducing the principle of virtual work expressed as follows:

$$\delta \mathbf{d}^T \mathbf{T} = \delta \mathbf{d}^T \mathbf{R} + \int_0^{h_n} \delta \mathbf{U}^T \mathbf{W} \, dz \tag{8.9}$$

Equation (8.9) states that the virtual work performed by the external tractions $\mathbf{T}$ is equal to that performed by the interfacial stresses $\mathbf{S}$ and the unbalanced internal body forces $\mathbf{W}$ caused by the assumption of the displacements. Applying the principle of virtual work to the element, performing lengthy and simple operations, we obtain the approximate system equation of dynamic equilibrium for the element as

$$\mathbf{F} = \mathbf{M}\ddot{\mathbf{d}} + \mathbf{K}_D \mathbf{d} \tag{8.10}$$

where

$$\mathbf{K}_D = -\mathbf{A}_1 \frac{\partial^2}{\partial x^2} - \mathbf{A}_2 \frac{\partial^2}{\partial x \partial y} - \mathbf{A}_3 \frac{\partial^2}{\partial y^2} + \mathbf{A}_4 \frac{\partial}{\partial x} + \mathbf{A}_5 \frac{\partial}{\partial y} + \mathbf{A}_6 \tag{8.11}$$

The force vector **F** in Eq. (8.2) is given by

$$\mathbf{F} = \mathbf{T} + \int_0^{h_n} \mathbf{N}^T \mathbf{f} \, dz \tag{8.12}$$

which is the vector of total external force consisting of nodal forces applied on the nodal surfaces of the element and distributed forces applied within the element. In Eq. (8.11), mass matrix **M** and matrices $\mathbf{A}_i$ $(i = 1,...,6)$ have been given in Chapter 5. From the above procedure, the original partial differential equation with respect to variables x, y, z, and t has been changed into an approximated partial differential Eq. (8.10) with respect only to variables x, y, and t.

Assembling the matrices of all the elements, we obtain a set of system equations of dynamic equilibrium for the entire laminate:

$$\mathbf{F}_t = \mathbf{K}_{Dt}\mathbf{d}_t + \mathbf{M}_t\ddot{\mathbf{d}}_t \tag{8.13}$$

where

$$\mathbf{K}_{Dt} = -\mathbf{A}_{1t}\frac{\partial^2}{\partial x^2} - \mathbf{A}_{2t}\frac{\partial^2}{\partial x \, \partial y} - \mathbf{A}_{3t}\frac{\partial^2}{\partial y^2} + \mathbf{A}_{4t}\frac{\partial}{\partial x} + \mathbf{A}_{5t}\frac{\partial}{\partial y} + \mathbf{A}_{6t} \tag{8.14}$$

where the subscript D denotes the matrix $\mathbf{K}_t$ is a differential operator matrix. Matrices $\mathbf{M}_t$ and $\mathbf{A}_{it}$ $(i = 1, 2,...,6)$ are obtained by assembling **M** and $\mathbf{A}_i$ $(i = 1, 2,...,6)$ for all the elements. The external force vector $\mathbf{F}_t$ is obtained by assembling the force vectors **F** for all the elements. Because quadratic shape functions are used and the three nodal surfaces in one element, the dimension of these matrices are $M \times M$ $(M = 6N + 3)$, where N is the number of the layer elements, and the dimension of these vectors is $M \times 1$.

8.3   Equation in Wavenumber Domain

The Fourier transform of spatial to wavenumber can be defined as follows:

$$\tilde{\mathbf{d}}_t(k_x, k_y, t) = \int_{-\infty}^{+\infty}\int_{-\infty}^{+\infty} \mathbf{d}_t(x, y, t)e^{ik_x x}e^{ik_y y} \, dx \, dy \tag{8.15}$$

The application of the Fourier transform to Eq. (8.13) leads to a set of system equations for the entire plate as follows:

$$\tilde{\mathbf{F}}_t = \mathbf{M}_t\ddot{\tilde{\mathbf{d}}}_t + \mathbf{K}_t\tilde{\mathbf{d}}_t \tag{8.16}$$

where $\tilde{\mathbf{F}}_t$, $\tilde{\ddot{\mathbf{d}}}_t$, and $\tilde{\mathbf{d}}_t$ are the Fourier transform of $\mathbf{F}_t$, $\ddot{\mathbf{d}}_t$, and $\mathbf{d}_t$, respectively. Matrix $\mathbf{K}_t$ is given by

$$\mathbf{K}_t = k_x^2 \mathbf{A}_{1t} + k_x k_y \mathbf{A}_{2t} + k_y^2 \mathbf{A}_{3t} - i k_x \mathbf{A}_{4t} - i k_y \mathbf{A}_{5t} + \mathbf{A}_{6t} \tag{8.17}$$

which is a constant matrix for given wavenumbers k_x and k_y.

8.4 Displacement in Wavenumber Domain

The model analysis is used to obtain the Fourier transform of the displacement vector $\tilde{\mathbf{d}}_t$. To this end, we need to solve the following eigenvalue equation of free wave motion

$$0 = [\mathbf{K}_t - \omega^2 \mathbf{M}_t] \mathbf{\Phi}^R \tag{8.18}$$

to obtain M wave modes with eigen-frequencies ω_m ($m = 1, 2, \dots, M$) and the corresponding right eigenvectors $\mathbf{\Phi}_m^R$. Since the $\mathbf{K}_t$ (see Eq. (8.18)) is a Hermitian matrix (Booth and Crampin, 1983), we have the following relationship between the left and the right eigenvector:

$$(\mathbf{\Phi}_m^L)^T = (\mathbf{\Phi}_m^R)^* \tag{8.19}$$

where the superscript T denotes the transpose of vector and the asterisk denotes the complex conjugate.

Expressing the transformed displacement vector $\tilde{\mathbf{d}}$ and the transformed force vector $\tilde{\mathbf{F}}$ in terms of the right eigenvectors, we have

$$\tilde{\mathbf{d}} = \sum_{m=1}^{M} \alpha_m \mathbf{\Phi}_m^R \tag{8.20}$$

and

$$\tilde{\mathbf{F}} = \sum_{m=1}^{M} \beta_m \mathbf{M} \mathbf{\Phi}_m^R \tag{8.21}$$

where α_m, β_m are functions of time. Substitution of Eqs. (8.20) and (8.21) into (8.16) leads to the dynamic equation for the mth wave mode:

$$\ddot{\alpha}_m + \omega_m^2 \alpha_m = \beta_m \tag{8.22}$$

with

$$\beta_m = \frac{\boldsymbol{\Phi}_m^L \tilde{\mathbf{F}}}{\boldsymbol{\Phi}_m^L \mathbf{M}_t \boldsymbol{\Phi}_m^R} \tag{8.23}$$

For a given force vector $\mathbf{F}$, we can obtain β_m from Eq. (8.23). Using initial conditions and solving Eq. (8.22) for α_m, the displacement vector $\tilde{\mathbf{d}}$ in the wavenumber domain can then be obtained using Eq. (8.20).

We assume that the force vector $\mathbf{F}_t$ has the following form

$$\mathbf{F}_t = \mathbf{F}_{0t} H(t) \tag{8.24}$$

where $H(t)$ is the Heaviside step function defined by

$$H(t) = \begin{cases} 1 & t > 0 \\ 0 & t \leq 0 \end{cases} \tag{8.25}$$

and $\mathbf{F}_{0t}$ is the constant amplitude vector of the applied force. Applying the Fourier transform to Eq. (8.24), we have

$$\tilde{\mathbf{F}}_t = \tilde{\mathbf{F}}_{0t} H(t) \tag{8.26}$$

From Eqs. (8.23), (8.22), and (8.26), we obtain

$$\ddot{\alpha}_m + \omega_m^2 \alpha_m = \frac{\boldsymbol{\Phi}_m^L \tilde{\mathbf{F}}_{0t} H(t)}{\boldsymbol{\Phi}_m^L \mathbf{M}_t \boldsymbol{\Phi}_m^R} \tag{8.27}$$

We assume that the laminate stays still before the application of the load. The initial conditions can then be written as

$$\mathbf{U}\big|_{t=0} = \frac{\partial \mathbf{U}}{\partial t}\bigg|_{t=0} = 0 \tag{8.28}$$

which gives the conditions for α_m, i.e.,

$$\alpha_m\big|_{t=0} = \dot{\alpha}_m\big|_{t=0} = 0 \tag{8.29}$$

Solving Eq. (8.27) for α_m under the initial condition of Eq. (8.29), we find the displacement vector in the Fourier transform domain as follows:

$$\tilde{\mathbf{d}}_t(k_x, k_y, t) = \sum_{m=1}^{M} \frac{\boldsymbol{\Phi}_m^L \tilde{\mathbf{F}}_{0t} \boldsymbol{\Phi}_m^R (1 - \cos \omega_m t)}{\omega_m^2 M_m} \tag{8.30}$$

where M_m is expressed by

$$M_m = \Phi_m^L \mathbf{M}_t \Phi_m^R \qquad (8.31)$$

which is the equivalent mass of the mth mode, and it is often termed the *equivalent mass in* vibration analysis.

8.5   Response in Space-Time Domain

8.5.1   Three-Dimensional Response

Performing the inverse Fourier transform, the displacement response in the space-time domain can be expressed by

$$\mathbf{d}_t(x,y,t) = \frac{1}{4\pi^2}\int_{-\infty}^{+\infty}\int_{-\infty}^{+\infty}\tilde{\mathbf{d}}_t(k_x,k_y,t)e^{-ik_x x}e^{-ik_y y}\,dk_x\,dk_y \qquad (8.32)$$

The integration in Eq. (8.32) can be carried out using two-dimensional (2D) fast Fourier transform (FFT) techniques. Since the square of ω_m is in the denominator of Eq. (8.30), the value of the fraction decreases rapidly with increasing order of the modes used in the superposition, and fairly good results can be obtained using only a few lowest modes in practical computations.

8.5.2   Two-Dimensional Response

The analysis and the formulas for 2D problems can be easily inferred from its 3D counterparts. In 2D problems, we assume that the load and the displacement field are independent of y-axis. Hence all expressions in the previous sections stay valid if we set $\frac{\partial}{\partial y} = 0$ and $k_y = 0$. The displacement response in the space-time domain can be obtained by performing the one-dimensional inverse Fourier transformation as follows:

$$\mathbf{d}_t^{2D}(x,t) = \frac{1}{2\pi}\int_{-\infty}^{+\infty}\tilde{\mathbf{d}}_t^{2D}(k_x,t)e^{-ik_x x}\,dk_x \qquad (8.33)$$

where

$$\tilde{\mathbf{d}}_t^{2D}(k_x,t) = \sum_{m=1}^{M}\frac{\Phi_m^L\tilde{\mathbf{F}}_{0t}^{2D}\Phi_m^R(1-\cos\omega_m t)}{\omega_m^2 M_m} \qquad (8.34)$$

where ω_m, Φ_m^R, Φ_m^L, M_m can be obtained from Eq. (8.18) with $k_y = 0$. Vector $\underset{\sim}{\mathbf{F}}_t^{2D}$ can be obtained in the same assembly procedure as for vector $\tilde{\mathbf{F}}_t$.

8.5.3   Computational Procedure

1. Divide the laminate into layer elements. In general, the number of plate elements N is taken to be greater than the number of layers of the laminated plate. A rule-of-thumb on the number of elements that should be used is that the thickness of the element should be less than one quarter of the wavelength of waves in the laminates, if quadratic shape functions are used.

2. For each element, calculate matrices $\mathbf{M}$ and $\mathbf{A}_i$ ($i = 1, 2,\ldots,6$) using matrix $\mathbf{c}$, ρ, θ, and h for the element, using the equations given in Chapter 5.

3. Assemble the matrices $\mathbf{M}$ and $\mathbf{A}_i$ for all the elements in the laminate to form global matrices $\mathbf{M}_t$ and $\mathbf{A}_{it}$ in the same way as with the finite element method.

4. For the step-impact force, obtain the frequencies and the corresponding left and right eigenvectors by solving Eq. (8.18) for various wavenumber k_x and k_y ($k_y = 0$, in the case of a 2D problem).

5. Using Eq. (8.30), obtain the displacement vector in the wavenumber domain, $\mathbf{d}_t(k_x,k_y,t)$. Finally, the displacement response can be obtained using Eq. (8.32) or Eq. (8.33) and the standard routines of numerical FFT or a quadrature discussed later in this chapter.

8.6   Response to Line Time-Step Load

The hybrid numerical method is applied to analyze transient waves in both isotropic plates and anisotropic laminated plates [C0/G+45/G–45]$_s$ excited by time-step loads. The results are presented and discussed in this section.

8.6.1   Settings of the Parameters

For a practical calculation, we must choose properly the number of plate elements N, the number of modes M, the ranges of wavenumbers, k_x and k_y, and the number of sampling points in the ranges of k_x and k_y, according to the required accuracy. The results with an accuracy shown in Figure 8.7 are obtained by choosing $N = 12$, $M = 10$, $-12\pi \le k_x H \le 12\pi$, $-12\pi \le k_y H \le 12\pi$ and 256 sampling points within the ranges. The accuracy of the results can be improved by increasing the number of elements, but the higher order modes over 10 seems to have little influence on the displacement response to the time-step excitation. If far-field and long-time responses are of interest, fewer elements can be used and the ranges of k_x and k_y can be reduced to obtain fairly good results. However, if the Rayleigh peak must be observed, more wave modes have to be used, and the integration must be carried out in larger ranges than $-8\pi \le k_x H \le 8\pi$ and $-8\pi \le k_y H \le 8\pi$ with sufficient sampling points.

Note that for general anisotropic materials, the eigen-functions (eigenvalue and eigenvector) are complex valued, and complex integrals have to be eva-luated. However, for anisotropic materials of 13 independent constants, the Hermitian stiffness matrix $\mathbf{K}$ can be changed to a real symmetrical one (see Kausel, 1986), and hence the eigen-functions are still real valued. Fiber-reinforced laminates fall into this case if the fiber is laid within the laminate plane. Therefore, the integration can be done in real number.

To compare with the results given by Scott and Miklowitz (1967), we use the same excitation force, i.e., a vertical line time-step load acting symmet-rically and simultaneously on both the lower and upper surfaces of the plate. It is given by

$$\mathbf{F}_t^{2D} = H(t)\delta(x)\mathbf{Q}_t \tag{8.35}$$

where $H(t)$ is the Heaviside step function of time, $\delta(x)$ is the Dirac delta function of x, and $\mathbf{Q}_t$ is a constant vector expressed by

$$\mathbf{Q}_t^T = \{0, 0, q_0, 0, \ldots, 0, 0, -q_0\} \tag{8.36}$$

where q_0 is constant. Applying the Fourier transformation to Eq. (8.35), we have

$$\tilde{\mathbf{F}}_t^{2D} = \int_{-\infty}^{\infty} \mathbf{F}_t^{2D} e^{ik_x x} dx = H(t)\mathbf{Q}_t \tag{8.37}$$

Hence, the vector $\tilde{\mathbf{F}}_{0t}^{2D}$ in Eq. (8.26) can be obtained as follows:

$$\tilde{\mathbf{F}}_{0t}^{2D} = \mathbf{Q}_t \tag{8.38}$$

Here only the responses of the displacement in the x-axis direction on the upper surface of the plates are presented and the following dimensionless parameters are introduced:

$$\bar{u} = u/u_0, \quad u_0 = Hq_0/c_0, \quad \bar{t} = t/t_0, \quad t_0 = H\sqrt{\rho_0/c_0}, \quad \bar{x} = x/H$$

For the isotropic plate, ρ_0, c_0, and t_0 are the density of the plate, the shear modulus of the plate, and the time for the shear wave to travel across the

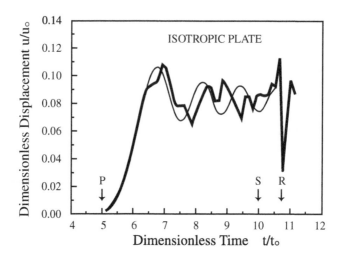

FIGURE 8.2

Comparison of time responses of u on the upper surface of an isotropic plate: $v = 1/3$, $x = 10H$; thick line: the present method, thin line: the method of head of the pulse approximation (Scott and Miklowitz, 1967). (From Liu, G. R., et al., *Journal of Vibration and Acoustics*, 113, 230, 1991a. With permission.)

thickness of the plate once, respectively. For composite laminates, ρ_0, c_0, and t_0 are the density of a carbon/epoxy layer in the laminate, the elastic coefficient c_{11} of this layer when the fiber orientation is in the x axis direction, and the time for the longitudinal wave (in the fiber orientation of the carbon/epoxy material) to travel once across the thickness of the laminate.

8.6.2 Results for an Isotropic Plate

Figure 8.2 shows the time response of the isotropic plate with Poisson's ratio of one-third obtained by the present method (thick line), together with those obtained by the method of the head of the pulse approximation (thin line) used by Scott and Miklowitz (1967) for comparison. The point of observation is at a distance of $10H$ from the position where the line load is applied. The letters P, S, and R mark the arrival times of the longitudinal, shear, and Rayleigh waves of the isotropic material, respectively. From Figure 8.2 we can see that the two curves rise almost at the same time after the arrival of the longitudinal wave, but when $\hat{t} \geq 6.4$ the two curves are noticeably different from each other. The curve obtained by the present method shows the sharp peak at R, but the curve obtained by the method of the head of the pulse approximation does not, due to the approximation made in the method.

Figure 8.3 shows the station responses of the same isotropic plate at time $\hat{t} = 5.124$ and $\hat{t} = 11.124$. For $\hat{t} = 5.124$, the arrival positions of the longitudinal, shear, and Rayleigh waves are marked by the smaller letters P, S, and R,

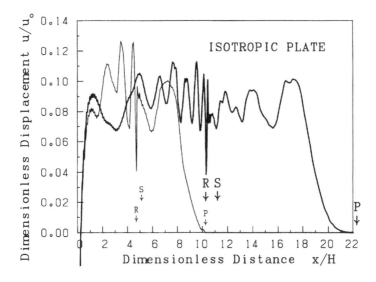

FIGURE 8.3

Station responses of u on the upper surface of an isotropic plate: $v = 1/3$, $u_0 = Hq_0/G$, $t_0 = H\sqrt{\rho_0/G}$; thick line: $\hat{t} = 5.124$, thin line: $\hat{t} = 11.124$. (From Liu, G. R., et al., *Journal of Vibration and Acoustics*, 113, 230, 1991a. With permission.)

respectively. For $\hat{t} = 11.124$ those positions are indicated by the larger letters P, S, and R, respectively. From this figure, we can see that after the arrival of the longitudinal wave, the responses rise speedily and at the positions of the Rayleigh waves, sharp peaks are observed.

Figure 8.4 shows the station and time responses of the isotropic plate. From this figure we can see how the displacement distribution on the upper surface of the isotropic plate changes with time.

8.6.3 Results for a Hybrid Laminate

Figure 8.5 shows the time and station responses of the displacement in the x direction on the surface of the laminate. From this figure we can see how the displacement distribution on the upper surface of the laminate changes with time. The station response at times $\hat{t} = 8.0$ and $\hat{t} = 14.0$ is plotted in Figure 8.6. For $\hat{t} = 8.0$, the smaller letters P, S, and R mark the arrival positions of the longitudinal, shear, and Rayleigh waves of the carbon/epoxy composite in the direction of fiber orientation, respectively. For $\hat{t} = 14.0$, those positions are indicated by the larger letters P, S, and R, respectively. From Figure 8.6 it can be seen that there are peaks near R and R, but the peaks are not very sharp and their positions differ from R and R. That is, the surface wave in the surface layer of the hybrid laminate is different from the Rayleigh wave in the pure carbon/epoxy composite. Another phenomenon that differs from the

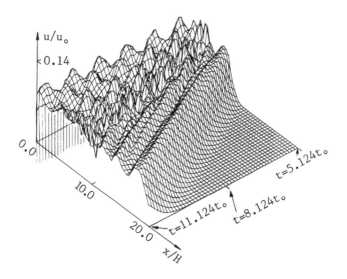

FIGURE 8.4

Station and time responses of u on the upper surface of an isotropic plate: $v = 1/3$, $u_0 = Hq_0/G$, $t_0 = H\sqrt{\rho_0/G}$. (From Liu, G. R., et al., *Journal of Vibration and Acoustics*, 113, 230, 1991a. With permission.)

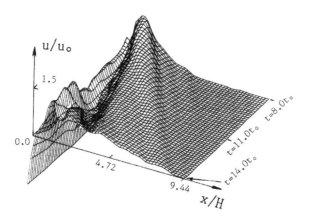

FIGURE 8.5

Station and time responses of u on the upper surface of a hybrid laminate $[C0/G+45/G-45]_s$ ($u_0 = Hq_0/c_{11}$, $t_0 = H\sqrt{\rho_0/c_{11}}$). (From Liu, G. R., et al., *Journal of Vibration and Acoustics*, 113, 230, 1991a. With permission.)

isotropic plate is that, after the arrival of the longitudinal wave, the responses rise very slowly. A reason is that for the hybrid laminate, the group velocity of the longitudinal wave of carbon/epoxy in the direction of fiber orientation is much larger than the group velocities of most of the other modes of Lamb waves.

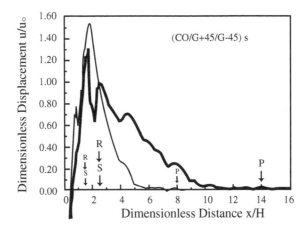

FIGURE 8.6
Station responses of u on the upper surface of a hybrid laminate $[C0/G+45/G-45]_s$: $u_0 = Hq_0/c_{11}$, $t_0 = H\sqrt{\rho_0/c_{11}}$, thick line: $\dot{t} = 14.0$, thin line: $\dot{t} = 8.0$. (From Liu, G. R., et al., *Journal of Vibration and Acoustics*, 113, 230, 1991a. With permission.)

8.7 Response to Point Time-Step Load

To compare with the results obtained using the exact method of generalized rays (Ceranoglu and Pao, 1981), we assume that the point time-step load $\mathbf{F}_t$ is applied, i.e.,

$$\mathbf{F}_t = H(t)\delta(x)\delta(y)\mathbf{Q}_t \tag{8.39}$$

where $H(t)$ is the Heaviside step function of time, $\delta(x)$ and $\delta(y)$ are the Dirac delta functions of x and y, respectively, and $\mathbf{Q}_t$ is a constant vector given by

$$\mathbf{Q}_t^T = \{0, 0, q_0, 0,\ldots,0, 0, 0\} \tag{8.40}$$

in which q_0 is constant. Equation (8.39) expresses a point step-impact force acting on the upper surface of a plate in the vertical direction.

After the application of the Fourier transformation to Eq. (8.39), we obtain

$$\tilde{\mathbf{F}}_t = \int_{-\infty}^{\infty}\int_{-\infty}^{\infty} \mathbf{F}_t e^{ik_x x} e^{ik_y y}\,dx\,dy = H(t)\mathbf{Q}_t \tag{8.41}$$

Hence the vector $\tilde{\mathbf{F}}_{0t}$ in Eq. (8.26) can be obtained as follows:

$$\tilde{\mathbf{F}}_{0t} = \mathbf{Q}_t \tag{8.42}$$

In this subsection, the responses of the vertical displacement w and the radial displacement u_r on the upper surface of the plates are considered and the following dimensionless parameters are employed:

$$\bar{w} = w/w_0, \quad w_0 = Hq_0/c_0, \quad \bar{u} = u/w_0$$

$$\bar{t} = t/t_0, \quad t_0 = H\sqrt{\rho_0/c_0}, \quad \bar{x} = x/H$$

$$\bar{y} = y/H, \quad \bar{r} = r/H, \quad r = \sqrt{x^2 + y^2}$$

8.7.1 Results for an Isotropic Plate

Figure 8.7 shows the time responses of the vertical displacement of the isotropic plate with Poisson's ratio of 0.21 obtained by the present method (thick line), together with those obtained by the method of generalized rays (thin line) for comparison. The point of observation is at a distance of $4H$ from the position where the point impact load is applied. The letters P, S, and R mark the arrival times of the longitudinal, shear, and Rayleigh waves of the isotropic material, respectively. From Figure 8.7, we can see that the present results are different from those obtained by the method of generalized rays earlier after the arrival of the longitudinal wave, where the responses are very small. When the response becomes stronger, the present results agree well with those obtained by the method of generalized rays. The Rayleigh peak is also observed, but the arrival of the Rayleigh peak obtained by the present method is a little faster than that obtained by the method of generalized rays. The reason is that in the present method, the plate is divided into layer elements and the displacement field within the elements is assumed. Hence, the plate can be considered to be subjected to some kind of constraint on the profile of

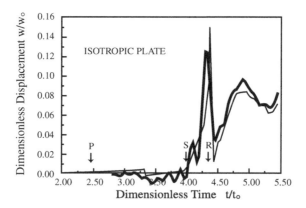

FIGURE 8.7

Time response of w on the upper surface of an isotropic plate: $v = 0.21$, $x = 4H$; thick line: the present method, thin line: the method of generalized rays. (From Liu, G. R., et al., *Journal of Vibration and Acoustics*, 113, 230, 1991a. With permission.)

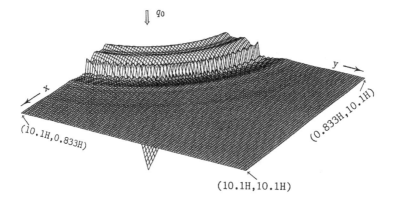

FIGURE 8.8
Station response of w on the upper surface of an isotropic plate: $v = 0.21$, $\bar{t} = 5.45$. (From Liu, G. R., et al., *Journal of Vibration and Acoustics*, 113, 230, 1991a. With permission.)

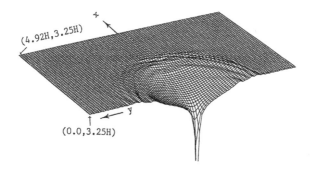

FIGURE 8.9
Station response of w on the upper surface of a hybrid laminate [C0/G+45/G–45]$_s$: $\bar{t} = 16.0$. (From Liu, G. R., et al., *Journal of Vibration and Acoustics*, 113, 230, 1991a. With permission.)

the displacements; the plate behaves a little more stiffly due to the constraint. Consequently, the surface wave is observed being a little faster. Further calculation of the response using a larger number of elements has confirmed that the frequencies of the wave modes decrease and that the arrival time of the Rayleigh peak is closer to that obtained by the method of generalized rays.

Figure 8.8 gives a quarter view of the distribution of the vertical displacement on the upper surface of the isotropic plate at $\bar{t} = 5.45$. Since the plate is isotropic, the vertical displacement field is axisymmetric and the response distribution of the vertical displacements forms the circles of waves.

8.7.2 Results for a Hybrid Laminate

Figure 8.9 shows a half view of the distribution of the vertical displacement on the upper surface of the hybrid laminated composite plate [C0/G+45/G–45]$_s$ at $\bar{t} = 16.0$. The waveform is no longer circular. For the hybrid laminate, the

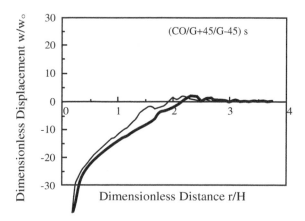

FIGURE 8.10
Comparison of station response of w on the upper surface of a hybrid laminate [C0/G+45/ G–45]$_s$: $\bar{t}$ = 16.0. thick line: on the x axis and thin line on the y axis. (From Liu, G. R., et al., *Journal of Vibration and Acoustics*, 113, 230, 1991a. With permission.)

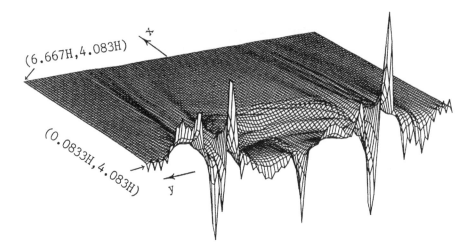

FIGURE 8.11
Station response of u_r on the upper surface of a hybrid laminate [C0/G+45/G–45]$_s$: $\bar{t}$ = 16.0. (From Liu, G. R., et al., *Journal of Vibration and Acoustics*, 113, 230, 1991a. With permission.)

anisotropy of the elastic coefficients corresponding to the vertical direction is less stiff. In fact, the elastic coefficients corresponding to the vertical direction are dependent mainly on those of the epoxy matrix. Hence the anisotropy of the responses of the vertical displacements is also weaker. To show the anisotropy of the responses of the vertical displacement quantitatively, the displacement distribution on both the x-axis and the y-axis are computed and plotted together in Figure 8.10.

Since the anisotropy of the elastic coefficients of the hybrid laminate corresponding to the x-y plane is strong, strong anisotropy of the in-plane displacement such as the radial displacement can be observed. Figure 8.11 shows a half view of the response distribution of the radial displacement on the upper surface of the hybrid laminate. The carbon fiber orientation of the upper layer of the hybrid laminate is in the x direction and, therefore, the plate is very stiff in this direction. Hence, the wave propagation in this direction is very fast, and the amplitudes of the radial displacement are very small. By contrast, the wave propagation in the y direction is much slower and the amplitudes of the radial displacement are much larger.

According to the results obtained in Chapter 5, the wave surfaces in the hybrid laminate are symmetrical with respect to the impact point (but not axisymmetrical). Hence, the other half views of Figures 8.8 and 8.11 can be obtained by using point symmetry.

8.8 Techniques for Inverse Fourier Integral

8.8.1 Problems

In using the fast Fourier transform technique (FFT) to carry out the integral in Eqs. (8.32) and (8.33), a large number of sampling points on the wavenumber axes k_x and k_y are needed to obtain good results, especially for long-time and far-field responses. The reason is that the kernel, $\mathbf{d}$, is a highly oscillatory function of k_x and k_y especially for large wavenumbers and for large time t. Figure 8.12 is an example of such an oscillatory function, which is for the displacement component w of hybrid laminate $[C90/G+45/G-45]_s$.

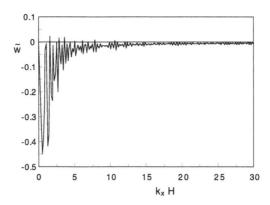

FIGURE 8.12
The integrand in the Fourier integral for displacement w on the upper surface for plate $[C90/G+45/G-45]_s$ subjected to a vertically applied time-step point load on the upper surface of the laminate ($\bar{t} = 20.0$, $k_y H = 1.0$).

From Eq. (8.32), it can be clearly seen that the term $\exp[-i(k_x x + k_y y)]$ is oscillatory, and the frequency of the oscillation is proportional to the observation distance x and/or y. Dealing with such a highly oscillatory integrand, we need a better technique of quadrature rather than relying on the use of a large number of even sampling points in the FFT. This section introduces a much more efficient integrated technique for handling this problem. This technique was originally proposed by Liu et al. in 1997 for 2D cases and Liu and Lam (1999) for 3D cases. The following section details the technique for 3D cases.

8.8.2   Technique

By introducing

$$k_x = k \cos\theta, \quad k_y = k \sin\theta \tag{8.43}$$

where θ is the direction of a plane wave with a wavenumber of k, the integration in Eq. (8.32) is changed into an integral in the polar coordinate system, i.e.,

$$\mathbf{d}_t(x, y) = \frac{1}{4\pi^2} \int_0^{2\pi} \int_0^\infty \tilde{\mathbf{D}}(k,\theta) e^{-is} \, dk \, d\theta \tag{8.44}$$

where

$$s = k(x \cos\theta + y \sin\theta) \tag{8.45}$$

and

$$\tilde{\mathbf{D}} = \sum_{m=1}^M \mathbf{\Psi}_m(k, \theta)(1 - \cos\omega_m t) \tag{8.46}$$

in which

$$\mathbf{\Psi}_m = \frac{\boldsymbol{\varphi}_m^L \mathbf{F}_{0t} \boldsymbol{\varphi}_M^R}{\omega_m^2 (\boldsymbol{\varphi}_m^L \mathbf{M}_t \boldsymbol{\varphi}_m^R)} k = \mathbf{\Phi}_m k \tag{8.47}$$

The eigenfrequencies and the corresponding eigenvectors can still be obtained as per normal, but the stiffness matrix $\mathbf{K}_t$ should be changed to

$$\mathbf{K}_t = k^2(\cos^2\theta \mathbf{A}_{1t} + \cos\theta \sin\theta \mathbf{A}_{2t} + \sin^2\theta \mathbf{A}_{3t}) - ik(\cos\theta \mathbf{A}_{4t} + \sin\theta \mathbf{A}_{5t}) + \mathbf{A}_{6t} \tag{8.48}$$

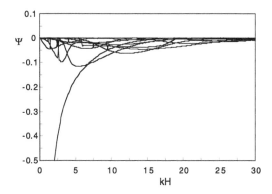

FIGURE 8.13
Lowest 10 modal factor functions for displacement w on the upper surface for plate [C90/G+45/ G–45]$_s$ ($\theta = \pi /4$).

Rewriting from Eq. (8.32) to Eq. (8.44) does not change the oscillatory nature of the integral kernel. However, the double infinite integration in Eq. (8.1) is changed to be a double integration over a finite range for θ and semi-infinite range for k. This is more convenient for the numerical evaluation of the integration.

It can be seen from Eq. (8.46) that a kernel for a wave mode is a product of a modal factor function Ψ_m and a time-related function $(1 - \cos \omega_m t)$. Modal factor function Ψ_m for the lowest ten wave modes in a hybrid laminate is examined and shown in Figure 8.13. It is found that Ψ_m has the following features:

- Function Ψ_m does not vibrate or vary rapidly except at the vicinity of origin, $k_x = 0$ and $k_y = 0$.
- For Φ_m of the lowest modes, the origin is a singular point, but for $\tilde{D}$ this singular point is removable and $\tilde{D}$ behaves well in the vicinity of the origin.

Hence, it is clearly seen that the source oscillation is the function $(1 - \cos \omega_m t)$, and the frequency of the oscillation is proportional to time t. It is also found that Ψ_m does not oscillate and vary rapidly with respect to θ because Ψ_m depends only on the anisotropy of the material for a given k, and the material properties do not change very rapidly with θ. Moreover, it has been found that ω_m does not vary rapidly with respect to k as well as θ. Taking advantage of the above-mentioned features of Ψ_m, ω_m, and $\tilde{D}$, the inverse integration in Eq. (8.44) is evaluated by the following procedure.

First, the integral in Eq. (8.44) is divided into two parts

$$d(x, y) = \frac{1}{4\pi^2}\int_0^{2\pi}\int_0^{\infty} \tilde{D}(k,\theta)e^{-iks}\, dk\, d\theta = I_1 + I_2 \qquad (8.49)$$

where d is any component of $\mathbf{d}$, $\tilde{D}$ is the component of $\tilde{\mathbf{D}}$ corresponding to d, and

$$I_1 = \frac{1}{4\pi^2}\int_0^{2\pi}\int_0^{k_a}\tilde{D}(k,\theta)e^{-iks}dk\,d\theta \tag{8.50}$$

$$I_2 = \frac{1}{4\pi^2}\int_0^{2\pi}\int_{k_a}^{\infty}\tilde{D}(k,\theta)e^{-is}dk\,d\theta$$

$$= \frac{1}{4\pi^2}\sum_{m=1}^{M}\int_0^{2\pi}\int_{k_a}^{\infty}\Psi_m(k,\theta)(1-\cos\omega_m t)e^{-is}\,dk\,d\theta$$

$$= \frac{1}{4\pi^2}\sum_{m=1}^{M}I_{2m} \tag{8.51}$$

where k_a is a small number, so that I_1 can be carried out using conventional evenly spaced quadrature routines. In dealing with I_2, the integration is evaluated for each wave mode, and

$$I_{2m} \approx \int_0^{2\pi}\int_{k_a}^{k_b}\Psi_m(k,\theta)(1-\cos\omega_m t)e^{-is}dk\,d\theta = I_{2m}^c - I_{2m}^t \tag{8.52}$$

where k_b is a very large number where Ψ_m vanishes. The truncation error caused by dropping the integral over k_b to ∞ will be discussed later. Integrals in Eq. (8.52) are further divided into two parts. One is independent of time t, and another is dependent on t as

$$I_{2m}^c = \int_0^{2\pi}\int_{k_a}^{k_b}\Psi_m(k,\theta)e^{-is}\,dk\,d\theta \tag{8.53}$$

$$I_{2m}^t = \int_0^{2\pi}\int_{k_a}^{k_b}\Psi_m(k,\theta)\cos(\omega_m t)e^{-is}\,dk\,d\theta \tag{8.54}$$

As mentioned before, Ψ_m does not vary rapidly in the range of $k \geq k_a$ and $0 \leq \theta \leq 2\pi$. Hence, we divide the region $k_a \leq k \leq k_b$ into $(L_k - 1)$ even intervals, i.e.,

$$k_a = k_1 < k_2 < \cdots < k_j < \cdots < k_{L_k} = k_b \tag{8.55}$$

and the region $0 \leq \theta \leq 2\pi$ is also divided into $(L_\theta - 1)$ even intervals as

$$0 = \theta_1 < \theta_2 < \cdots < \theta_j < \cdots < \theta_{L_\theta} = 2\pi \tag{8.56}$$

For sufficiently large L_k and L_θ, Ψ_m, s, and ω_m can be approximately replaced by a plane in each rectangular range of $k_i \le k \le k_{i+1}$ and $\theta_j \le \theta \le \theta_{j+1}$, i.e.,

$$\Psi_{mij}(k, \theta) = a_{mij}^{(1)} \theta + a_{mij}^{(2)} k + a_{mij}^{(3)} \tag{8.57}$$

$$s_{ij}(k, \theta) = b_{mij}^{(1)} \theta + b_{mij}^{(2)} k + b_{mij}^{(3)} \tag{8.58}$$

$$\omega_{mij}(k, \theta) = c_{mij}^{(1)} \theta + c_{mij}^{(2)} k + c_{mij}^{(3)} \tag{8.59}$$

where in Eq. (8.57), $a_{mij}^{(1)}$, $a_{mij}^{(2)}$, and $a_{mij}^{(3)}$ are determined by the following procedure:

1. The inclination of the plane with respect to θ equals that of the line passing through two points of $(\frac{\Psi_{m,i,j+1} + \Psi_{m,i,j}}{2}, \theta_{i+1})$ and $(\frac{\Psi_{m,i+1,j+1} + \Psi_{m,i+1,j}}{2}, \theta_{i+1})$, thus

$$a_{mij}^{(1)} = \frac{(\Psi_{m,i+1,j+1} + \Psi_{m,i+1,j}) - (\Psi_{m,i,j+1} + \Psi_{m,i,j})}{2(\theta_{i+1} - \theta_i)} \tag{8.60}$$

2. The inclination of the plane with respect to k equals that of the line passing through two points of $(\frac{\Psi_{m,i,j} + \Psi_{m,i+1,j}}{2}, k_j)$ and $(\frac{\Psi_{m,i,j+1} + \Psi_{m,i+1,j+1}}{2}, k_{j+1})$, we have

$$a_{mij}^{(2)} = \frac{(\Psi_{m,i,j+1} + \Psi_{m,i+1,j+1}) - (\Psi_{m,i,j} + \Psi_{m,i+1,j})}{2(k_{j+1} - k_j)} \tag{8.61}$$

3. The plane of approximation passes through the point of $(\frac{\theta_{i+1} - \theta_i}{2}, \frac{k_{j+1} - k_j}{2}, \frac{\Psi_{m,i,j+1} + \Psi_{m,i+1,j+1} + \Psi_{m,i,j} + \Psi_{m,i+1,j}}{4})$; hence we obtain

$$a_{mij}^{(3)} = (\Psi_{m,i,j+1} + \Psi_{m,i+1,j+1} + \Psi_{m,i,j} + \Psi_{m,i+1,j})$$
$$/4 - a_{mij}^{(1)}(\theta_{i+1} - \theta_i) - a_{mij}^{(2)}(k_{i+1} - k_i) \tag{8.62}$$

In Eqs. (8.58) and (8.59), b_{mij} s and c_{mij} s can be obtained in exactly the same way as obtaining a_{mij} s. Integrations in Eqs. (8.53) and (8.54) can then be expressed, respectively, as

$$I_{2m}^c = \sum_{i=1}^{L_\theta} \sum_{j=1}^{L_k} \int_{k_j}^{k_{j+1}} \int_{\theta_i}^{\theta_{i+1}} (a_{mij}^{(1)} \theta + a_{mij}^{(2)} k + a_{mij}^{(3)}) e^{-i(b_{mij}^{(1)} \theta + b_{mij}^{(2)} k + b_{mij}^3)} dk \, d\theta$$

$$= \sum_{i=1}^{L_\theta} \sum_{j=1}^{L_k} I_{mij}^c \tag{8.63}$$

$$I_{2m}^t = \sum_{i=1}^{L_\theta} \sum_{j=1}^{L_k} \int_{k_j}^{k_{j+1}} \int_{\theta_i}^{\theta_{i+1}} (a_{mij}^{(1)}\theta + a_{mij}^{(2)}k + a_{mij}^{(3)}) \cos[((c_{mij}^{(1)}\theta + c_{mij}^{(2)}k + c_{mij}^{(3)})t]$$

$$\times e^{-i(b_{mij}^{(1)}\theta + b_{mij}^{(2)}k + b_{mij}^{(3)})} dk\, d\theta = \sum_{i=1}^{L_\theta}\sum_{j=1}^{L_k} I_{mij}^t \qquad (8.64)$$

The integrals of I_{mij}^c and I_{mij}^t in the foregoing two equations can then be carried out analytically. Taking advantage of the symmetry of the integrand, the foregoing two equations can be further simplified.

1. For loads in the z direction, to obtain displacement components u and v, or for load in either x or y direction, to obtain the displacement component w, the displacement must be antisymmetric, and the integration in Eqs. (8.63) and (8.64) can be changed to

$$I_{mij}^c = \int_{k_j}^{k_{j+1}} \int_{\theta_i}^{\theta_{i+1}} (a_{mij}^{(1)}\theta + a_{mij}^{(2)}k + a_{mij}^{(3)})\ \sin(b_{mij}^{(1)}\theta + b_{mij}^{(2)}k + b_{mij}^{(3)})\ dk\, d\theta$$

$$\qquad\qquad (8.65)$$

$$I_{2m}^t = \frac{1}{2}\int_{k_j}^{k_{j+1}} \int_{\theta_i}^{\theta_{i+1}} (a_{mij}^{(1)}\theta + a_{mij}^{(2)}k + a_{mij}^{(3)})$$

$$\times \{\sin[(c_{mij}^{(1)}t + b_{mij}^{(1)})\theta + (c_{mij}^{(2)}t + b_{mij}^{(2)})k + (c_{mij}^{(3)}t + b_{mij}^{(3)})]$$

$$-\sin[(c_{mij}^{(1)}t - b_{mij}^{(1)})\theta + (c_{mij}^{(2)}t - b_{mij}^{(2)})k + (c_{mij}^{(3)}t - b_{mij}^{(3)})]\}dk\, d\theta \quad (8.66)$$

2. For loads in the z direction, to obtain displacement component w, or for load in either x or y direction, to obtain the displacement components u and v, the displacement must be symmetric, and the integration in Eqs. (8.63) and (8.64) can be changed to

$$I_{mij}^c = \int_{k_j}^{k_{j+1}} \int_{\theta_i}^{\theta_{i+1}} (a_{mij}^{(1)}\theta + a_{mij}^{(2)}k + a_{mij}^{(3)})\cos(b_{mij}^{(1)}\theta + b_{mij}^{(2)}k + b_{mij}^{(3)})dk\, d\theta$$

$$\qquad\qquad (8.67)$$

$$I_{2m}^t = \frac{1}{2}\int_{k_j}^{k_{j+1}} \int_{\theta_i}^{\theta_{i+1}} (a_{mij}^{(1)}\theta + a_{mij}^{(2)}k + a_{mij}^{(3)})$$

$$\times \{\cos[(c_{mij}^{(1)}t + b_{mij}^{(1)})\theta + (c_{mij}^{(2)}t + b_{mij}^{(2)})k + (c_{mij}^{(3)}t + b_{mij}^{(3)})]$$

$$+\cos[(c_{mij}^{(1)}t - b_{mij}^{(1)})\theta + (c_{mij}^{(2)}t - b_{mij}^{(2)})k + (c_{mij}^{(3)}t - b_{mij}^{(3)})]\}dk\, d\theta \quad (8.68)$$

Again, the integrations in Eqs. (8.65) to (8.68) can be carried out analytically and easily implemented into a computer program.

8.8.3 Application

The new technique has been implemented into the 3D HNM program. Responses of an isotropic plate loaded on the upper surface are computed by the modified HNM program. The results are compared with those obtained by an exact method. Response of a hybrid laminate [C90/G+45/G−45]$_s$ subjected to point loads on the upper surface of the plate is also computed using the modified HNM. The time histories of the loads are the time-step and one cycle of a sine function. The Poisson's ratio of the material for the isotropic plate is 0.25.

Figure 8.14 shows the time history of the displacement w on the surface of the isotropic plate subjected to a time-step point load. All the parameters are set exactly the same as in the case discussed in Figure 8.7. The load is on the upper surface at $x = 0$ and $y = 0$, and in the thickness direction of the plate. The observation position is at $x = 4.0H$, and $y = 0$. The dotted line is obtained using the present modified HNM, and the solid line is obtained using the method of generalized rays, an exact method (Ceranoglu and Pao, 1981). In the HNM, the integration for θ needs to be done only over 0 to $\pi/2$, due to the isotropy of the material. In the HNM, twelve elements and twenty modes have been used. Parameters used in Eqs. (8.53) and (8.54) are $k_a = \pi$ and $k_b = 11\pi$. The sampling points needed in the modified HNM is 70 for k and 51 for θ, or 3570 in total. The original HNM program using FFT has also been used for the same computation, and the sampling points are 512 for both k_x and k_y, or 262144 total, to get the same accuracy of results. This is about 70 times the sampling point needed for the modified HNM. In addition, the accuracy of the results obtained using the present integral scheme (Figure 8.14) is significantly improved compared to that using FFT (Figure 8.7).

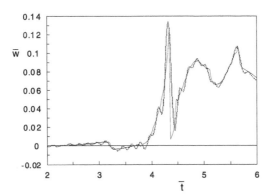

FIGURE 8.14

Time histories of the vertical displacement w at $x = 4H$ on the upper surface of an isotropic plate subjected to a time-step point load acting on the upper surface at $x = 0$, $y = 0$. The dotted line is obtained by the present technique, and the solid line is obtained by an exact method (Ceranoglu and Pao, 1981).

A difference can still be observed in Figure 8.14 between results obtained using the HNM and the exact method. Errors in the results by the HNM are caused by three factors: the number of the elements N, the number of mode M, and truncating point k_b. As a time-step load is considered, the exciting source contains frequencies of infinity. On the other hand, due to the nature of the HNM, it cannot take frequencies at infinity into account, even though using larger N, M, and/or k_b will get better results. The purpose to obtain the responses for the time-step load is to obtain the response for a load of any-time history. For a practical case, the frequency of a load is finite, and hence the HNM is useful for these kinds of loads. The results for a load of any-time history can obtained from results for the time-step load by the following equation.

$$d(t) = \sum_{i=1}^{L_t} [d^H(t - \tau_i) - d^H(t - \tau_{i+1})] f\left(\frac{\tau_{i+1} + \tau_i}{2}\right) \tag{8.69}$$

where the superscript H stands for results for the time-step load, and

$$0 = \tau_1 < \tau_2 < \cdots < \tau_i < \cdots < \tau_{L_t} = t_d \tag{8.70}$$

where t_d is the duration of the loading.

8.9   Response to Transient Load of Arbitrary Time Function

We consider now a point-transient load with a time-history function of one cycle sine function defined by

$$f(t) = \sin(2\pi t/t_d), \quad 0 < t < t_d \tag{8.71}$$

The frequency spectrum for $\bar{t}_d = 2.0$ is shown in Figure 8.15. Responses of the vertical displacement on the upper surface of the isotropic plate subjected to the point load with a time history of Eq. (8.71) are computed using the HNM with the present integral scheme, and results are shown in Figure 8.16. The load acts in the vertical direction and on the upper surface at $x = 0$ and $y = 0$. The solid line is obtained by using $N = 12$, $M = 20$, and $k_b = 11\pi$, and the dotted line is obtained by doubling the values of these parameters, ($N = 24$, $M = 40$, and $k_b = 22\pi$). However, these two curves are almost coincident and hard to distinguish. The reasons are as follows. From Figure 8.15, it is found that the components with a dimensionless frequency higher than 15.0 are very small. Therefore, the highest frequency is considered approximately 15.0. At this highest frequency, the wavelength of the shear wave in the

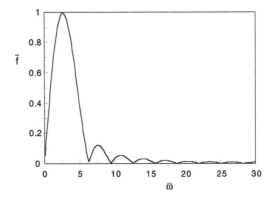

FIGURE 8.15
Frequency spectrum of a point load with a time history of one cycle of sine function ($\bar{t}_d = 2.0$).

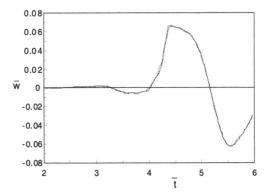

FIGURE 8.16
Time histories of the vertical displacement w at $x = 4H$ on the upper surface of the plate [C90/G+45/G–45]$_s$ subjected to a point load acting on the upper surface at $x = 0$, $y = 0$. The time history of the load is one cycle sine function ($\bar{t}_d = 2.0$). The dotted line: $N = 24$, $M = 40$, and $k_b = 22\pi$. Solid line: $N = 12$, $M = 20$, and $k_b = 11\pi$.

material of the plate is about $0.42H$. For this frequency, $N = 12$, $M = 20$, and $k_b = 11\pi$ is sufficient to obtain good results, because the thickness of one element is far less than a quarter of the shear wavelength, and frequencies for wave modes higher than 20 and for $k > k_b$ are higher than the highest frequency of the load. This means the present method can be expected to give a very accurate transient response.

Overviews of the wave field excited by a point load with time variation of one cycle of sine function given by Eq. (8.71), are shown in Figure 8.17 for the isotropic plate and in Figure 8.18 for the hybrid laminate [C90/G+45/G–45]$_s$. The observation time is $6.0\bar{t}$, and the displacement in the radial direction u_r on the upper surface of the plate are computed in a rectangular range of about $5H$ by $10H$. From Figure 8.17, it can be seen again that the

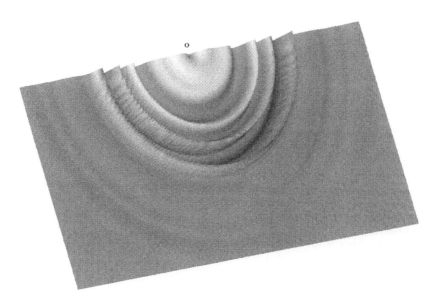

FIGURE 8.17
Overview of the displacement field for u_r. Isotropic plate excited by a point load on the upper surface at $x = 0$ and $y = 0$. The time history of the load is one cycle sine function ($\dot{t}_d = 1.0$, $\dot{t} = 6.0$).

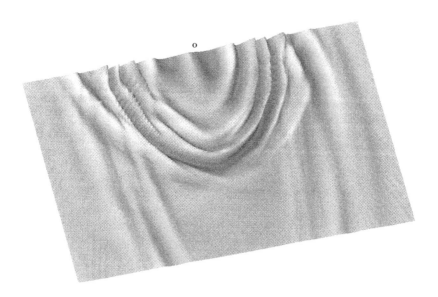

FIGURE 8.18
Overview of the displacement field for u_r. Laminate $[C90/G+45/G–45]_s$ excited by a point load on the upper surface at $x = 0$ and $y = 0$. The time history of the load is one cycle sine function ($\dot{t}_d = 1.0$, $\dot{t} = 6.0$).

displacement field forms cycles due to the isotropy of the material. However, for the laminated plate, the displacement field depends on the direction of observation point to the point load.

8.10 Remarks

The formulation of the hybrid numerical method is presented to analyze the displacement response of anisotropic laminates excited by transient loads. This method is straightforward and easy to use. Moreover, this method can be used for the analysis of arbitrary anisotropic laminated plates with many layers subjected to a line or point load.

In the present method, the FEM and the Fourier transform are combined to obtain a transient response of anisotropic laminates. Hence, the merits of the FEM and the Fourier transform are incorporated. With the use of the Fourier transform together with the modal analysis approach, the displacement in the wavenumber domain can be obtained analytically.

A technique for handling the oscillatory integrand in Eqs. (8.32) and (8.33) has also been presented to achieve significantly better performance. The HNM method is well suited for computing the displacement response in composite laminated plates. In terms of performance, the HNM is much more economic than the exact method introduced in Chapter 5 for transient response analysis. The application of the HNM method is not limited to composite materials. It can also be used to compute the seismic waves in a stratified anisotropic medium—which is a very important problem in geophysics—the transient waves in ultrasonic devices, and so on. In short, the HNM is applicable wherever waves propagate in stratified, anisotropic media.

Finally, comparing the present method with the method of normal modes, it can be found that the two methods express the displacements in the Fourier transform domain with a series of eigenmodes. The method of normal modes uses the exact expressions of the eigenmodes for isotropic plates, which exact modes could be very cumbersome to obtain. The present method uses the eigenmodes obtained using layer element method. Thereby the present method can be easily used for 3D problems of anisotropic laminated structures with a very large number of layers.

9

Waves in Functionally Graded Plates

9.1 Introduction

Laminates made of layers of composite materials have material property discontinuity between plies. Such discontinuity in material property could produce singularity of interlaminar stresses and consequently rip off the laminates. Plates made from functionally graded materials (FGM) have no interlaminar interfaces as the material property varies continuously through the thickness. FGMs have, therefore, found many applications in engineering, especially for thermal protection systems for rocket and spacecraft engines.

Chapter 2 used an analytical method to deal with harmonic and transient waves in functionally graded material (FGM). The problem has been idealized as a one-dimensional one, and the solution in the frequency domain is obtained in the form of confluent hypergeometric functions that require a numerical approach to evaluate. The analytical derivation is tedious, and its application is very limited. Chapter 5 introduced a layer element method for investigating characteristics of Lame waves in anisotropic laminates. Chapter 8 introduced the HNM for analyzing transient waves in anisotropic laminated plates excited by transient loadings. This chapter extends the layer element and the HNM to the case of FGM plates. The layer element and HNM can be, in fact, directly applied to FGM plates for 1D, 2D, and 3D analysis. However, as demonstrated in Chapter 1 for 1D cases, a significant number of elements is needed to achieve accurate results, as the element used in the layer element method and HNM must be homogeneous. Therefore, our task here is to introduce sets of formulation for the layer element and HNM methods allowing the material property changes continuously in the thickness direction of the element in a given function. The present formulation was originally proposed by Liu (1991).

Upon the establishment of the formulation, several numerical examples are presented to illustrate and investigate the characteristics of waves in FGM plates. This includes the frequency spectra, group velocity, and the mode shapes, as well as the transient response of FGM plates excited by transient loads. The results obtained for FGM plates are compared with those for isotropic homogeneous plates to reveal important features for waves in FGM plates. Very interesting findings that are useful in developing surface acoustic wave (SAW) devices are presented.

9.2 Dynamic System Equation

Consider a FGM plate with a varying material property in the thickness direction. Without loss of generality, the material property in the plane of the plate is assumed anisotropic. The total thickness of the plate is denoted by H. The plate is divided into N layer elements. The thickness of an element is denoted by h that can be different from element to element, as shown in Figure 9.1. The elastic coefficient matrix and the mass density on the lower, middle, and upper surfaces of the element are denoted by $\mathbf{c}^l = (c_{ij})^l$, ρ^l, $\mathbf{c}^m = (c_{ij})^m$, ρ^m, $\mathbf{c}^u = (c_{ij})^u$ ($i, j = 1,...,6$), and ρ^u, respectively, where superscript u and l stand for the upper and lower surfaces of an element. These material properties can also be different from element to element. For FGM plates, it is usually the case that $\mathbf{c}^u$ and ρ^u of an element equal $\mathbf{c}^l$ and ρ^l of the element above, since the material properties vary continuously in FGM plates. The dynamic equilibrium equation for an element is given in matrix form as follows:

$$\mathbf{W} = \rho(z)\ddot{\mathbf{U}} - \mathbf{L}^T\mathbf{c}(z)\mathbf{L}\mathbf{U} - \mathbf{f} = 0 \tag{9.1}$$

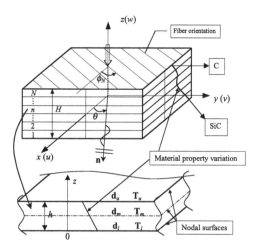

FIGURE 9.1
FGM plate and the isolated element.

where the dots are the differentiation with respect to time, superscript T stands for the transposed matrix, and $\mathbf{f}$ is the external body force vector in the element. The foregoing equation is basically the same as that in the previous chapter, except that the material properties $\mathbf{c}$ and ρ are now functions of coordinate z.

9.3 Dispersion Relation

Consider a harmonic plane wave of wavenumber k propagating in the plane of the plate in the direction of θ with respect to x-axis. Following exactly the same procedure for the layer element formulation described in Chapter 5, with special attention to the fact that the material property is now a function of z, we have

$$\mathbf{F} = [\mathbf{K} - \omega^2 \mathbf{M}]\mathbf{d} \tag{9.2}$$

where $\mathbf{F}$ and $\mathbf{d}$ are the same as those obtained in Chapter 8. The stiffness matrix $\mathbf{K}$ is found to be

$$\mathbf{K} = [\mathbf{A}_1^l + \mathbf{A}_1^d]k^2\cos^2\theta + [\mathbf{A}_2^l + \mathbf{A}_2^d]k^2\cos\theta\sin\theta + [\mathbf{A}_3^l + \mathbf{A}_3^d]k^2\sin^2\theta$$
$$+ i[\mathbf{A}_4^l + \mathbf{A}_4^d]k\cos\theta + i[\mathbf{A}_5^l + \mathbf{A}_5^d]k\sin\theta + [\mathbf{A}_6^l + \mathbf{A}_6^d] \tag{9.3}$$

and the mass matrix becomes

$$\mathbf{M} = \mathbf{M}^l + \mathbf{M}^d \tag{9.4}$$

In Eqs. (9.3) and (9.4), the matrices $\mathbf{M}^l$ and $\mathbf{A}_i^l$ ($i = 1,\dots,6$) are the same as the matrices $\mathbf{M}$ and $\mathbf{A}_i$ ($i = 1,\dots,6$) given in Chapter 5, except that the density and the elastic coefficients on the lower surface of the element should be used. The matrices $\mathbf{M}^d$ and $\mathbf{A}_i^d$ ($i = 1,\dots,6$) are the additional matrices for the variation of the density and the elastic coefficients in the thickness direction of the element. By assuming that the density and elastic coefficients are linear functions of z within the element, the matrices $\mathbf{M}^d$ and $\mathbf{A}_i^d$ ($i = 1,\dots,6$) can be obtained as follows:

$$\mathbf{M}^d = \frac{\rho^d h}{60}\begin{bmatrix} \mathbf{I} & 0 & -\mathbf{I} \\ 0 & 16\mathbf{I} & 4\mathbf{I} \\ -\mathbf{I} & 4\mathbf{I} & 7\mathbf{I} \end{bmatrix} \tag{9.5}$$

$$\mathbf{A}_1^d = \frac{h}{30} \begin{bmatrix} 0.5\mathbf{D}_1^d & 0 & -0.5\mathbf{D}_1^d \\ 0 & 8\mathbf{D}_1^d & 2\mathbf{D}_1^d \\ -0.5\mathbf{D}_1^d & 2\mathbf{D}_1^d & 3.5\mathbf{D}_1^d \end{bmatrix} \tag{9.6}$$

$$\mathbf{A}_2^d = \frac{h}{30} \begin{bmatrix} 0.5\mathbf{D}_2^d & 0 & -0.5\mathbf{D}_2^d \\ 0 & 8\mathbf{D}_2^d & 2\mathbf{D}_2^d \\ -0.5\mathbf{D}_2^d & 2\mathbf{D}_2^d & 3.5\mathbf{D}_2^d \end{bmatrix} \tag{9.7}$$

$$\mathbf{A}_3^d = \frac{h}{60} \begin{bmatrix} 0.5\mathbf{D}_3^d & 0 & -0.5\mathbf{D}_3^d \\ 0 & 8\mathbf{D}_3^d & 2\mathbf{D}_3^d \\ -0.5\mathbf{D}_3^d & 2\mathbf{D}_3^d & 3.5\mathbf{D}_3^d \end{bmatrix} \tag{9.8}$$

$$\mathbf{A}_4^d = \frac{1}{15} \begin{bmatrix} 2\mathbf{D}_4^d - 2\mathbf{D}_7^d & -4\mathbf{D}_4^d - \mathbf{D}_7^d & 2\mathbf{D}_4^d + 0.5\mathbf{D}_7^d \\ 6\mathbf{D}_4^d - \mathbf{D}_7^d & 8\mathbf{D}_4^d - 8\mathbf{D}_7^d & -14\mathbf{D}_4^d - \mathbf{D}_7^d \\ 0.5\mathbf{D}_7^d - 3\mathbf{D}_4^d & 16\mathbf{D}_4^d - \mathbf{D}_7^d & 13\mathbf{D}_7^d - 2\mathbf{D}_4^d \end{bmatrix} \tag{9.9}$$

$$\mathbf{A}_5^d = \frac{1}{15} \begin{bmatrix} 2\mathbf{D}_5^d - 2\mathbf{D}_8^d & -4\mathbf{D}_5^d - \mathbf{D}_8^d & 2\mathbf{D}_5^d - 0.5\mathbf{D}_8^d \\ 6\mathbf{D}_5^d - \mathbf{D}_8^d & 8\mathbf{D}_5^d - 8\mathbf{D}_8^d & -14\mathbf{D}_5^d - \mathbf{D}_8^d \\ 0.5\mathbf{D}_8^d - 3\mathbf{D}_5^d & 16\mathbf{D}_5^d - \mathbf{D}_8^d & 13\mathbf{D}_8^d - 13\mathbf{D}_5^d \end{bmatrix} \tag{9.10}$$

$$\mathbf{A}_6^d = \frac{1}{6h} \begin{bmatrix} 3\mathbf{D}_6^d & -4\mathbf{D}_6^d & \mathbf{D}_6^d \\ -4\mathbf{D}_6^d & 16\mathbf{D}_6^d & -12\mathbf{D}_6^d \\ \mathbf{D}_6^d & -12\mathbf{D}_6^d & 11\mathbf{D}_6^d \end{bmatrix} \tag{9.11}$$

Here $\mathbf{D}_i^d = \mathbf{D}_i^u - \mathbf{D}_i^l$ $(i = 1,\ldots,8)$, $\rho^d = \rho^u - \rho^l$. The matrices $\mathbf{D}_i^u$ can be obtained by replacing c_{ij} in $\mathbf{D}_i$ with c_{ij}^u. The matrices $\mathbf{D}_i^l$ can be obtained by replacing c_{ij} in $\mathbf{D}_i$ with c_{ij}^l. Matrices $\mathbf{D}_i$ have been given in Chapter 5.

Assembling the matrices of all the elements, we obtain the dynamic system equation for FGM plates:

$$\mathbf{F}_t = [\mathbf{K}_t - \omega^2 \mathbf{M}_t] \mathbf{d}_t \tag{9.12}$$

where

$$\mathbf{K}_t = [\mathbf{A}_{1t}^l + \mathbf{A}_{1t}^d]k^2\cos^2\theta + [\mathbf{A}_{2t}^l + \mathbf{A}_{2t}^d]k^2\cos\theta\,\sin\theta + [\mathbf{A}_{3t}^l + \mathbf{A}_{3t}^d]k^2\sin^2\theta$$
$$+ i[\mathbf{A}_{4t}^l + \mathbf{A}_{4t}^d]k\cos\theta + i[\mathbf{A}_{5t}^l + \mathbf{A}_{5t}^d]k\sin\theta + [\mathbf{A}_{6t}^l + \mathbf{A}_{6t}^d] \qquad (9.13)$$

In these equations, the subscript t indicates that the matrices are for the entire FGM plate, and the matrices $\mathbf{d}_t$, $\mathbf{M}_t$, $\mathbf{A}_{it}^l$, and $\mathbf{A}_{it}^d$ ($i = 1,...,6$) are the results of assembling the contributions of $\mathbf{d}$, $\mathbf{M}$, $\mathbf{A}_i^l$, and $\mathbf{A}_i^d$ ($i = 1,...,6$) for all the elements. The size of $\mathbf{d}_t$ is $M \times 1$ and the sizes of $\mathbf{M}_t$, $\mathbf{A}_{it}^l$, and $\mathbf{A}_{it}^d$ are $M \times M$. When the number of the element is N, we have $(2N + 1)$ nodal surfaces and $M = 3 \times (2N + 1)$.

The vectors $\mathbf{d}_t$ and $\mathbf{F}_t$ represent, respectively, the nodal displacement amplitudes and the external forces applied at the nodal surfaces of the plate. The wave modes of harmonic waves in the FGM plate are determined by considering free wave motion and setting the external force vector $\mathbf{F}_t$ in Eq. (9.12) to zero, i.e.,

$$0 = [\mathbf{K}_t - \omega^2\mathbf{M}_t]\mathbf{d}_t \qquad (9.14)$$

The elements of $\mathbf{K}_t$ are functions of the wavenumber k. When the wavenumber k is given, the eigen-frequency ω and its corresponding eigenvectors can be obtained from Eq. (9.14). Consequently Eq. (9.14) gives the dispersion relation for harmonic plane waves propagating in the FGM plate. The eigenvectors obtained represent the shape of the wave modes in the FGM plate.

9.4 Group Velocity

The matrix $\mathbf{K}_t$ is a Hermitian matrix for a given real wavenumber. Hence, the frequency obtained from Eq. (9.14) must be real, and the left and right eigenvectors are complex conjugate. For the mth mode, we find the group velocity in the same manner discussed in Chapter 5.

$$c_{gm} = \frac{\boldsymbol{\psi}_m^L \mathbf{K}_{t,k} \boldsymbol{\psi}_m^R}{2\omega_m \boldsymbol{\psi}_m^L \mathbf{M}_t \boldsymbol{\psi}_m^R} \qquad (9.15)$$

where

$$\mathbf{K}_{t,k} = 2k[\mathbf{A}_{1t}^l + \mathbf{A}_{1t}^d]\cos^2\theta + 2k[\mathbf{A}_{2t}^l + \mathbf{A}_{2t}^d]\cos\theta\,\sin\theta + 2k[\mathbf{A}_{3t}^l + \mathbf{A}_{3t}^d]\sin^2\theta$$
$$+ i[\mathbf{A}_{4t}^l + \mathbf{A}_{4t}^d]\cos\theta + i[\mathbf{A}_{5t}^l + \mathbf{A}_{5t}^d]\sin\theta \qquad (9.16)$$

9.5   Response Analysis

Following exactly the same procedure for the HNM formulation described in Chapter 8, with special attention to the fact that the material property is now a function of z, we shall have

$$\mathbf{F}_t = \mathbf{M}\ddot{\mathbf{d}}_t + \mathbf{K}_{Dt}\mathbf{d}_t \qquad (9.17)$$

where

$$\mathbf{K}_{Dt} = -[\mathbf{A}_{1t}^l + \mathbf{A}_{1t}^d]\frac{\partial^2}{\partial x^2} - [\mathbf{A}_{2t}^l + \mathbf{A}_{2t}^d]\frac{\partial^2}{\partial x \partial y} - [\mathbf{A}_{3t}^l + \mathbf{A}_{3t}^d]\frac{\partial^2}{\partial y^2}$$

$$+ [\mathbf{A}_{4t}^l + \mathbf{A}_{4t}^d]\frac{\partial}{\partial x} + [\mathbf{A}_{5t}^l + \mathbf{A}_{5t}^d]\frac{\partial}{\partial y} + [\mathbf{A}_6^l + \mathbf{A}_6^d] \qquad (9.18)$$

In Eq. (9.17), vector $\mathbf{F}_t$ can be obtained by assembling the matrices $\mathbf{F}$ of all the elements in the FGM plate. The application of the Fourier transform to Eq. (9.17) leads to the following dynamic system equations in the wavenumber domain:

$$\tilde{\mathbf{F}}_t = \mathbf{M}_t\ddot{\tilde{\mathbf{d}}}_t + \mathbf{K}_t\tilde{\mathbf{d}}_t \qquad (9.19)$$

where

$$\mathbf{K}_t = k_x^2[\mathbf{A}_{1t}^l + \mathbf{A}_{1t}^d] + k_x k_y[\mathbf{A}_{2t}^l + \mathbf{A}_{2t}^d] + k_y^2[\mathbf{A}_{3t}^l + \mathbf{A}_{3t}^d]$$

$$-ik_x[\mathbf{A}_{4t}^l + \mathbf{A}_{4t}^d] - ik_y[\mathbf{A}_{5t}^l + \mathbf{A}_{5t}^d] + [\mathbf{A}_{6t}^l + \mathbf{A}_{6t}^d] \qquad (9.20)$$

With Eq. (9.17), we can readily obtain responses of FGM plates to various loads in the similar way to those discussed in Chapter 8 for anisotropic laminates.

9.6   Two-Dimensional Problem

The formulas for 2D problems can be easily inferred from the 3D counterparts. In 2D problems, we assume that the load is independent of y, such as a line load that is constant along the y-axis, which leads to the independency of all the field variables on y. In this case, the stiffness matrix becomes much simpler as

$$\mathbf{K}_t = k_x^2[\mathbf{A}_{1t}^l + \mathbf{A}_{1t}^d] + ik_x[\mathbf{A}_{4t}^l + \mathbf{A}_{4t}^d] + [\mathbf{A}_{6t}^l + \mathbf{A}_{6t}^d] \qquad (9.21)$$

The contents of the matrices $\mathbf{A}_{it}^l$ and $\mathbf{A}_{it}^d$ ($i = 1, 4, 6$) remain the same as their 3D counterparts.

9.7 Computational Procedure

We use a SiC-C plate as an example to describe the procedure for calculating the frequency spectra, group velocities, and mode shapes of the waves in FGM plates. The plate is made by combining two materials, SiC and C, using a chemical vapor deposition (CVD) technique (Sasaki et al., 1989). The distribution of the content of SiC or C is plotted in Figure 9.2. The material properties of the SiC monolith and C monolith are given in Table 9.1. The Young's modulus E, the shear modulus G, and Poisson's ratio v are obtained by using the method given by Kerner (1956) and are also shown in Figure 9.2. The distribution of the density of the SiC-C plate in the thickness direction shown in Figure 9.2 is obtained by following equation of rule of mixture.

$$\rho = \rho_c v_c + \rho_{sic} v_{sic} \qquad (9.22)$$

where ρ_c and v_c are, respectively, the density and volume fraction of the C monolith, and ρ_{sic} and v_{sic} are those of the SiC monolith, respectively. Natural frequencies, shapes of wave modes, dispersion relations, and transient

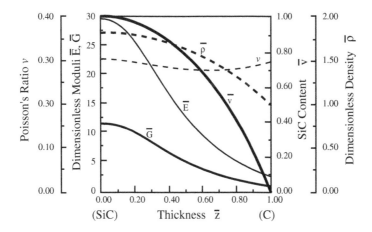

FIGURE 9.2
Distributions of material properties along the thickness of a SiC-C plate.

TABLE 9.1

Material Properties of SiC and C Monolith
Materials

Materials Constants	E (GPa)	v	ρ (g/cm^3)
CVD-SiC	320	0.3	3.22
CVD-C	28	0.3	1.78

responses excited by time-step impact and time-pulse loads are computed
for the SiC-C plate. The computational procedure is summarized briefly
below.

1. First, divide the plate into N plate elements in the thickness direction of the plate and acquire the values of material properties, such
   as mass density, Young's moduli, shear moduli, and Poisson's
   ratios on the lower and upper surfaces of the elements, using the
   distribution curve shown in Figure 9.2.

2. Calculate the mass density ρ^u and ρ^l, as well as the elastic coefficient
   matrices on the lower and upper surfaces of the element $c^l = (c_{ij})^l$
   and $c^u = (c_{ij})^u$ ($i, j = 1,...,6$). The matrices M^l, A_i^d, M^d, and A_i^d ($i =$
   $1,...,6$) can be formed using Eqs. (9.4) to (9.11).

3. Then, assembly matrices M_t, A_{it}^l, and A_{it}^d using the matrices M^l,
   A_i^l, M^d, and A_i^d for all the elements in the FGM plate.

4. In the characteristic analysis, the matrix K_t can be obtained from
   Eq. (9.13) for a given direction of wave propagation θ (when the
   plate is isotropic, the characteristics are independent of θ, we use
   $\theta = 0$) and for a given wavenumber k. Solving Eq. (9.14), the eigen-
   frequency ω_m and the corresponding eigenvector ψ_m can be
   obtained. The group velocities can be computed using Eq. (9.15).

5. In the analysis of the 3D displacement responses, we discretize,
   appropriately, the ranges wavenumbers of $[-K_x\ K_x]$ and $[-K_y\ K_y]$,
   using the integral technique described in Chapter 8. Build the
   matrix K_t for various discrete points of (k_{xi}, k_{yi}) using Eq. (9.20).
   Eigen-frequencies ω_m and eigenvector ψ_m as well as displacement
   responses of the plate can be obtained following exactly the procedures in Chapter 5.

9.8 Dispersion Curves

Figure 9.3 shows the frequency spectra for the lowest fifteen wave modes
of the SiC-C plate. The group velocities of these wave modes are shown in
Figure 9.5. For comparison, the frequency spectra and the group velocity

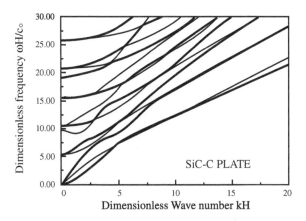

FIGURE 9.3
Frequency spectra for the SiC-C plate (thick line: odd modes; thin line: even modes).

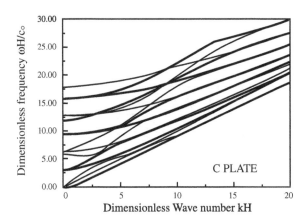

FIGURE 9.4
Frequency spectra for the C plate (thick line: antisymmetric modes; thin line: symmetric modes).

spectra for the C monolith plate are shown in Figures 9.4 and 9.6, respectively. Comparing Figures 9.3 and 9.4, it can be found that the lowest symmetric mode of the C plate has no dispersion, but all of the modes of the SiC-C plate are observed to have dispersion. This property is similar to that of laminates discussed in Chapter 5. Comparison of Figures 9.5 and 9.6 reveals that the group velocities of the SiC-C plate approach constant values more quickly than those of the C plate with increasing wavenumber. This indicates that the dispersions of the lower modes of the SiC-C plate are weaker than those of the C plate.

The shape of wave modes is very important for understanding the properties of waves in plates. We calculated the lowest fifteen mode shapes of

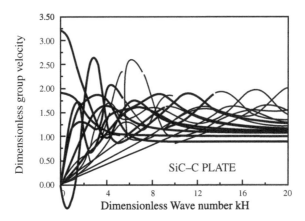

FIGURE 9.5
Group velocity spectra for the SiC-C plate (thick line: modes 1–8; thin line: modes 9–15).

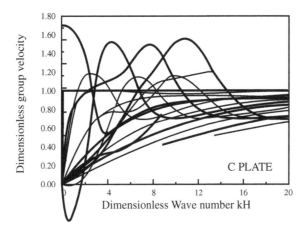

FIGURE 9.6
Group velocity spectra for the C plate (thick line: antisymmetric modes; thin line: symmetric modes).

the SiC-C plate for wavenumbers 6.0 and 25.0 and show them in Figures 9.7 and 9.9, respectively. In these figures, the upper surface is C monolith and the lower surface is SiC monolith. From these figures, note that the displacement component v is not coupled with the displacement components u and w, since the SiC-C plate is isotropic in the plane of the plate. Furthermore, it can be found that the displacements on the upper surface are larger than those on the lower surface when the wavenumber is smaller ($k = 6.0$). The displacements are completely concentrated only on the upper surface of the SiC-C plate, when the wavenumber is very large ($k = 25.0$). The lowest fifteen

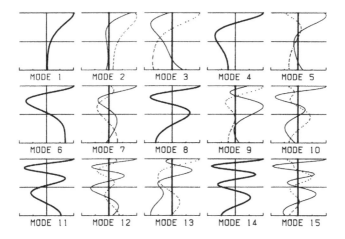

FIGURE 9.7
Mode shapes of harmonic waves in the SiC-C plate ($\bar{k}$ = 6.0; thin line: u; dotted line: w; thick line: v).

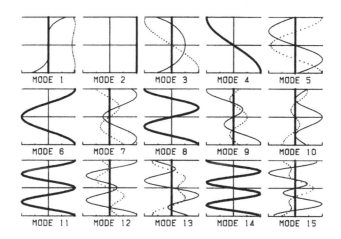

FIGURE 9.8
Mode shapes of harmonic waves in the C plate ($\bar{k}$ = 6.0; thin line: u; dotted line: w; thick line: v).

mode shapes of the C plate for wavenumber 6.0 and 25.0 are shown in Figures 9.8 and 9.10, respectively. Comparing Figures 9.7 and 9.8, it is clear that the displacements are more concentrated on the upper surface of the SiC-C plate than those on the surfaces of the C plate, when k = 6.0. When k = 25.0 (Figures 9.9 and 9.10), the displacements of the lowest fourteen modes of the SiC-C plate are strongly concentrated on the upper surface, but the displacements of only the lowest two modes of the C plate are concentrated on both surfaces. Examining the mode shapes of the displacement v (the SH

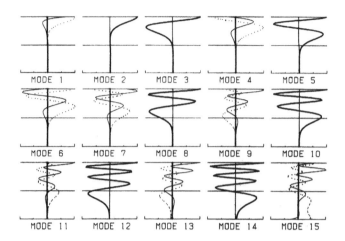

FIGURE 9.9

Mode shapes of harmonic waves in the SiC-C plate ($\bar{k} = 25.0$; thin line: u; dotted line: w; thick line: v).

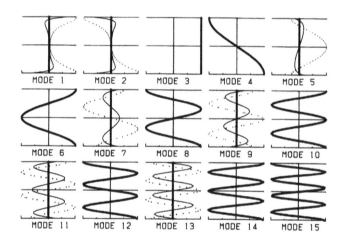

FIGURE 9.10

Mode shapes of harmonic waves in the C plate ($\bar{k} = 25.0$; thin line: u; dotted line: w; thick line: v).

wave modes in plates), reveals that the mode shapes of the C plate are independent of wavenumber. The mode shape of SH waves in the SiC-C plate depends on wavenumber and the displacements become more concentrated on the upper surface of the plate with increasing wavenumber. The reason is that the SiC monolith that is on the lower surface of the SiC-C plate is much harder than the C monolith on the upper surface (refer to Table 9.1). From this property, it can be inferred that the responses of surface waves on the upper surface of the SiC-C plate should be stronger than those of the C plate.

9.9 Transient Response to Line Time-Step Loads

9.9.1 Results for Vertical Transient Loads

Only line transient loads that are constant along the y-axis are considered here, which means that we consider only 2D cases. It is assumed that the loads are applied on the upper surface of plates. First, consider a line time-step load vertical to the plane of the plate. In this case, we have

$$q_x = 0, \quad q_y = 0, \quad q_z = 1 \tag{9.23}$$

where q_x, q_y, and q_z represent the ratios of the components of the force in the x, y, and z directions, respectively. Figure 9.11 shows the responses of the displacement u on the upper surface of the SiC-C plate at $\bar{t} = 5.0$, 7.0, and 10.0. From this figure, a surface wave with velocity of $0.92c_0$ is observed (e.g., at $\bar{x} \approx 9.2$ and $\bar{t} = 10.0$). This velocity is close to the velocity of the Rayleigh wave of C monolith ($0.9274c_0$) and is in agreement with the velocity of the first mode shown in Figure 9.7. The responses on $\bar{z} = 0$, 0.5, and 1.0 at $\bar{t} = 5.0$ are shown in Figure 9.13. From this figure it can be seen that a suddenly changing displacement, as in the case of $\bar{z} = 1.0$ and $\bar{x} \approx 4.6$, is not observed at $\bar{z} = 0$, 0.5. The displacement responses of the C plate excited by the same load are shown in Figures 9.12 and 9.14. Comparison of Figures 9.11 and 9.12 shows that the response of the surface wave on the upper surface of the SiC-C plate is stronger than that on the upper surface of the C plate. From Figures 9.13 and 9.14 it can be seen that the absolute values of the displacements in the upper and lower surfaces of the C plate are approximately equal.

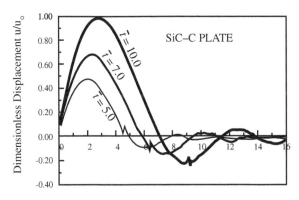

FIGURE 9.11
Responses of u on the upper surface of the SiC-C plate excited by a vertical time-step load ($\bar{z} = 1.0$).

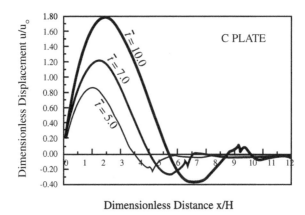

FIGURE 9.12

Responses of u on the upper surface of the C plate excited by a vertical time-step load ($\bar{z} = 1.0$).

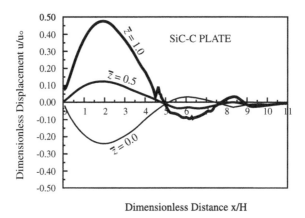

FIGURE 9.13

Responses of u in the SiC-C plate excited by a vertical time-step load ($\bar{t} = 5.0$).

The absolute values of the displacement on the upper surface of the SiC-C plate is about twice that on the lower surface since the lower surface of the SiC-C plate is harder than the upper surface. Figure 9.15 shows the responses of the displacement w on the upper surface of the SiC-C plate at $\bar{t} = 5.0, 7.0$, and 10.0. From this figure, the surface wave with velocity of $0.92c_0$ is also observed, but the responses of the surface wave are weaker than that of the surface wave observed in Figure 9.11. The responses for $\bar{z} = 0, 0.5$, and 1.0 at $\bar{t} = 5.0$ are shown in Figure 9.17 where it can be seen that a suddenly changing displacement, as in the case of $\bar{z} = 1.0$ and $\bar{x} \approx 4.6$, is not observed in the cases of $\bar{z} = 0$ and 0.5. The displacement responses of the C plate under the same load are shown in Figures 9.16 and 9.18. Comparing Figures 9.15, 9.16,

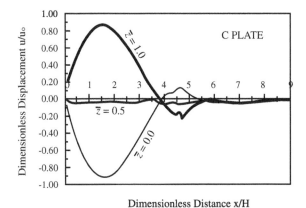

FIGURE 9.14
Responses of u in the C plate excited by a vertical time-step load ($\bar{t} = 5.0$).

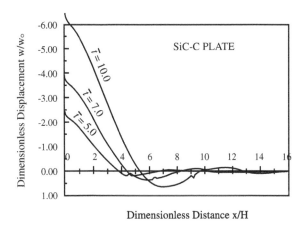

FIGURE 9.15
Responses of w on the upper surface of the SiC-C plate excited by a vertical time-step load ($\bar{z} = 1.0$).

and 9.17 with 9.18, it can be seen that the response of the surface wave on the upper surface of the SiC-C plate is relatively stronger than that on the upper surface of the C plate. It can be seen from Figure 9.14 that at the places far from the load and the surface wave, the displacement u on the two surfaces of the C plate possesses opposite sign, and their absolute values are approximately equal. In addition, the displacement on the middle surface is close to zero. Furthermore, Figure 9.18 shows that the values of the displacement w on the three surfaces are approximately equal. These results indicate that the deformation of the monolith plate under a vertical time-step transient load is dominated by a bending deformation, and the plate theory based on

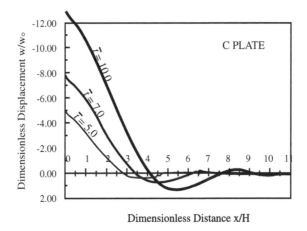

FIGURE 9.16
Responses of w on the upper surface of the C plate excited by a vertical time-step load ($\bar{z} = 1.0$).

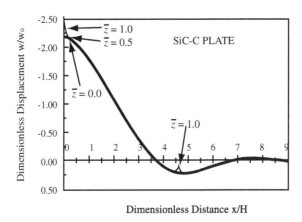

FIGURE 9.17
Responses of w in the SiC-C plate excited by a vertical time-step load ($\bar{t} = 5.0$).

Kirchhoff's assumption is suitable for the C plate to obtain approximate results. Study of Figures 9.13 and 9.17 has also revealed that the plate theory is also suitable for the SiC-C plate to obtain approximated results at places far from the load and the surface wave.

When two identical transient loads act symmetrically and simultaneously on both the lower and upper surfaces of a plate, the displacement responses excited by the two loads compensate each other and there will not be bending deformation. Methods based on the theory of bending plates will not work because the results obtained will simply be zero.

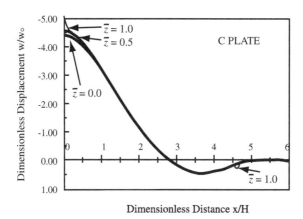

FIGURE 9.18
Responses of w in the C plate excited by a vertical time-step load ($\bar{t} = 5.0$).

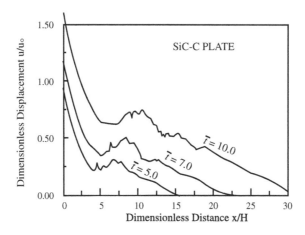

FIGURE 9.19
Responses of u on the upper surface of the SiC-C plate excited by a shear time-step load in the x-direction ($\bar{z} = 1.0$).

9.9.2 Results for a Shear Load in the x Direction

Consider now a shear force applied on the upper surface of the plate. We shall have

$$q_x = 1, \quad q_y = 0, \quad q_z = 0 \tag{9.24}$$

Figure 9.19 shows the responses of the displacement u on the upper surface of the SiC-C plate excited by the shear force. From this figure, a surface wave with velocity of $0.92c_0$ is observed. The responses on $\bar{z} = 0$, 0.5, and 1.0 at $\bar{t} = 0.5$ are

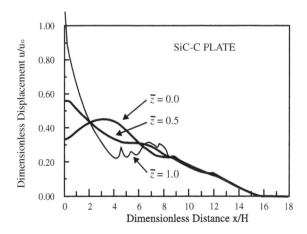

FIGURE 9.20
Responses of u in the SiC-C plate excited by a shear time-step load in the x-direction ($\bar{t} = 5.0$).

shown in Figure 9.20 in which it can be seen that the displacements on the lower, middle, and upper surfaces are approximately equal at a position far from the load ($\bar{x} > 9.0$).

Comparing Figures 9.15 and 9.19, it can be found that the response of the displacement w on the upper surface of the plate excited by the vertical load arrives at $\bar{x} \approx 9.0$ and $\bar{t} = 5.0$. On the other hand, the response of the displacement u by the shear load in the x direction arrives at $\bar{x} \approx 16.0$ and $\bar{t} = 5.0$. This fact indicates that the response of the displacement u by the shear load in the x direction contains more components of the longitudinal wave than that of the displacement w by the vertical load.

9.9.3 Results for a Shear Load in the y Direction

To observe the SH waves in the FGM plate, the responses of the displacement v excited by a shear force in the direction y-axis are computed. In this case, we have

$$q_x = 0, \quad q_y = 1, \quad q_z = 0 \qquad (9.25)$$

Figure 9.21 shows the responses of the displacement v on the upper surface of the SiC-C plate, and a surface wave with velocity of $1.03c_0$ is observed (e.g., at $\bar{x} = 10.3$ and $\bar{t} = 5.0$). This is close to the velocity of the shear wave of C monolith (c_0) and is in agreement with the velocity of the third mode shown in Figure 9.9. The responses on $\bar{z} = 0, 0.5$, and 1.0 at $\bar{t} = 5.0$ are shown in Figure 9.23 in which it can be seen that a suddenly changing displacement, as in the case of $\bar{z} = 1.0$ and $\bar{x} \approx 5.0$, is not observed when $\bar{z} = 0$ and 0.5. The displacement responses of the C plate excited by the same load are shown in Figures 9.22 and 9.24. Comparison of Figures 9.21 and 9.22 shows that the

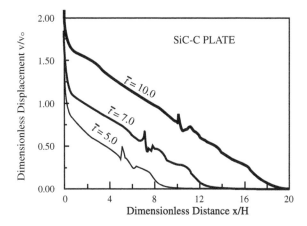

FIGURE 9.21
Responses of v on the upper surface of the SiC-C plate excited by a shear time-step load in the y-direction ($\bar{z} = 1.0$).

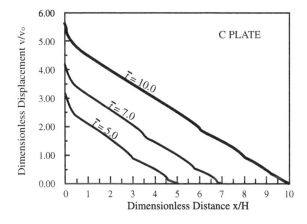

FIGURE 9.22
Responses of v on the upper surface of the C plate excited by a shear time-step load in the y-direction ($\bar{z} = 1.0$).

strong response of the surface wave on the upper surface of the SiC-C plate is observed, but no suddenly changing displacements are observed on the surfaces of the C plate. This fact, which has been also shown in Figures 9.7–9.10, indicates that in the C plate the SH waves are not concentrated on the surfaces, but in the SiC-C plate there are SH waves that are concentrated on the upper surface of the SiC-C plate. Figures 9.23 and 9.24 show that the values of the displacements in the upper, middle, and lower surfaces of the SiC-C plate are approximately equal at places that are far from the load, and the surface wave and that of the C plate is approximately equal at places far from the load.

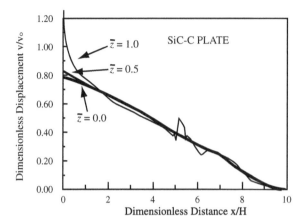

FIGURE 9.23

Responses of v in the SiC-C plate excited by a shear time-step load in the y-direction ($\bar{t} = 5.0$).

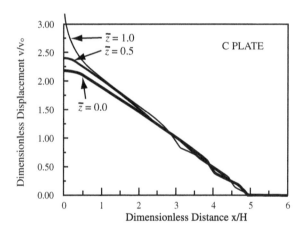

FIGURE 9.24

Responses of v in the C plate excited by a shear time-step load in the y-direction ($\bar{t} = 5.0$).

9.9.4 Results for a Line Time-Pulse Load

To observe the surface waves due to the transient load with high frequency, the displacement responses of the SiC-C plate excited by a line time-pulse load are investigated. As an example, the response of the displacement u on the upper surface of the SiC-C plate excited by a shear transient load in the x direction ($q_x = 0$, $q_y = 0$, $q_z = 1$) are shown in Figure 9.25. The responses for C plate are shown in Figure 9.26 for comparison. From Figure 9.25 the surface wave is again observed with velocity of $0.92c_0$, which is close to that of the surface wave of the C plate shown in Figure 9.26 and is in agreement with

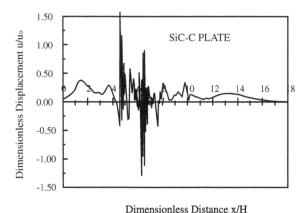

FIGURE 9.25
Responses of u on the upper surface of the SiC-C plate excited by a shear time-pulse load in the x-direction ($\bar{t} = 5.0$, $\bar{z} = 1.0$).

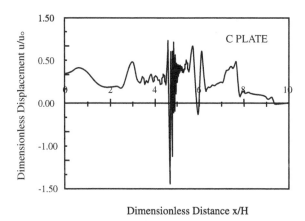

FIGURE 9.26
Responses of u on the upper surface of the C plate excited by a shear time-pulse load in the x-direction ($\bar{t} = 5.0$, $\bar{z} = 1.0$).

the first mode shown in Figure 9.9. After that, a faster surface wave, which does not appear on the surface of the C plate and whose velocity is close to those of modes 4, 6, 7, and 9, is observed.

It is confirmed that the displacement responses of the surface waves of the plate excited by the time-pulse transient load are much stronger than those excited by the time-step transient load. This fact seems to suggest that excitations of higher frequency generate more wave modes of surface waves in plates.

9.10 Remarks

The formulation of the dispersion relation and the group velocity of the harmonic plane waves in the FGM plates and the displacement responses of the FGM plates excited by the time-step transient loads and the time-pulse ones are presented in this chapter, by extending the layer element and HNM methods for anisotropic laminated plates. In the present method, the variations of material properties in the thickness direction of the element have been taken into consideration in the formulation. By assuming linear variation on material properties in the thickness direction within elements, detailed formulae for mass and stiffness matrices have been presented. These matrices can be used directly to compute the natural frequencies, wave modes, and transient responses of FGM plates.

For the material with 13 elastic coefficients, the matrix $\mathbf{K}_t$ can be changed into a real symmetric matrix for real wavenumbers and the formulae can be simplified. Furthermore, the symmetry of the displacements in the plane of the plate can also be used to reduce the effort of computation in the numerical inverse Fourier transforms.

The present formulae can be directly used for anisotropic laminated plates without any difficulty by simply omitting the additional matrices $\mathbf{M}^d$ and $\mathbf{A}_i^d$ ($i = 1,\ldots,6$). Therefore, the laminated plate can actually be treated as a special case of FGM plate.

From the numerical examples, the following conclusions can be drawn.

- The present method is effective, powerful, and easy to use for analyzing both harmonic waves and transient waves in FGM plates.
- For large wavenumbers, the displacements of the lowest modes are concentrated on the softer surface of the SiC-C plate.
- The responses of the surface waves on the soft surface of the SiC-C plate are relatively stronger than those on the surfaces of the homogeneous C plate.
- The responses of the surface waves excited by time-pulse transient loads are much stronger than those excited by time-step transient loads. This implies that excitations of higher frequency generate more surface wave modes and, therefore, stronger surface waves.
- SH waves are propagating as surface waves on the softer surface of the SiC-C plate. This feature of FGM plates is very important and may be used in surface acoustic wave (SAW) devices.

Finally, there are two factors in determining the number of elements to be used in the analysis. First, the number of elements should be sufficient for accurately modeling the material variation in the FGM plate. More elements have to be used if the FGM plates have a larger gradient of material property in the thickness direction. Second, the number of the elements should also be sufficient in modeling the variation of wave field variables (displacement and stresses) in the thickness direction of the FGM plates. More elements have to be used if higher wave modes or transient responses of higher frequencies are required. For many practical cases, the gradient of FGM plates is not very large. Hence, the number of linear variation elements required to model the variation of the FGM plate should be smaller than the number of elements required for modeling the variation of the wave field in the thickness direction of the plates. This is especially true when higher wave modes and response of higher frequency are of interest. For those cases, the number of elements should be determined by the requirement of wave-field modeling. Elements with a higher order for material variation can increase the modeling accuracy in material variation. These elements can be developed following the procedure described in this chapter.

10

Waves in Anisotropic Functionally Graded Piezoelectric Plates

10.1 Introduction

Chapter 2 introduced an analytic method to deal with one-dimensional wave propagation problems in plates of functionally graded materials (FGM). The previous chapter introduced a layer element method and a hybrid numerical method and investigated 2D problems of harmonic and transient waves propagating in FGM plates.

FGMs can be used not only in thermal-protection systems for structures of space-planes but also in electronic and chemical fields as well as many other fields. The results obtained in the last chapter for the FGM plate opens a new avenue of developing high performance surface acoustic wave (SAW) devices using the concept of FGMs. Chapter 9 showed that strong SH surface waves and other surface waves can be generated on the softer surface of FGM plates if functionally graded piezoelectric materials (FGPM) can be made and used as the substrate of the SAW devices. Fortunately, techniques for fabricating the FGPM have also been developed (e.g., Kawai and Miyazaki et al., 1990). Hence, it is possible to fabricate FGPM substrates for SAW devices. However, there are still many problems to solve in wave propagation analysis: what characteristics do the waves in the FGPM plates have? In the design process, how can one simulate the displacement responses of the FGPM plate excited by electrodes and the electrostatic potential responses excited by mechanical loads? The purpose of this chapter is to provide an answer to these questions. It introduces a procedure to extend the layer element method and HNM for analyzing the characteristics and transient responses in FGPM plates, so as to simulate surface waves propagating in FGPM SAW devices.

For the characteristic analysis of harmonic waves in a piezoelectric medium, investigations have been undertaken. A good review is given by Lewis (1985). For waves in homogeneous piezoelectric plates, the finite element method (e.g., Koshiba and Suzuki, 1978) and the method of transmission line model (Hikita and Koshiba et al., 1973) are used for investigating the characteristics of Lamb waves in piezoelectric plates. Moreover, Lamb waves in a thin $LiTaO_3$ plate have been investigated by an experimental method combined with theoretical calculations (Toda and Mizutani, 1988). For waves in laminated piezoelectric structures, the finite element method (e.g., Pualey and Dong, 1975) and a matrix formulation (Ono and Wasa et al. 1977) have also been used. Shiosaki and Mikamura et al. (1986) have investigated Love waves in a three-layered piezoelectric structure, using an experimental method with exact theoretical calculations. However, all of these studies deal with the homogeneous piezoelectric plates or laminated piezoelectric plates of which material properties were homogeneous in the thickness of the layers, where only the characteristics of harmonic waves were considered.

In this chapter, the layer element method and hybrid numerical method proposed in the last chapter for investigating waves in FGM plates is extended to FGPM plates. The present method was originally proposed by Liu (1991). The method was published later in a paper authored by Liu and Tani (1994). In using this method, it is considered that the material properties of the FGPM plates vary continuously in the thickness direction, but are homogeneous, anisotropic in the plane of the plate. First, the FGPM plate is divided into layer elements. In an element, we assume that the displacement and the electrostatic potential fields are expressed by second order polynomials. The variational principle is used to develop a set of approximate governing system equations. The dispersion relations and the mode shapes of harmonic waves in the plates are obtained using the mechanical and electrical boundary conditions. The group velocities of the plane waves are formulated with the aid of the Rayleigh quotient. The method of Fourier transforms in conjunction with the modal analysis is used to determine the responses of displacements and electrostatic potentials. The present method developed for FGPM plates is directly applicable to homogenous and laminated piezoelectric plates as a special case of FGPM plates.

Examples are presented for calculating the dispersion relations, group velocities, electromechanical coupling constants, and mode shapes of harmonic plane waves propagating in a hypothetical FGPM plate and a homogeneous z-x $LiTaO_3$ plate. The displacement and electrostatic potential responses of the FGPM plate excited by mechanical loads and electrodes are also calculated and discussed in comparison with those of a homogeneous z-x $LiTaO_3$ plate.

10.2   Basic Equations

Consider a FGPM plate with a thickness H as shown in Figure 10.1. The plate is divided into N layer elements. The thickness of an element is denoted by h which can vary from element to element. In an element, the constitutive relationships for piezoelectric and elastic materials are given in

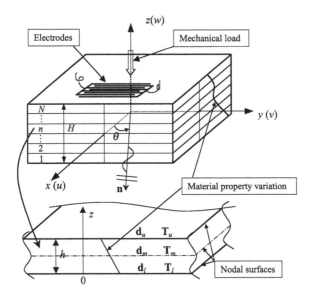

FIGURE 10.1
FGPM plate and an isolated layer element with linear variation on material property.

matrix form of

$$\boldsymbol{\sigma} = \mathbf{c}(z)\boldsymbol{\varepsilon} - \mathbf{e}^T(z)\mathbf{E} \qquad (10.1)$$

and

$$\mathbf{D} = \mathbf{g}(z)\mathbf{E} + \mathbf{e}(z)\boldsymbol{\varepsilon} \qquad (10.2)$$

where

$$\boldsymbol{\sigma}^T = \{\sigma_{xx} \quad \sigma_{yy} \quad \sigma_{zz} \quad \sigma_{yz} \quad \sigma_{xz} \quad \sigma_{xy}\} \qquad (10.3)$$

$$\boldsymbol{\varepsilon}^T = \{\varepsilon_{xx} \quad \varepsilon_{yy} \quad \varepsilon_{zz} \quad \varepsilon_{yz} \quad \varepsilon_{xz} \quad \varepsilon_{xy}\} \qquad (10.4)$$

$$\mathbf{D}^T = \{D_x \quad D_y \quad D_z\} \qquad (10.5)$$

$$\mathbf{E}^T = \{E_x \quad E_y \quad E_z\} \qquad (10.6)$$

are, respectively, the stress tensor, strain tensor, electric displacement vector, and electric field vector, while $\mathbf{c} = (c_{ij})$ $(i, j = 1,...,6)$, $\mathbf{e} = (e_{ij})$ $(i = 1, 2, 3;$ $j = 1,...,6)$, and $\mathbf{g} = (g_{ij})$ $(i, j = 1, 2, 3)$ are, respectively, the elastic, piezoelectric, and dielectric constant matrices of the material of the element.

The relationship between the displacement vector $\mathbf{U}$ and strain tensor $\boldsymbol{\varepsilon}$ is given by

$$\boldsymbol{\varepsilon} = \mathbf{L}\mathbf{U} \qquad (10.7)$$

where

$$\mathbf{U}^T = \{u \quad v \quad w\} \tag{10.8}$$

The operator matrix $\mathbf{L}$ has been defined in Eq. (4.5).

The Maxwell equations in the quasistatic approximation are

$$\mathbf{E} = -\mathrm{grad}\,\varphi = -\mathbf{L}_\varphi \varphi \tag{10.9}$$

where

$$\mathrm{div}\,\mathbf{D} = 0 \tag{10.10}$$

where φ is the electrostatic potential, while $\mathbf{L}_\varphi$ is given in the matrix form by

$$\mathbf{L}_\varphi^T = \left\{ \frac{\partial}{\partial x} \quad \frac{\partial}{\partial y} \quad \frac{\partial}{\partial z} \right\} \tag{10.11}$$

The operator matrix $\mathbf{L}_\varphi$ can be written in a form similar to Eq. (4.3) as follows:

$$\mathbf{L}_\varphi = \mathbf{L}_{\varphi x} \frac{\partial}{\partial x} + \mathbf{L}_{\varphi y} \frac{\partial}{\partial y} + \mathbf{L}_{\varphi z} \frac{\partial}{\partial z} \tag{10.12}$$

where $\mathbf{L}_{\varphi x}$, $\mathbf{L}_{\varphi y}$, $\mathbf{L}_{\varphi z}$ can be obtained by inspection of Eq. (10.11) as

$$\mathbf{L}_{\varphi x}^T = \{1 \quad 0 \quad 0\} \tag{10.13}$$

$$\mathbf{L}_{\varphi y}^T = \{0 \quad 1 \quad 0\} \tag{10.14}$$

$$\mathbf{L}_{\varphi z}^T = \{0 \quad 0 \quad 1\} \tag{10.15}$$

This rewriting of Eqs. (10.11) and (10.12) simplifies the derivation of formulae and thereby the electromechanical coupling and capacitance matrices will have the same form as the stiffness matrix for FGM plates.

The functional for piezoelectric materials can be written by

$$\Pi = \frac{1}{2}\left[\int_0^{h_n} (\boldsymbol{\varepsilon}^T \boldsymbol{\sigma} - \mathbf{E}^T \mathbf{D} + \rho_n \mathbf{U}^T \ddot{\mathbf{U}})\, dz - \mathbf{d}^T \mathbf{F} - \boldsymbol{\varphi}^T \mathbf{D}_z \right] \tag{10.16}$$

where **d** and **F** are the nodal displacement and external traction vectors, i.e.,

$$\mathbf{d}^T = \{\mathbf{d}_l^T \quad \mathbf{d}_m^T \quad \mathbf{d}_u^T\} \tag{10.17}$$

and

$$\mathbf{F}^T = \{\mathbf{F}_l^T \quad \mathbf{F}_m^T \quad \mathbf{F}_u^T\} \tag{10.18}$$

where $\mathbf{d}_i^T = \{v \ u \ w\}_i$, $\mathbf{F}_i^T = \{F_x \ F_y \ F_z\}_i$ ($i = l, m, u$). In Eq. (10.16), $\boldsymbol{\varphi}$ and $\mathbf{D}_z$ are the nodal electrostatic potential vector and electric displacement vector in the z direction, given as follows:

$$\boldsymbol{\varphi}^T = \{\phi_l \quad \phi_m \quad \phi_u\} \tag{10.19}$$

and

$$\mathbf{D}_z^T = \{(D_z)_l \quad (D_z)_m \quad (D_z)_u\} \tag{10.20}$$

10.3   Approximated Governing Equations

We approximate the displacement **U** and the electrostatic potential φ within an element in the form of

$$\mathbf{U}(x, y, z, t) = \mathbf{N}_d(z)\mathbf{d}(x, y, t) \tag{10.21}$$

and

$$\varphi(x, y, z, t) = \mathbf{N}_\phi(z)\boldsymbol{\varphi}(x, y, t) \tag{10.22}$$

where **d** and $\boldsymbol{\varphi}$ are functions of the axes x, y, and time t, while $\mathbf{N}_d$ and $\mathbf{N}_\phi$ are the matrices of shape functions for mechanical displacement,

$$\mathbf{N}_d(z) = [(1 - 3\bar{z} + 2\bar{z}^2)\mathbf{I} \quad 4(\bar{z} - \bar{z}^2)\mathbf{I} \quad (2\bar{z}^2 - \bar{z})\mathbf{I}] \tag{10.23}$$

and for electrostatic potential,

$$\mathbf{N}_\phi(z) = [(1 - 3\bar{z} + 2\bar{z}^2) \quad 4(\bar{z} - \bar{z}^2) \quad (2\bar{z}^2 - \bar{z})] \tag{10.24}$$

in which $\bar{z} = z/h$.

Using Eqs. (10.7) and (10.21), as well as Eqs. (10.9) and (10.22), we obtain

$$\varepsilon = \mathbf{B}_d \mathbf{d} \tag{10.25}$$

and

$$\mathbf{E} = -\mathbf{B}_\phi \boldsymbol{\varphi} \tag{10.26}$$

where

$$\mathbf{B}_d = \mathbf{L}_x \mathbf{N}_d \frac{\partial}{\partial x} + \mathbf{L}_y \mathbf{N}_d \frac{\partial}{\partial y} + \mathbf{L}_z \frac{d\mathbf{N}_d}{dz} \tag{10.27}$$

in which matrices $\mathbf{L}_x$, $\mathbf{L}_y$, and $\mathbf{L}_z$ are constant matrices given by Eqs. (4.6)–(4.8). The operator matrix $\mathbf{B}_\phi$ in Eqs. (10.25) and (10.26) is given by

$$\mathbf{B}_\phi = \mathbf{L}_{\phi x} \mathbf{N}_\phi \frac{\partial}{\partial x} + \mathbf{L}_{\phi y} \mathbf{N}_\phi \frac{\partial}{\partial y} + \mathbf{L}_{\phi z} \frac{d\mathbf{N}_\phi}{dz} \tag{10.28}$$

Substitution of Eqs. (10.25) and (10.26) into Eqs. (10.1) and (10.2), respectively, leads to

$$\boldsymbol{\sigma} = \mathbf{c}(z) \mathbf{B}_d \mathbf{d} + \mathbf{e}^T(z) \mathbf{B}_\phi \boldsymbol{\varphi} \tag{10.29}$$

and

$$\mathbf{D} = -\mathbf{g}(z) \mathbf{B}_\phi \boldsymbol{\varphi} + \mathbf{e}(z) \mathbf{B}_d \mathbf{d} \tag{10.30}$$

Applying the variational principle to Eq. (10.16), we have

$$\delta \Pi = 0 \tag{10.31}$$

The stationary point of the functional gives a set of partial differential equations with respect to x, y, and t as follows.

$$\mathbf{T} = \mathbf{M}\ddot{\boldsymbol{\psi}} + \mathbf{K}_D \boldsymbol{\psi} \tag{10.32}$$

where

$$\mathbf{K}_D = \begin{bmatrix} \mathbf{A}_D & \mathbf{C}_D \\ \mathbf{C}_D^T & -\mathbf{G}_D \end{bmatrix} \tag{10.33}$$

$$\mathbf{M} = \begin{bmatrix} \mathbf{M}_s & 0 \\ 0 & 0 \end{bmatrix} \tag{10.34}$$

$$\boldsymbol{\psi}^T = \{\mathbf{d}^T \quad \boldsymbol{\varphi}^T\} \tag{10.35}$$

and

$$\mathbf{T}^T = \{\mathbf{F}^T \quad \mathbf{D}_z^T\} \tag{10.36}$$

In Eq. (10.33)

$$\mathbf{A}_D = -\mathbf{A}_1 \frac{\partial^2}{\partial x^2} - \mathbf{A}_2 \frac{\partial^2}{\partial x \partial y} - \mathbf{A}_3 \frac{\partial^2}{\partial y^2} + \mathbf{A}_4 \frac{\partial}{\partial x} + \mathbf{A}_5 \frac{\partial}{\partial y} + \mathbf{A}_6 \tag{10.37}$$

where the subscript D stands for a differential operator matrix, and $\mathbf{A}_i = \mathbf{A}_i^l + \mathbf{A}_i^d$, $\mathbf{M}_s = \mathbf{M}_s^l + \mathbf{M}_s^d$, while $\mathbf{A}_i^l$, $\mathbf{A}_i^d$, $\mathbf{M}_s^l$, and $\mathbf{M}_s^d$ are the same as the matrices given in Chapter 9, by assuming that the elastic constants change linearly in the thickness of the element.

The matrices $\mathbf{C}_D$ and $\mathbf{G}_D$ in Eq. (10.33) can be obtained using exactly the same form of Eq. (10.37) and substituting $\mathbf{A}_i$ with $\mathbf{C}_i = \mathbf{C}_i^l + \mathbf{C}_i^d$ and $\mathbf{G}_i = \mathbf{G}_i^l + \mathbf{G}_i^d$, respectively. By assuming the piezoelectric property varies linearly in the thickness direction of the element, matrices $\mathbf{C}_i^l$, $\mathbf{C}_i^d$, $\mathbf{G}_i^l$, and $\mathbf{G}_i^d$ can be derived and are given by

$$\mathbf{G}_i^l = \frac{h g_i^l}{30} \begin{bmatrix} 4 & 2 & -1 \\ 2 & 16 & 2 \\ -1 & 2 & 4 \end{bmatrix} \tag{10.38}$$

$$\mathbf{G}_i^d = \frac{h g_i^d}{60} \begin{bmatrix} 1 & 0 & -1 \\ 0 & 16 & 4 \\ -1 & 4 & 7 \end{bmatrix} \tag{10.39}$$

with $i = 1, 2, 3.$ $g_1^l = g_{11}^l$, $g_2^l = 2g_{12}^l$, $g_3^l = g_{22}^l$, $g_4^l = g_{13}^l$, $g_5^l = g_{23}^l$, $g_6^l = g_{33}^l$,
$g_i^d = g_i^u - g_i^l$.

$$\mathbf{G}_i^l = \frac{g_i^l}{3} \begin{bmatrix} 0 & -4 & 1 \\ 4 & 0 & -4 \\ -1 & 4 & 0 \end{bmatrix} \tag{10.40}$$

$$\mathbf{G}_i^d = \frac{hg_i^d}{30} \begin{bmatrix} 0 & -10 & 5 \\ 10 & 0 & 2 \\ -5 & -2 & 0 \end{bmatrix} \tag{10.41}$$

with $i = 4, 5.$

$$\mathbf{G}_6^l = \frac{g_6^l}{3h} \begin{bmatrix} 7 & -8 & 1 \\ -8 & 16 & -8 \\ 1 & -8 & 7 \end{bmatrix} \tag{10.42}$$

$$\mathbf{G}_6^d = \frac{g_6^d}{6h} \begin{bmatrix} 3 & -4 & 1 \\ -4 & 16 & -12 \\ 1 & -12 & 11 \end{bmatrix} \tag{10.43}$$

$$\mathbf{C}_i^l = \begin{bmatrix} e_i^l \mathbf{n}_1 \\ e_i^l \mathbf{n}_2 \\ e_i^l \mathbf{n}_3 \end{bmatrix} \tag{10.44}$$

where

$$\mathbf{n}_1 = \frac{h}{30} \{4 \quad 2 \quad -1\} \tag{10.45}$$

$$\mathbf{n}_2 = \frac{h}{30} \{2 \quad 16 \quad 2\} \tag{10.46}$$

$$\mathbf{n}_3 = \frac{h}{30} \{-1 \quad 2 \quad 4\} \tag{10.47}$$

$$\mathbf{C}_i^d = \begin{bmatrix} e_i^d \mathbf{q}_1 \\ e_i^d \mathbf{q}_2 \\ e_i^d \mathbf{q}_3 \end{bmatrix} \tag{10.48}$$

where

$$\mathbf{q}_1 = \frac{h}{60}\{1 \quad 0 \quad -1\} \tag{10.49}$$

$$\mathbf{q}_2 = \frac{h}{60}\{0 \quad 16 \quad 4\} \tag{10.50}$$

$$\mathbf{q}_3 = \frac{h}{60}\{-1 \quad 4 \quad 7\} \tag{10.51}$$

with $i = 1, 2, 3$.

$$e_1^l = [e_{11}^l \quad e_{16}^l \quad e_{15}^l]^T, \quad e_2^l = [e_{21}^l + e_{14}^l \quad e_{26}^l + e_{12}^l \quad e_{25}^l + e_{16}^l]^T \tag{10.52}$$

$$e_3^l = [e_{24}^l \quad e_{22}^l \quad e_{26}^l]^T, \quad e_i^d = e_i^u - e_i^l \tag{10.53}$$

$$\mathbf{C}_i^l = \begin{bmatrix} -e_i^l \mathbf{r}_1 + e_i'^l \mathbf{r}_1' \\ -e_i^l \mathbf{r}_2 + e_i'^l \mathbf{r}_2' \\ -e_i^l \mathbf{r}_3 + e_i'^l \mathbf{r}_3' \end{bmatrix} \tag{10.54}$$

where

$$\mathbf{r}_1 = \frac{1}{6}\{-3 \quad 4 \quad -1\}, \quad \mathbf{r}_1' = \frac{1}{6}\{-3 \quad -4 \quad 1\} \tag{10.55}$$

$$\mathbf{r}_2 = \frac{1}{6}\{-4 \quad 0 \quad 4\}, \quad \mathbf{r}_2' = \frac{1}{6}\{4 \quad 0 \quad -4\} \tag{10.56}$$

$$\mathbf{r}_3 = \frac{1}{6}\{1 \quad -4 \quad 3\}, \quad \mathbf{r}_3' = \frac{1}{6}\{-1 \quad 4 \quad 3\} \tag{10.57}$$

$$\mathbf{C}_i^d = \begin{bmatrix} -e_i^d \mathbf{p}_1 + e_i'^d \mathbf{p}_1' \\ -e_i^d \mathbf{p}_2 + e_i'^d \mathbf{p}_2' \\ -e_i^d \mathbf{p}_3 + e_i'^d \mathbf{p}_3' \end{bmatrix} \tag{10.58}$$

where

$$\mathbf{p}_1 = \frac{1}{30}\{-2 \quad 4 \quad -2\}, \quad \mathbf{p}_1' = \frac{1}{30}\{-2 \quad -6 \quad 3\} \tag{10.59}$$

$$\mathbf{p}_2 = \frac{1}{30}\{-6 \quad -8 \quad 14\}, \quad \mathbf{p}_2' = \frac{1}{30}\{4 \quad -8 \quad -16\} \tag{10.60}$$

$$\mathbf{p}_3 = \frac{1}{30}\{3 \quad -16 \quad 13\}, \quad \mathbf{p}_3' = \frac{1}{30}\{-2 \quad 14 \quad 13\} \tag{10.61}$$

with $i = 4, 5$.

$$e_4^l = [e_{31}^l \quad e_{36}^l \quad e_{35}^l]^T, \quad e_4'^l = [e_{15}^l \quad e_{14}^l \quad e_{13}^l]^T \tag{10.62}$$

$$e_5^l = [e_{34}^l \quad e_{32}^l \quad e_{36}^l]^T, \quad e_5'^l = [e_{25}^l \quad e_{24}^l \quad e_{23}^l]^T \tag{10.63}$$

$$e_6^l = [e_{35}^l \quad e_{34}^l \quad e_{33}^l]^T, \quad e_i'^d = e_i'^u - e_i'^l \tag{10.64}$$

$$\mathbf{C}_6^l = \begin{bmatrix} e_6^l \mathbf{s}_1 \\ e_6^l \mathbf{s}_2 \\ e_6^l \mathbf{s}_3 \end{bmatrix} \tag{10.65}$$

where

$$\mathbf{s}_1 = \frac{1}{3h}\{7 \quad -8 \quad 1\} \tag{10.66}$$

$$\mathbf{s}_2 = \frac{1}{3h}\{-8 \quad 16 \quad -8\} \tag{10.67}$$

$$\mathbf{s}_3 = \frac{1}{3h}\{1 \quad -8 \quad 7\} \tag{10.68}$$

$$\mathbf{C}_6^d = \begin{bmatrix} e_6^d \mathbf{t}_1 \\ e_6^d \mathbf{t}_2 \\ e_6^d \mathbf{t}_3 \end{bmatrix} \tag{10.69}$$

where

$$\mathbf{t}_1 = \frac{1}{6h}\{3 \quad -4 \quad 1\} \tag{10.70}$$

$$\mathbf{t}_2 = \frac{1}{6h}\{-4 \quad 16 \quad -12\} \tag{10.71}$$

$$\mathbf{t}_3 = \frac{1}{6h}\{1 \quad -12 \quad 11\} \tag{10.72}$$

Assembling matrices of all the elements, we obtain the dynamic differential equation for the entire FGPM plate:

$$\mathbf{T}_t = \mathbf{M}_t \ddot{\boldsymbol{\psi}}_t + \mathbf{K}_{Dt} \boldsymbol{\psi}_t \tag{10.73}$$

$$\mathbf{K}_{Dt} = \begin{bmatrix} \mathbf{A}_{Dt} & \mathbf{C}_{Dt} \\ \mathbf{C}_{Dt}^T & -\mathbf{G}_{Dt} \end{bmatrix} \tag{10.74}$$

$$\mathbf{M}_t = \begin{bmatrix} \mathbf{M}_{st} & 0 \\ 0 & 0 \end{bmatrix} \tag{10.75}$$

$$\boldsymbol{\psi}_t^T = \{ \mathbf{d}_t^T \quad \boldsymbol{\varphi}_t^T \} \tag{10.76}$$

$$\mathbf{A}_{Dt} = -\mathbf{A}_{1t} \frac{\partial^2}{\partial x^2} - \mathbf{A}_{2t} \frac{\partial^2}{\partial x \partial y} - \mathbf{A}_{3t} \frac{\partial^2}{\partial y^2} + \mathbf{A}_{4t} \frac{\partial}{\partial x} + \mathbf{A}_{5t} \frac{\partial}{\partial y} + \mathbf{A}_{6t} \tag{10.77}$$

For matrices $\mathbf{C}_{Dt}$ and $\mathbf{G}_{Dt}$, we have exactly the same form of equation as Eq. (10.77), but $\mathbf{A}_{it}$ is replaced by $\mathbf{C}_{it}$ and $\mathbf{G}_{it}$, respectively. The matrices $\mathbf{A}_{it}$, $\mathbf{C}_{it}$, $\mathbf{G}_{it}$, $\mathbf{M}_{st}$, $\mathbf{d}_t$, and $\boldsymbol{\varphi}_t$ are the matrices for the entire FGPM plate and are, respectively, the results of assembling the contributions of the matrices $\mathbf{A}_i$, $\mathbf{C}_i$, $\mathbf{G}_i$, $\mathbf{M}_s$, $\mathbf{d}$, and $\boldsymbol{\varphi}$ for all the elements. The dimension of these matrices $\mathbf{K}_{Dt}$ and $\mathbf{M}_t$ is $M \times M$, where $M = 4 \times (2N + 1)$, and the dimension of these vectors $\boldsymbol{\psi}_t$ and $\mathbf{T}_t$ is $M \times 1$.

10.4 Equations in Transform Domain

We introduce the Fourier transformations with respect to the horizontal coordinates x and y as follows:

$$\tilde{\boldsymbol{\psi}}_t(k_x, k_y, t) = \int_{-\infty}^{+\infty} \int_{-\infty}^{+\infty} \boldsymbol{\psi}_t(x, y, t) e^{-ik_x x} e^{-ik_y y} \, dx \, dy \tag{10.78}$$

The application of the Fourier transformations indicated by Eq. (10.78) to Eq. (10.73), leads to following governing equation in the wavenumber domain.

$$\tilde{\mathbf{T}}_t = \mathbf{M}_t \ddot{\tilde{\boldsymbol{\psi}}}_t + \mathbf{K}_t \tilde{\boldsymbol{\psi}}_t \tag{10.79}$$

where $\tilde{\mathbf{T}}_t$, $\ddot{\boldsymbol{\psi}}_t$, and $\ddot{\boldsymbol{\psi}}_t$ are the transforms of $\mathbf{T}_t$, $\boldsymbol{\psi}_t$, and $\boldsymbol{\psi}_t$, respectively. Matrix $\mathbf{K}_t$ is given by

$$\mathbf{K}_t = \begin{bmatrix} \mathbf{A}_t & \mathbf{C}_t \\ \mathbf{C}_t^T & -\mathbf{G}_t \end{bmatrix} \tag{10.80}$$

where $\mathbf{A}_t$ is the stiffness matrix given by

$$\mathbf{A}_t = k_x^2 \mathbf{A}_{1t} + k_x k_y \mathbf{A}_{2t} + k_y^2 \mathbf{A}_{3t} + i k_x \mathbf{A}_{4t} + i k_y \mathbf{A}_{5t} + \mathbf{A}_{6t} \tag{10.81}$$

The above equation stays valid also for the electromechanical coupling matrix, $\mathbf{C}_t$ and the capacitance matrix, $\mathbf{G}_t$, if $\mathbf{A}$ is replaced by $\mathbf{C}$ and $\mathbf{G}$, respectively. The matrices $\mathbf{A}_{it}$ and $\mathbf{G}_{it}$ are symmetric when $i = 1, 2, 3, 6$ and are anti-symmetric when $i = 4, 5$. Hence, the matrix $\mathbf{K}_t$ is a Hermitian matrix for given real wavenumbers.

There are mechanical boundary conditions (M.B.C.) and electrical boundary conditions (E.B.C.) for wave propagation analysis of piezoelectric plates. Here, the following four boundary conditions are considered.

For investigating the characteristics of waves in FGPM plates with electrically open and shorted boundary conditions, we have

$$\mathbf{F}_t = 0, \quad \mathbf{D}_z = 0 \tag{10.82}$$

$$\mathbf{F}_t = 0, \quad \mathbf{D}_z = 0, \quad \phi_{ti} = 0. \tag{10.83}$$

For investigating the responses of the displacements and electrostatic potentials of FGPM plates excited by mechanical loads and electrodes, we use

$$\mathbf{F}_t = \hat{\mathbf{F}}_t, \quad \mathbf{D}_z = 0 \tag{10.84}$$

$$\mathbf{F}_t = 0, \quad \mathbf{D}_z = 0, \quad \phi_{ti} = \phi_e. \tag{10.85}$$

In Eqs. (10.83) and (10.85), the subscript i denotes the ith node of the plate. Equations (10.83) and (10.85) are the forced electrical boundary conditions. These conditions can be handled by introducing the method given by Payne and Irons (see Zienkiewicz, 1989) for the forced mechanical boundary conditions. For example, if the ith node of the plate excited by electrodes with $\phi_{ti} = \phi_e$, it can be satisfied by multiplying the element $\mathbf{G}_{ii}$ in matrix $\mathbf{G}$ by a very large value of α and replacing the element D_{zi} in vector $\mathbf{D}_z$ by $-\alpha \mathbf{G}_{ii} \phi_e$.

10.5 Characteristics of Waves in FGPM Plates

The imposition of Eq. (10.82) or (10.83) in Eq. (10.79), and consideration of free wave motion of harmonic plane waves in the FGPM plates, leads to the following eigenvalue equation.

$$0 = \mathbf{K}_t \tilde{\mathbf{\psi}}_t - \omega^2 \mathbf{M}_t \tilde{\mathbf{\psi}}_t \tag{10.86}$$

For a plane wave propagating in the direction of θ with respect to the x-axis (see Figure 10.1), the wavenumber variables in Eq. (10.81) become

$$k_x = k \cos\theta, \quad k_y = k \sin\theta \tag{10.87}$$

where k is the wavenumber of the plane wave propagating in the direction of θ.

Solving Eq. (10.86) for given wavenumber k, $M = 4 \times (2N + 1)$ eigenfrequencies ω_m and the corresponding eigenvectors can be obtained. The relationship between ω_m and k gives the dispersion relation of harmonic plane waves in FGPM plates. The Rayleigh quotient for the mth eigenvalue can be written as

$$\omega_m^2 = \frac{\mathbf{\psi}_m^L \mathbf{K}_t \mathbf{\psi}_m^R}{\mathbf{\psi}_m^L \mathbf{M}_t \mathbf{\psi}_m^R} \tag{10.88}$$

where superscripts L and R stand for the left and right eigenvectors. Those eigenvectors have the following relationship for $\mathbf{K}_t$, a Hermitian matrix:

$$\mathbf{\psi}_m^L = (\mathbf{\psi}_m^R)^T \tag{10.89}$$

According to Eq. (10.76), the eigenvectors can be written by

$$\mathbf{\psi}_m^L = \{\mathbf{\Lambda}_m^L \quad \mathbf{\varphi}_m^L\} \tag{10.90}$$

$$\mathbf{\psi}_m^R = \{(\mathbf{\Lambda}_m^R)^T \quad (\mathbf{\varphi}_m^R)^T\}^T \tag{10.91}$$

where $\mathbf{\Lambda}_m^L$ and $\mathbf{\varphi}_m^L$ are the mth left eigenvectors for displacement and electrostatic potential, respectively, and $\mathbf{\Lambda}_m^R$ and $\mathbf{\varphi}_m^R$ are the mth right eigenvectors for displacement and electrostatic potential, respectively.

The phase velocity of the mth mode c_{pm} can be expressed by

$$c_{pm} = \omega_m / k \tag{10.92}$$

With differentiation of Eq. (10.92) with respect to k and application of the orthogonality condition of the left and right eigenvectors, the group velocity of the mth mode c_{gm} can be expressed by

$$c_{gm} = \frac{d\omega_m}{dk} = \frac{\boldsymbol{\psi}_m^L \frac{\partial \mathbf{K}_t}{\partial k} \boldsymbol{\psi}_m^R}{2\omega_m \boldsymbol{\psi}_m^L \mathbf{M}_t \boldsymbol{\psi}_m^R} \tag{10.93}$$

The eigenvalue equation, Eq. (10.86), can be solved directly. However, it is convenient to change Eq. (10.86) into the following condensed (dimension-reduced) equation:

$$0 = \mathbf{K}_{st} \tilde{\mathbf{d}}_t - \omega^2 \mathbf{M}_{st} \tilde{\mathbf{d}}_t \tag{10.94}$$

where

$$\mathbf{K}_{st} = \mathbf{A}_t + \mathbf{C}_t \mathbf{G}_t^{-1} \mathbf{C}_t^T \tag{10.95}$$

When the Fourier transformation of the displacement or the eigenvector of the displacement is obtained, the Fourier transform of the electrostatic potential or the eigenvector of the electrostatic potential can be obtained using the following relations:

$$\tilde{\boldsymbol{\varphi}}_t = \mathbf{G}_t^{-1} \mathbf{C}_t^T \tilde{\mathbf{d}}_t \quad \boldsymbol{\varphi}_m = \mathbf{G}_t^{-1} \mathbf{C}_t^T \boldsymbol{\Lambda}_m \tag{10.96}$$

For obtaining the group velocity of the mth mode, it is convenient to use the following equation:

$$c_{gm} = \frac{\boldsymbol{\Lambda}_m^L \frac{\partial \mathbf{A}_t}{\partial k} \boldsymbol{\Lambda}_m^R + 2\mathrm{Re}\left(\boldsymbol{\Lambda}_m^L \frac{\partial \mathbf{C}_t}{\partial k} \boldsymbol{\varphi}_m^R \right) - \boldsymbol{\varphi}_m^L \frac{\partial \mathbf{G}_t}{\partial k} \boldsymbol{\varphi}_m^R}{2\omega_m \boldsymbol{\psi}_m^L \mathbf{M}_t \boldsymbol{\psi}_m^R} \tag{10.97}$$

Finally, the electromechanical coupling constant of the mth mode is given by

$$K_{pm}^2 = \frac{2|c_{pm(open)} - c_{pm(short)}|}{c_{pm(open)}} \tag{10.98}$$

where $c_{pm(open)}$ and $c_{pm(short)}$ are the phase velocities of the mth mode obtained, respectively, by using the electrically open boundary condition, Eq. (10.82), and the electrically shorted boundary condition, Eq. (10.83).

The characteristic wave surfaces can also be obtained for FGPM plates using the same procedure as described in Chapter 5.

10.6 Transient Response Analysis

Changing Eq. (10.79) into the following form:

$$\tilde{\mathbf{F}}_t + \tilde{\mathbf{F}}_{te} = \mathbf{M}_{st}\,\ddot{\tilde{\mathbf{d}}}_t + \mathbf{K}_{st}\tilde{\mathbf{d}}_t \tag{10.99}$$

where the first term in the left side of the equation expresses the mechanical load, and the second term expresses the equivalent load generated by electrodes that can be obtained by

$$\tilde{\mathbf{F}}_{te} = \mathbf{C}_t\mathbf{G}_t^{-1}\tilde{\mathbf{D}}_z \tag{10.100}$$

Using the method of modal analysis in Eq. (10.99) and the initial condition that the plate stands stationary before the excitation and action of the load, the displacement in the wavenumber domain can be obtained for various types of transient excitations (see Chapter 8). The formulae of the displacement in the wavenumber domain for the time-step, time-pulse, and periodic mechanical loads are as follows:

A. Time-step excitation with a time function of the Heaviside function $H(t)$:

$$\tilde{\mathbf{d}}_t(k_x, k_y, t) = \sum_{m=1}^{M} \frac{\mathbf{\Lambda}_m^L \tilde{\mathbf{F}}_{0t}\mathbf{\Lambda}_m^R(1 - \cos \omega_m t)}{\omega_m^2 M_m} \tag{10.101}$$

B. Time-pulse excitation with a time function of the Dirac delta function $\delta(t)$:

$$\tilde{\mathbf{d}}_t(k_x, k_y, t) = \sum_{m=1}^{M} \frac{\mathbf{\Lambda}_m^L \tilde{\mathbf{F}}_{0t}\mathbf{\Lambda}_m^R \sin \omega_m t}{\omega_m M_m} \tag{10.102}$$

C. Periodic excitation with a time function of $\exp(-i\omega_\phi t)$:

$$\tilde{\mathbf{d}}_t(k_x, k_y, t) = \sum_{m=1}^{M} \frac{\mathbf{\Lambda}_m^L \tilde{\mathbf{F}}_{0t}\mathbf{\Lambda}_m^R \exp(-i\omega_f t)}{(\omega_m^2 - \omega_f^2)M_m} \tag{10.103}$$

From the above equations, it is found that the Fourier transformation of the displacement becomes infinity when $\omega_f = \omega_m(k_x, k_y)$, due to the resonance of the external excitation with the natural frequency of the wave modes.

10.7 Interdigital Electrodes Excitation

It is considered here that the plate is excited by interdigital electrodes. For convenience, the interdigital electrodes are assumed to be very long in the y direction; therefore, the displacement and electrical fields are independent of y. The electrodes are applied on the ith nodal plane and the distribution of the electrostatic potential on the plane, $\bar{\phi}_e$, is assumed to be

$$
\bar{\phi}_e = \begin{cases}
\delta_{n_e}, & \left(n_e - \dfrac{1}{2}\right)d_e - \dfrac{h_e}{2} < x < \left(n_e - \dfrac{1}{2}\right)d_e + \dfrac{h_e}{2} \\[2mm]
& n_e = 1,\ldots,N_e \\[3mm]
& 0 < x < \dfrac{(d_e - h_e)}{2};\ x > \left(N_e - \dfrac{1}{2}\right)d_e + \dfrac{h_e}{2} \\[3mm]
0, & \left(n_e - \dfrac{1}{2}\right)d_e + \dfrac{h_e}{2} < x < \left(n_e + \dfrac{1}{2}\right)d_e - \dfrac{h_e}{2} \\[2mm]
& n_e = 1,\ldots,N_e - 1
\end{cases}
\tag{10.104}
$$

where

$$
\delta_{n_e} = \begin{cases}
+1, & n = \text{odd} \\
-1, & n = \text{even}
\end{cases}
\tag{10.105}
$$

and d_e, h_e, and N_e are the pitch of electrodes (distance between the neighboring electrodes), the width of an electrode and the number of the electrodes, respectively, as shown in Figure 10.2. In Eq. (10.85), ϕ_e is given by

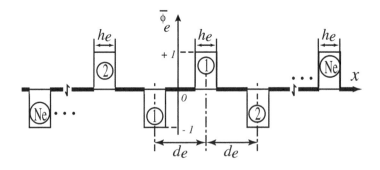

FIGURE 10.2
Distribution of electrostatic potential on a nodal plane of a plate generated by interdigital electrodes. (From Liu, G. R. and Tani, J., *ASME Journal of Vibration and Acoustics,* 116, 440, 1994. With permission.)

$\phi_e = p_0 \bar{\phi}_e$, and p_0 is a constant expressing the value of the electrostatic potential. The electrodes are arranged anti-symmetrically with respect to the ϕ_e axis; hence, we have, $\phi_e(-x) = -\phi_e(x)$.

The application of the Fourier transform to ϕ_e leads to

$$\tilde{\phi}_e = \frac{2ip_0}{k_x} \cos \frac{k_x h_e}{2} \sum_{n_e=1}^{N_e} \cos \frac{(2n_e - 1)k_x d_e}{2} \tag{10.106}$$

or

$$\tilde{\phi}_e = \frac{i4p_0}{k_x} \cos \frac{k_x h_e}{2} \left(\frac{\cos N_e k_x d_e}{\cos \frac{k_x d_e}{2}} - \tan^2 \frac{k_x d_e}{2} \right) \tag{10.107}$$

The boundary condition in the wavenumber domain for this case is $\tilde{\phi}_{ti} = \tilde{\phi}_e$. It can be imposed by modifying matrices $\mathbf{G}_t$ and $\mathbf{D}_z$ in the following manner: (1) multiplying the element $\mathbf{G}_{ii}$ in matrix $\mathbf{G}_t$ by a very large value of α, and (2) replacing the element D_{zi} in vector $\mathbf{D}_z$ by $-\alpha \mathbf{G}_{ii} \tilde{\phi}_e$. The equivalent load generated by the electrodes can then be obtained with the help of Eq. (10.100) and using the modified matrices $\mathbf{G}_t$ and $\mathbf{D}_z$. The displacement in the wavenumber domain excited by the interdigital electrodes can be obtained for given excitation of electrostatic potential using Eqs. (10.101), (10.102), and (10.103), respectively, for time-step, time-pulse, and periodic time functions. Finally, the electrostatic potentials in the plate in the wavenumber domain can be obtained using Eq. (10.96), after the displacement in the wavenumber domain is obtained.

From Eq. (10.107), it can seen that $\tilde{\phi}_e$ becomes infinity when $k_x d_e = \pi$. This fact indicates that the waves with wavelength of $2d_e$ will be excited most strongly by the interdigital electrodes with a pitch of d_e.

10.8 Displacement and Electrostatic Potential Response

Applying the inverse Fourier transform to $\tilde{\mathbf{d}}_t$ and $\tilde{\psi}_t$, the displacement and electrostatic potential responses in the space-time domain can be expressed, respectively, by

$$\mathbf{d}_t(x, y, t) = \left(\frac{1}{2\pi} \right)^2 \int_{-\infty}^{+\infty} \int_{-\infty}^{+\infty} \tilde{\mathbf{d}}_t(k_x, k_y, t) e^{ik_x x} e^{ik_y y} dk_x dk_y \tag{10.108}$$

and

$$\phi_t(x, y, t) = \left(\frac{1}{2\pi}\right)^2 \int_{-\infty}^{+\infty}\int_{-\infty}^{+\infty} \tilde{\phi}_t(k_x, k_y, t)e^{ik_x x}e^{ik_y y}\, dk_x\, dk_y \qquad (10.109)$$

The integration in Eqs. (10.108) and (10.109) can be carried out using the 2D fast Fourier transform (FFT) techniques or the integral quadrature schemes described in Chapter 8.

10.9 Computation Procedure

As an example, the layer element method is applied to analyze characteristics of harmonic waves in a hypothetical FGPM plate, and the HNM method is applied to analyze the transient waves in the FGPM plate. The dispersion relations, group velocities, electromechanical coupling constants, and mode shapes of waves propagating in a hypothetical FGPM plate are first examined using the layer element method. Displacement and electrostatic potential responses of the plate excited by mechanical loads and electrodes are calculated using the HNM and compared with those obtained using the same code for a homogeneous z-x LiTaO$_3$ plate. The hypothetical FGPM plate is designed as follows.

The material properties on the upper surface ($z = H$) of the hypothetical FGPM plate are the same as those of the homogeneous z-x LiTaO$_3$ plate, i.e.,

$$c^U = c, \quad e^U = e, \quad g^U = g, \quad \rho^U = \rho \qquad (10.110)$$

where c, e, and g are, respectively, the elastic, piezoelectric, and dielectric constant matrices of the z-x LiTaO$_3$ plate. All these material properties are given in Table 10.1.

The material properties on the lower surface ($z = 0$) of the FGPM plate are assumed to be

$$c^L = 12c, \quad e^L = 0.1e, \quad g^L = 0.5g, \quad \rho^L = 1.8\rho \qquad (10.111)$$

In the thickness direction, the material properties are assumed to vary in the following quadratic fashion:

$$\left.\begin{aligned}
c(z) &= c^U + (c^L - c^U)(1 - z/H)^2 \\
e(z) &= e^U + (e^L - e^U)(1 - z/H)^2 \\
g(z) &= g^U + (g^L - g^U)(1 - z/H)^2 \\
\rho(z) &= \rho^U + (\rho^L - \rho^U)(1 - z/H)^2
\end{aligned}\right\} \qquad (10.112)$$

TABLE 10.1

Material Properties of z-x LiTaO$_3$ Plate

$$
\mathbf{c} = \begin{bmatrix}
22.98 & 0.440 & 0.812 & -0.104 & 0.000 & 0.000 \\
0.440 & 2.298 & 0.812 & 0.104 & 0.000 & 0.000 \\
0.812 & 0.812 & 2.798 & 0.000 & 0.000 & 0.000 \\
-0.104 & 0.104 & 0.000 & 0.968 & 0.000 & 0.000 \\
0.000 & 0.000 & 0.000 & 0.000 & 0.968 & -0.104 \\
0.000 & 0.000 & 0.000 & 0.000 & -0.104 & 0.929
\end{bmatrix} \times 10^{11} (\text{N/m}^2)
$$

$$
\mathbf{e} = \begin{bmatrix}
0.00 & 0.00 & 0.00 & 0.00 & 2.72 & -1.67 \\
-1.67 & 1.67 & 0.00 & 2.72 & 0.00 & 0.00 \\
-0.38 & -0.38 & 1.09 & 0.00 & 0.00 & 0.00
\end{bmatrix} \times (\text{C/m}^2)
$$

$$
\mathbf{g} = \begin{bmatrix}
42.6 & 0.00 & 0.00 \\
0.00 & 42.6 & 0.00 \\
0.00 & 0.00 & 42.8
\end{bmatrix} \times \varepsilon_0, \quad \rho_0 = 7456 \ \text{kg/m}^2
$$

Source: Toda and Mizutani, 1988.

The computation procedure can be summarized as follows:

1. Dividing the plate into N layer elements in the thickness direction of the plate, calculating the elastic, piezoelectric, dielectric constant matrices, and densities on the lower and upper surfaces of each element.

2. Form matrices $\mathbf{M}_s^l$, $\mathbf{M}_s^d$, $\mathbf{A}_i^l$, $\mathbf{A}_i^d$, $\mathbf{C}_i^l$, $\mathbf{C}_i^d$, $\mathbf{G}_i^l$, $\mathbf{G}_i^d$ ($i = 1,...,6$) for all the elements, which are then used to assemble the global matrices $\mathbf{M}_{st}$, $\mathbf{A}_{it}$, $\mathbf{C}_{it}$, $\mathbf{G}_{it}$.

3. In the characteristic analysis, the matrix $\mathbf{K}_{st}$ can be obtained from Eq. (10.95), for given wavenumber k, and the direction of wave propagation θ of a harmonic plane wave. Solving Eq. (10.94), M eigenfrequencies ω_m and the corresponding eigenvectors $\mathbf{\Lambda}_m$ can be obtained. The phase velocities, group velocities, and electrome-chanical coupling constants can be obtained using Eqs. (10.92), (10.93), and (10.98), respectively.

4. In the analysis of the 3D displacement responses, we divide appro-priately the integral ranges of wavenumbers of $[-K_x, K_x]$ and $[-K_y, K_y]$. Building the matrix $\mathbf{K}_{st}$ for various discrete points of (k_{xi}, k_{yi}) using Eq. (10.95), ω_m and $\mathbf{\Lambda}_m$ can be obtained by solving Eq. (10.94) using standard routines for eigenvalue analysis. The displacement in the wavenumber domain $\tilde{\mathbf{d}}_t(k_x, k_y, t)$ can be obtained using Eq. (10.101), (10.102), or (10.103). The electrostatic potential in the Fourier trans-forms domain $\tilde{\phi}_t(k_x, k_y, t)$ can be obtained using Eq. (10.96). Finally, the displacement and electrostatic potential responses can be obtained by carrying out the integration of Eqs. (10.108) and (10.109), respectively. In the analysis of 2D problems, k_x and k_y are

given by Eq. (10.87) and the integration is only for k. FFT or the integral technique described in Chapter 8 may be used. In this chapter, the integration in Eqs. (10.108) and (10.109) is carried out using standard numerical FFT routines or the integral scheme described in Chapter 8.

In this section, the following dimensional parameters are employed.

$$\bar{t} = t/t_0, \quad t_0 = H/c_0, \quad c_0 = \sqrt{c_{66}/\rho_0}$$

$$\bar{x} = x/H, \quad \bar{z} = z/H, \quad \bar{k} = k_x H, \quad \bar{c}_g = c_g/c_0$$

$$\bar{u} = u/u_0, \quad \bar{v} = v/v_0, \quad \bar{w} = w/w_0$$

$$u_0 = Hf_0/c_{66}, \quad v_0 = Hf_0/c_{66}, \quad w_0 = Hf_0/c_{66}.$$

$$\bar{\varphi} = \varphi/\varphi_0, \quad \varphi_0 = (e_s H f_0)/(g_s c_{66})$$

where H is the thickness of the plates, ρ_0 and c_{66} are the density and the elastic constant c_{66} of the z-x LiTaO$_3$ plate, respectively. c_0 is the shear wave velocity of the z-x LiTaO$_3$ plate and t_0 is the time for the shear wave to travel once across the plate thickness. For the mechanical loads, $e_s = C/m^2$, $g_s = 10^{-10}$ F/m, $f_0 = q_0$ and for the electrode excitation, $f_0 = e_s p_0/H$. Only plane waves of $\theta = 0$ are considered here. The plates are excited by line mechanical loads, a line electrode, and interdigital electrodes, which are on the upper surface of the plates.

10.10   Dispersion Curves

Figures 10.3 and 10.4 show the frequency spectra of the lowest ten modes of the FGPM and LiTaO$_3$ plates, respectively. The thick lines are the dispersion curves for the boundary condition of electrically open on the upper surface of the plates, and the thin lines are those for the boundary condition of electrically shorted. Since the piezoelectric constants are very small, they have little effect on the dispersion curves. Hence, the thick and thin lines of each mode are so close to each other that one almost cannot distinguish them. This implies that the piezoelectric effects on the phase velocity of the waves in the piezoelectric plate with material property given in Table 10.1 are very small.

The shape (displacement distribution along the thickness of the plates) of wave modes is very important for understanding the properties of harmonic waves propagating in plates. We calculate the lowest 15 mode shapes of the FGPM and z-x LiTaO$_3$ plates, for two cases of wavenumbers $\bar{k} = 3.927$ and 25.0. The wavelengths corresponding to these two wavenumbers are approximately 1.6H and 0.25H. The results are shown in Figures 10.5–10.10. Results for the FGPM plate are plotted in Figures 10.5, 10.7, and 10.9, and those for

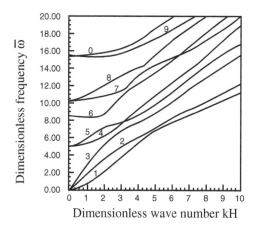

FIGURE 10.3
Frequency spectra of the FGPM plate (thick line: open; thin line: short).

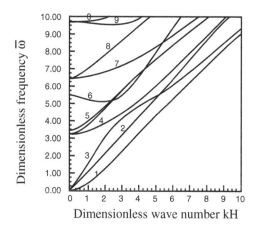

FIGURE 10.4
Frequency spectra of the LiTaO$_3$ plate (thick line: open; thin line: short).

the LiTaO$_3$ plate are plotted in Figures 10.6, 10.8, and 10.10. The mode shapes of wave modes are shown in Figures 10.4 and 10.6 for the smaller wavenumber of $\bar{k}$ = 3.927 and in Figures 10.7 and 10.8 for the larger wavenumber of $\bar{k}$ = 25.0. Comparing Figures 10.5 and 10.6, it can be seen that the displacements are more concentrated on the upper surface of the FGPM plate than those on the surfaces of the LiTaO$_3$ plate, when $\bar{k}$ = 3.927. Comparing Figures 10.7 and 10.8, it can be seen that the displacements of the lowest 14 modes of the FGPM plate are completely concentrated on the upper surface, but the displacements of only the two lowest modes of the LiTaO$_3$ plate are concentrated on the surfaces, when $\bar{k}$ = 25.0. The electrostatic potential shapes are shown in Figures 10.9 and 10.10. From these figures it can be seen

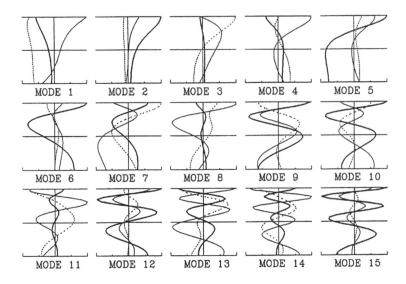

FIGURE 10.5
Displacement distribution in the thickness direction for wave modes in the FGPM plate ($\bar{k} = 3.927$; thin line: u; dotted line: w; thick line: v). (From Liu, G. R. and Tani, J., *ASME Journal of Vibration and Acoustics*, 116, 440, 1994. With permission.)

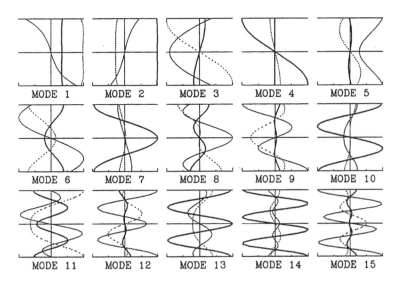

FIGURE 10.6
Displacement distribution in the thickness direction for wave modes in the LiTaO$_3$ plate ($\bar{k} = 3.927$; thin line: u; dotted line: w; thick line: v). (From Liu, G. R. and Tani, J., *ASME Journal of Vibration and Acoustics*, 116, 440, 1994. With permission.)

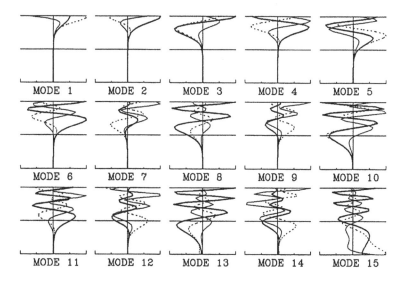

FIGURE 10.7
Displacement distribution in the thickness direction for wave modes in the FGPM plate ($\bar{k}$ =25.0; thin line: u; dotted line: w; thick line: v).

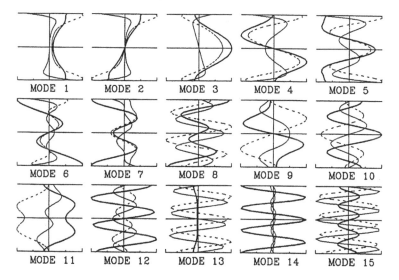

FIGURE 10.8
Displacement distribution in the thickness direction for wave modes in the LiTaO$_3$ plate ($\bar{k}$ = 25; thin line: u; dotted line: w; thick line: v).

that the electrostatic potentials are more concentrated on the upper surface of the FGPM plate than those on the two surfaces of the LiTaO$_3$ plate, when $\bar{k}$ = 3.927. It is confirmed that the electrostatic potentials become more concentrated on the upper surface of the FGPM plate for larger wavenumbers or smaller wavelength.

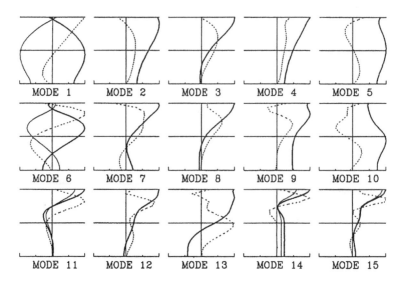

FIGURE 10.9
Electrostatic potential distribution in the thickness direction for wave modes in the FGPM plate
($\bar{k} = 3.927$; thin line: E_x; dotted line: E_z; thick line: φ).

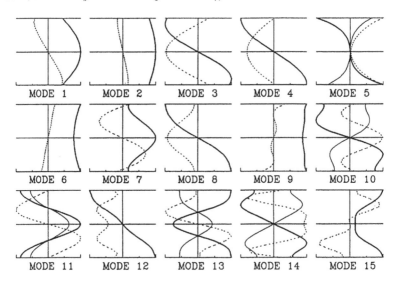

FIGURE 10.10
Electrostatic potential distribution in the thickness direction for wave modes in the LiTaO$_3$ plate
($\bar{k} = 3.927$; thin line: E_x; dotted line: E_z; thick line: φ).

Table 10.2 shows the dimensionless group velocities of the lowest six modes of the FGPM and LiTaO$_3$ plates for $\bar{k} = 25.0$, while the upper surface of the plates are electrically open and shorted. From Table 10.2 it can be seen that the group velocity of each mode for the electrically open boundary

TABLE 10.2

Dimensionless Group Velocities of the Lowest 6 Modes ($kH = 25.0$)

Plate	E.B.C.	M_1	M_2	M_3	M_4	M_5	M_6
FGPM	Open	0.885	1.048	0.990	1.139	1.055	1.116
	Short	0.884	1.039	0.987	1.140	1.058	1.116
LiTaO$_3$	Open	0.918	0.914	0.929	0.896	0.917	0.997
	Short	0.917	0.915	0.929	0.899	0.928	0.991

Source: Liu, G. R. and Tani, J., *ASME Journal of Vibration and Acoustics*, 116, 440, 1994. With permission.

condition is close to that for the electrically shorted boundary condition. For the LiTaO$_3$ plate, the group velocities of M_1 and M_2 are close to the velocity of the Rayleigh wave on the surface of a half space of the z-x LiTaO$_3$ material. For the FGPM plate, the group velocity of M_1 is close to the velocity of the Rayleigh wave, but the group velocity of M_2 is close to the velocity of the shear wave in the unbounded solid of z-x LiTaO$_3$ material.

Because of the anisotropy of the plates, the three displacement components u, v, and w are coupled. Hence, in the LiTaO$_3$ plate, there is neither pure SH wave nor pure SH surface wave, which has been observed in a SiC-C plate in the previous chapter. However, the second mode in the piezoelectric plates can be regarded as a pseudo-SH surface wave (pseudo-SHSW) for the following two reasons:

- The displacement component v of this mode is larger than the two components u and w, and
- The group velocity of the second mode is close to the velocity of lowest mode of SH waves (Table 10.2).

A feature of the pseudo-SHSW is that the penetration depth into the FGPM plate is as small as that of the Rayleigh wave. The pseudo-SHSW is different from the famous Bleustein-Gulyaev waves (Bleustein, 1968; Gulyaev, 1969) that are pure SH surface waves, in a homogeneous piezoelectric medium, where the displacement components u and w are decoupled from v and the potential ϕ. The penetration depth of the Bleustein-Gulyaev waves into the medium is much larger than the wavelength (Lewis, 1985). It may be mentioned that for isotropy FGPM plates or some kinds of anisotropic FGPM plate (e.g., crystals in *class* 6 mm), pure SH surface waves can be obtained in certain propagation directions of waves.

Figures 10.11 and 10.12 show the curves of the electromechanical coupling constants of waves in the FGPM and LiTaO$_3$ plates, respectively. Comparing these two figures, it can be seen that, for a given wavenumber, the coupling constants of some modes of the FGPM plate are smaller than those of the LiTaO$_3$ plate. For a given mode, the coupling constants of the

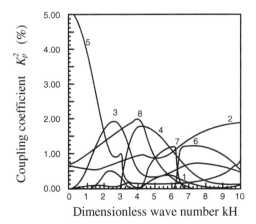

FIGURE 10.11
Electromechanical coupling constants for the FGPM plate (thick line: odd modes; thin line: even modes).

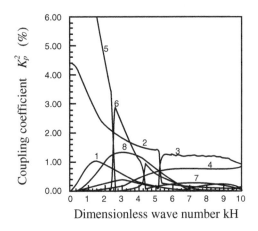

FIGURE 10.12
Electromechanical coupling constants for the LiTaO₃ plate (thick line: odd modes; thin line: even modes).

FGPM plate are larger in some regions and smaller in others compared to those of the LiTaO₃ plate. For example, for the second mode, the coupling constants of the FGPM plate are smaller than those of LiTaO₃ plate when $\bar{k}$ is small and are larger than those of LiTaO₃ plate when $\bar{k}$ is large.

For an infinite piezoelectric material body, it is well known that the electromechanical coupling constant K^2 is proportional to the square of the piezoelectric constant. For a piezoelectric material plate, this is not exact but approximately true. From Eqs. (10.110) to (10.112), we can see that only the piezoelectric constants on the upper surface of the FGPM plate equal those of the LiTaO$_3$ plate. The piezoelectric constants decrease rapidly with increasing depth into the FGPM plate. When $\bar{k}$ is small, the penetration depth of the displacement and potential field into the FGPM plate is large. Hence, the coupling constants of the FGPM plate are smaller than those of the LiTaO$_3$ plate, because the piezoelectric constants in the inner part of the FGPM plate are much smaller compared to those of the LiTaO$_3$ plate. When $\bar{k}$ is large, the penetration depth into the FGPM plate is small and the piezoelectric constants of the FGPM plate are only a little smaller compared to those of the LiTaO$_3$ plate near the upper surface. On the other hand, the effect of the electrical boundary condition of the FGPM plate on the phase velocities (hence on the coupling constants) is larger than that of the LiTaO$_3$ plate, because the electric potential is concentrated on the upper surface of the FGPM plate. Consequently, the coupling constants of the FGPM plate are larger than those of the LiTaO$_3$ plate when $\bar{k}$ is large.

10.11 Excitation of Time-Step Shear Force in y Direction

The responses of the displacement and the electrostatic potential excited by the time-step mechanical shear force in the y-axis are computed. In this case, we have

$$q_x = 0, \quad q_y = 1, \quad q_z = 0 \tag{10.113}$$

Figure 10.13 shows the responses of the displacement u of the FGPM plate. The arrival positions of wave modes M$_i$ are marked by arrows. From this figure, it can be seen that the displacements on the upper surface ($z = 1.0H$) of the FGPM plate change suddenly at the arrival positions of M$_1$ and M$_2$. The sudden change in response is not observed at $z = 0$ and $0.5H$, indicating that two strong surface waves with velocities of about 0.89 and 1.05, respectively, are excited by the shear force. It can also be seen that the response of surface wave M$_2$ is stronger than that of the surface wave M$_1$. The responses of the displacement u of the LiTaO$_3$ plate are shown in Figure 10.14 which shows that the displacements on the upper and lower surfaces of the plate are almost the same. Two weak surface waves are also observed at the arrival position of M$_1$ and M$_2$, and the position of M$_6$. Comparing Figures 10.13 and 10.14, it can be found that the responses of the surface waves of the FGPM plate are stronger than those of the LiTaO$_3$ plate. Figures 10.15 and 10.16 show the responses of the electrostatic potential of the FGPM and LiTaO$_3$ plates, respectively. From Figure 10.15 it is seen again that the surface waves appear at the arrival positions of M$_1$ and M$_2$, and that the surface wave of M$_2$ is stronger than that of M$_1$. These surface waves are so weak that one cannot confirm their appearance.

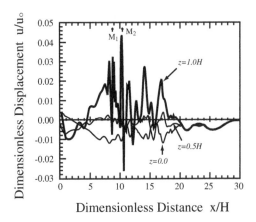

FIGURE 10.13
Responses of displacement u of the FGPM plate excited by a shear time-step load in the y-direction ($\dot{t} = 10$). (From Liu, G. R. and Tani, J., *ASME Journal of Vibration and Acoustics*, 116, 440, 1994. With permission.)

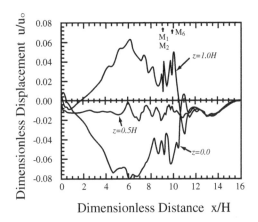

FIGURE 10.14
Responses of displacement u of the LiTaO$_3$ plate excited by a shear time-step load in the y-direction ($\dot{t} = 10$).

10.12   Excitation of a Line Electrode

The responses of the displacement excited by a line electrode on the upper surface of the plates are computed. It is assumed that the electrode is long enough, so that the displacement and electrical fields can be considered independent of y. The electrode generates a time-step electrostatic potential, and the plates are assumed electrically shorted on the upper surface.

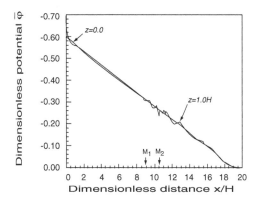

FIGURE 10.15
Electrostatic potential responses of the FGPM plate excited by a shear time-step load in the y-direction ($\bar{t} = 10$).

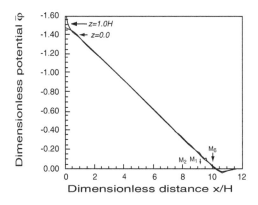

FIGURE 10.16
Electrostatic potential responses of the LiTaO$_3$ plate excited by a shear time-step load in the y-direction ($\bar{t} = 10$).

The potential on the upper surface of the plate is expressed by

$$\phi_e = p_0 \delta(x) H(t) \tag{10.114}$$

Figures 10.17 and 10.18 show the responses of the displacement v of the FGPM and LiTaO$_3$ plates, respectively. From Figure 10.17 it can be seen that the displacement on the lower surface is very small. On the upper surface two strong surface waves are observed at the arrival positions of M$_1$ and M$_2$. In this case, the surface wave of M$_2$ is much stronger than the surface wave of M$_1$. From Figure 10.18, it can be seen that the displacements on both surfaces of the LiTaO$_3$ plate are almost the same. Comparing Figures 10.17 and 10.18, it can be found that the surface wave M$_2$ of FGPM plate is stronger than any of the surface waves observed in the LiTaO$_3$ plate.

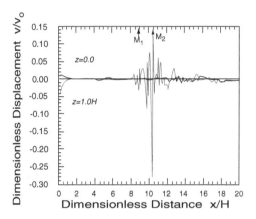

FIGURE 10.17
Responses of displacement v of the FGPM plate excited by a line electrode ($\dot{t} = 10$).

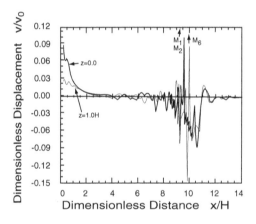

FIGURE 10.18
Responses of displacement v of the LiTaO$_3$ plate excited by a line electrode ($\dot{t} = 10$; thin line: upper surface; thick line: middle surface; thicker line: lower surface).

10.13 Excitation of Interdigital Electrodes

The responses of the plates excited by interdigital electrodes on the upper surface of the plates are computed. The same configuration has been also used by Toda and Mizutani (1988). The parameters of the interdigital electrode are

$$N_e = 10, \quad d_e = 0.8H, \quad h_e = d_e/2 \tag{10.115}$$

The distribution of the potential on the upper surface of the plates is expressed by Eq. (10.104). The Fourier transform of the potential of the interdigital electrodes are given in Eq. (10.107). From Eq. (10.107), it can be found that the waves with wavenumber of $\bar{k} = \pi H/d_e \approx 3.927$ or wavelength of $1.6H$ will be largely excited. The displacement shapes of the lowest 15 wave modes for $\bar{k} = 3.927$ are shown in Figures 10.5 and 10.6, when the upper surface of the plate is electrically open. We have confirmed that the displacement shapes of the wave modes for the electrically open boundary condition is very similar to those for the electrically short boundary condition. We found that the eigenfrequencies of the second modes of the FGPM and LiTaO$_3$ plates are $5.944\omega_0$ and $4.066\omega_0$, respectively, when $\bar{k} = 3.927$. Hence, the excitation frequency of the electrodes should be $5.944\omega_0$ for the FGPM plate, and $4.066\omega_0$ for the LiTaO$_3$ plate. The electromechanical coupling constants of the FGPM and LiTaO$_3$ plates are 0.93% and 1.6%, respectively.

Applying a periodic excitation using the interdigital electrodes, the response can be obtained using Eq. (10.103). Note that the integrand given in Eq. (10.103) will be singular at the location where $\omega_m(k_x, k_y) = \omega_f$. We have seen similar cases in Chapters 1 and 4. For 2D problems of point excitation, the inverse Fourier integral can be evaluated analytically using Cauchy's theorem in a manner similar to that described in Chapter 1. Alternatively, the seminumerical method (SNM) introduced in Chapter 12 can be used. Here we use numerical integration together with the complex path technique (Chapter 4) to carry out the integrals, due to the complexity caused by the excitation of interdigital electrodes. Figures 10.19 and 10.20 show, respectively, the distribution of the displacement amplitudes for the FGPM plate and the LiTaO$_3$ plate. These figures clearly show that waves with wavelength of $1.6H$ are indeed excited. Comparing Figures 10.5 and 10.19 reveals that the second modes of the FGPM plate are excited. The displacement v on the upper surface of the FGPM plate is the largest followed by those at the middle and

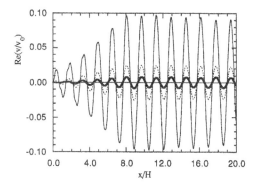

FIGURE 10.19

Responses of displacement v of the FGPM plate excited by interdigital electrodes ($\omega_f = 5.944\omega_0$; thin line: upper surface; thick line: middle surface; thicker line: lower surface). (From Liu, G. R. and Tani, J., *ASME Journal of Vibration and Acoustics*, 116, 440, 1994. With permission.)

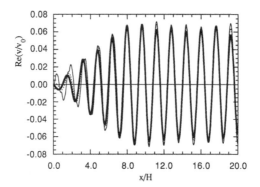

FIGURE 10.20
Responses of displacement v of the LiTaO$_3$ plate excited by interdigital electrodes ($\omega_f = 4.066\omega_0$; thin line: upper surface; thick line: middle surface; thicker line: lower surface). (From Liu, G. R. and Tani, J., *ASME Journal of Vibration and Acoustics*, 116, 440, 1994. With permission.)

lower surfaces. The displacements at these three surfaces possess the same sign. Comparing Figures 10.6 and 10.20 reveals also the fact that the second modes of the LiTaO$_3$ plate are excited. The displacement v on the upper, middle, and lower surfaces of the LiTaO$_3$ are almost the same in terms of both amplitudes and signs. The displacements at these three surfaces possess the same sign. Comparing Figures 10.19 and 10.20, we can see that the three curves of displacement amplitudes on the lower, middle, and upper surfaces of the LiTaO$_3$ almost overlap. For the FGPM plate, however, the displacement amplitudes on the upper surface are much larger compared to those on the middle and lower surfaces. The displacement amplitudes on the upper surfaces are about 12 times those on the lower surface of the FGPM plate. This means the waves in the FGPM plates appears as a surface wave.

In general, if the coupling constant is large, large displacement responses excited by electrodes or the large potential responses excited by mechanical load should be observed. However, the above example shows that the coupling constant of the LiTaO$_3$ plate is much larger than that of the FGPM plate, but the displacement amplitude on the upper surface of the FGPM plate is larger than that of the LiTaO$_3$ plate. This fact indicates that the effect of the wave modes with strong concentration of the displacement and electrical field on the upper surface is very significant.

10.14   Remarks

The layer element method and the hybrid numerical method have been extended for solving wave propagation problems for FGPM plates. The formulation of the dispersion relation, the energy propagation velocities, and the electromechanical coupling constants of waves in FGPM plates have been presented. Formulae for displacement and electrostatic potential responses of FGPM plates excited by mechanical loads and the electrodes have been derived and provided. Using the present method, the material properties can vary in an arbitrary manner in the thickness of the plate, as it is divided into layer elements. By assuming linear variation of material property in the thickness direction within elements, the stiffness, electromechanical coupling, capacitance, and mass matrices have been obtained and provided in this chapter.

The present formulae can be directly used for anisotropic laminated piezo-electric plates without any difficulty by simply omitting the additional matrices $\mathbf{M}^d$, $\mathbf{A}_i^d$, $\mathbf{C}_i^d$, and $\mathbf{G}_i^d$. It is also applicable for nonpiezoelectric plates or laminates by simply omitting matrices $\mathbf{C}_i$ and $\mathbf{G}_i$.

The frequency spectra, energy velocities, electromechanical coupling constants, mode shapes of waves in a hypothetical FGPM plate, and the responses of the FGPM plate excited by mechanical impact loads and electrodes have been computed. From these numerical examples, the following conclusions can be drawn:

- The displacements are more concentrated on the softer surface of the FGPM plate than those on the two surfaces of the homogeneous $LiTaO_3$ plate.

- For large wavenumbers, the displacements of the lowest modes are strongly concentrated on the softer surface of the FGPM plate; i.e., the waves in the FGPM appear as surface waves.

- The responses of the surface waves on the soft surface of the FGPM plate are relatively stronger compared to those on the surfaces of the homogeneous z-x $LiTaO_3$ plate.

- There is a strong pseudo-SH surface wave propagating on the surface of the FGPM plate, and the penetration depth of the wave into the plate is as small as that of the Rayleigh wave. For large wavenumbers, the high electromechanical coupling effect and the strong displacement concentration effect can be obtained. This feature of FGPM plates is very important and can be used in designing high performance surface acoustic wave (SAW) devices.

11

Strip Element Method for Stress Waves in Anisotropic Solids

11.1 Introduction

Stress wave propagation in linearly elastic solids is one of the classic problems in solid mechanics. There are exact solutions available for unbounded isotropic or anisotropic media in textbooks, such as Achenbach (1973) and Payton (1983). For anisotropic laminated media however, it is very difficult to obtain an analytical solution; instead, numerical methods, such as a finite element method and boundary element method, are usually used. The finite element method generally requires a large memory in computation. Techniques have been developed over the years, such as bandwidth storage (Meyer, 1973, 1975) and sky-line storage techniques (Bathe et al., 1976; Felippa, 1975; Jennings, 1966). The boundary element method (Beskos, 1987) is more efficient in terms of storage requirement than the finite element method for many problems, and it has been successfully used for a great variety of problems of wave propagation. However, the boundary element method is difficult to apply for anisotropic and inhomogeneous solids because Green's functions for such a media have not been found in a simple form.

This chapter presents a strip element method (SEM) for analyzing stress waves in anisotropic and/or inhomogeneous solids of layered structure. The SEM was originally presented by Liu and Achenbach (1994). This technique has a clear advantage over the finite element method in terms of storage requirements. Formulation and validation for the SEM method is detailed in this chapter. Applications of the SEM to anisotropic laminated plates are presented in following chapters.

11.2 System Equations

Consider a domain $\mathbf{D}_f$ defined by $x = \pm a$, $-\infty < y < \infty$, which includes the domain of problem definition $\mathbf{D}_p$ with boundaries S_1, S_2, S_3, and S_4, as shown in Figure 11.1. We consider a 2D problem, and the field is independent of the z coordinate. We assume also that the anisotropy of the material is confined

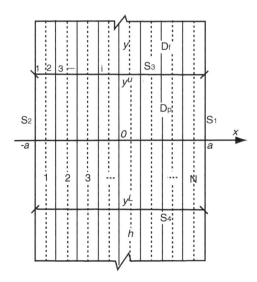

FIGURE 11.1
Problem domain and coordinate system. (From Liu, G. R. and Achenbach, J. D., *ASME Journal of Applied Mechanics*, 61, 270, 1994. With permission.)

within the *x-y* plane. Therefore, there are only two displacement components, *u*, *v*. Hence the degree of freedom at any point is two. Problems in the frequency domain are considered in this section, but the time harmonic term exp($-i\omega t$), where ω is the angular frequency, has been suppressed for simplicity. First, we make use of the oportunity to provide a set of system equations for 2D anisotropic solids.

11.2.1   Strain-Displacement Relation

The strain-displacement relations are given in matrix form by

$$\boldsymbol{\varepsilon} = \mathbf{L}\mathbf{U} \tag{11.1}$$

where

$$\boldsymbol{\varepsilon} = \begin{Bmatrix} \varepsilon_{xx} \\ \varepsilon_{yy} \\ \gamma_{xy} \end{Bmatrix} \tag{11.2}$$

is the vector of strains, and

$$\mathbf{u} = \begin{Bmatrix} u \\ v \end{Bmatrix} \tag{11.3}$$

is the vector of displacements. Here u and v are the displacement components in the x- and y-directions, respectively. The differential operator matrix **L** is given by

$$\mathbf{L} = \begin{bmatrix} \dfrac{\partial}{\partial x} & 0 \\ 0 & \dfrac{\partial}{\partial y} \\ \dfrac{\partial}{\partial y} & \dfrac{\partial}{\partial x} \end{bmatrix} \tag{11.4}$$

The operator matrix **L** can be rewritten in a concise form of

$$\mathbf{L} = \mathbf{L}_x \frac{\partial}{\partial x} + \mathbf{L}_y \frac{\partial}{\partial y} \tag{11.5}$$

where the matrices $\mathbf{L}_x$ and $\mathbf{L}_y$ are constant matrices given by

$$\mathbf{L}_x = \begin{bmatrix} 1 & 0 \\ 0 & 0 \\ 0 & 1 \end{bmatrix} \tag{11.6}$$

$$\mathbf{L}_y = \begin{bmatrix} 0 & 0 \\ 0 & 1 \\ 1 & 0 \end{bmatrix} \tag{11.7}$$

11.2.2 Stress-Strain Relation

The stress-strain relations are also given in matrix form by

$$\boldsymbol{\sigma} = \mathbf{c}\boldsymbol{\varepsilon} \tag{11.8}$$

where

$$\boldsymbol{\sigma} = \begin{Bmatrix} \sigma_{xx} \\ \sigma_{yy} \\ \tau_{xy} \end{Bmatrix} \tag{11.9}$$

is the vector of stresses, and

$$\mathbf{c} = \begin{bmatrix} c_{11} & c_{12} & c_{13} \\ c_{12} & c_{22} & c_{23} \\ c_{13} & c_{23} & c_{33} \end{bmatrix} \tag{11.10}$$

is the matrix of the off-principal-axis stiffness coefficients of the anisotropic material whose expressions in terms of engineering constants can be found in a book by Vinson and Sierakowski (1987).

11.2.3   Equation of Motion

The equation of motion, or dynamic equilibrium equation, for the anisotropic media in the absence of body force is given in matrix form of

$$\rho\ddot{U} - L^T\sigma = 0 \tag{11.11}$$

where the dot represents the differentiation with respect to time, ρ is the mass density of the material, and the superscript T stands for the transpose. Substituting Eqs. (11.1) and (11.8) into Eq. (11.11), we obtain the equation of motion in terms of displacement:

$$\rho\ddot{U} - L^T cLU = 0 \tag{11.12}$$

where

$$L^T cL = D_{xx}\frac{\partial^2}{\partial x^2} + 2D_{xy}\frac{\partial^2}{\partial x\,\partial y} + D_{yy}\frac{\partial^2}{\partial y^2} \tag{11.13}$$

in which D_{xx}, D_{yy}, and D_{xy} can be easily found to be

$$D_{xx} = L_x^T cL_x = \begin{bmatrix} c_{11} & c_{13} \\ c_{13} & c_{33} \end{bmatrix} \tag{11.14}$$

$$D_{yy} = L_y^T cL_y = \begin{bmatrix} c_{33} & c_{23} \\ c_{23} & c_{22} \end{bmatrix} \tag{11.15}$$

$$D_{xy} = \frac{1}{2}[L_x^T cL_y + L_y^T cL_x] = \frac{1}{2}\begin{bmatrix} 2c_{13} & c_{33} + c_{12} \\ c_{33} + c_{12} & 2c_{23} \end{bmatrix} \tag{11.16}$$

The stresses acting on a plane of $y = $ constant can be written in terms of displacement vector as

$$R_y = L_y^T cLU = L_y^T cL_x\frac{\partial U}{\partial x} + L_y^T cL_y\frac{\partial U}{\partial y} = D'_{yx}\frac{\partial U}{\partial x} + D_{yy}\frac{\partial U}{\partial y} \tag{11.17}$$

and the stresses acting on a plane of $x = $ constant can be written as

$$\mathbf{R}_x = \mathbf{L}_x^T \mathbf{cLU} = \mathbf{L}_x^T \mathbf{cL}_x \frac{\partial \mathbf{U}}{\partial x} + \mathbf{L}_x^T \mathbf{cL}_y \frac{\partial \mathbf{U}}{\partial y} = \mathbf{D}_{xx} \frac{\partial \mathbf{U}}{\partial x} + \mathbf{D}'_{xy} \frac{\partial \mathbf{U}}{\partial y} \quad (11.18)$$

where

$$\mathbf{D}'_{xy} = \mathbf{L}_x^T \mathbf{cL}_y = \begin{bmatrix} c_{13} & c_{12} \\ c_{33} & c_{23} \end{bmatrix} \quad (11.19)$$

$$\mathbf{D}'_{yx} = \mathbf{L}_y^T \mathbf{cL}_x = \begin{bmatrix} c_{13} & c_{33} \\ c_{12} & c_{23} \end{bmatrix} = [\mathbf{D}'_{xy}]^T \quad (11.20)$$

11.2.4 Strip Element Method Equation

The domain $\mathbf{D}_f$ is divided into N infinite strip elements. The displacement field within an element is assumed in the form of

$$\mathbf{U}(x, y) = \mathbf{N}(x)\mathbf{V}(y)\exp(-i\omega t) \quad (11.21)$$

where

$$\mathbf{N}(x) = [(1 - 3\bar{x} + 2\bar{x}^2)\mathbf{I} \quad 4(\bar{x} - \bar{x}^2)\mathbf{I} \quad (-\bar{x} + 2\bar{x}^2)\mathbf{I}] \quad (11.22)$$

$$\mathbf{V}^T = \{\mathbf{V}_L^T \quad \mathbf{V}_M^T \quad \mathbf{V}_R^T\} \quad (11.23)$$

are, respectively, the shape function matrix and the vector of displacement amplitudes. In Eq. (11.23), $\mathbf{V}_L$, $\mathbf{V}_M$, and $\mathbf{V}_R$ denote the displacement amplitude vectors on the left, middle, and right node lines of the element. The displacement amplitude vectors are functions of y.

Assume that the external harmonic forces take the form of

$$\mathbf{T} = \bar{\mathbf{T}} \exp(-i\omega t) \quad (11.24)$$

where $\bar{\mathbf{T}}$ is the amplitude vector. Following the general procedure described in previous chapters, we obtain the following set of approximate differential equations for an element

$$\bar{\mathbf{T}} = -\mathbf{A}_2 \frac{\partial^2 \mathbf{V}}{\partial y^2} + \mathbf{A}_1 \frac{\partial \mathbf{V}}{\partial y} + \mathbf{A}_0 \mathbf{V} - \omega^2 \mathbf{MV} \quad (11.25)$$

where $\mathbf{A}_i$ ($i = 0, 1, 2$) and $\mathbf{M}$ are given by

$$\mathbf{A}_0 = \frac{1}{3h}\begin{bmatrix} 7\mathbf{D}_{xx} & -8\mathbf{D}_{xx} & \mathbf{D}_{xx} \\ -8\mathbf{D}_3 & 16\mathbf{D}_{xx} & -8\mathbf{D}_{xx} \\ \mathbf{D}_{xx} & -8\mathbf{D}_{xx} & 7\mathbf{D}_{xx} \end{bmatrix} \tag{11.26}$$

$$\mathbf{A}_1 = \frac{1}{3}\begin{bmatrix} 3(\mathbf{D}_{xy} - \mathbf{D}'_{xy}) & -4\mathbf{D}_{xy} & \mathbf{D}_{xy} \\ 4\mathbf{D}_{xy} & 0 & -4\mathbf{D}_{xy} \\ -\mathbf{D}_{xy} & 4\mathbf{D}_{xy} & -3(\mathbf{D}_{xy} - \mathbf{D}'_{xy}) \end{bmatrix} \tag{11.27}$$

$$\mathbf{A}_2 = \frac{h}{30}\begin{bmatrix} 4\mathbf{D}_{yy} & 2\mathbf{D}_{yy} & -\mathbf{D}_{yy} \\ 2\mathbf{D}_{yy} & 16\mathbf{D}_{yy} & 2\mathbf{D}_{yy} \\ -\mathbf{D}_{yy} & 2\mathbf{D}_{yy} & 4\mathbf{D}_{yy} \end{bmatrix} \tag{11.28}$$

$$\mathbf{M} = \frac{\rho h}{30}\begin{bmatrix} 4\mathbf{I} & 2\mathbf{I} & -\mathbf{I} \\ 2\mathbf{I} & 16\mathbf{I} & 2\mathbf{I} \\ -\mathbf{I} & 2\mathbf{I} & 4\mathbf{I} \end{bmatrix} \tag{11.29}$$

In the expression of the matrix $\mathbf{M}$, $\mathbf{I}$ is a 2×2 identity matrix.

Assembling all the strip elements in domain $\mathbf{D}_f$, a system of approximate differential equations for the whole domain $\mathbf{D}_f$ is obtained.

$$\bar{\mathbf{T}}_t = -\mathbf{A}_{2t}\frac{\partial^2 \mathbf{V}_t}{\partial y^2} + \mathbf{A}_{1t}\frac{\partial \mathbf{V}_t}{\partial y} + \mathbf{A}_{0t}\mathbf{V}_t - \omega^2 \mathbf{M}_t \mathbf{V}_t \tag{11.30}$$

where the subscript t denotes matrices or vectors for the whole domain. The matrices $\mathbf{A}_{it}$ ($i = 0, 1, 2$), $\mathbf{M}_t$ and the vectors $\mathbf{T}_t$, $\mathbf{V}_t$ can be obtained by assembling the corresponding matrices and vectors of all the elements. The dimension of matrices $\mathbf{A}_{it}$ and $\mathbf{M}_t$ is $M \times M$, where $M = 4N + 2$ with N being the number of the elements.

Equation (11.30) represents a set of second-order differential equations with constant coefficients. It can be solved exactly. Assuming

$$\mathbf{V}_t = \mathbf{d}_t \exp(iky) \tag{11.31}$$

$$\bar{\mathbf{T}}_t = \mathbf{P}_t \exp(iky) \tag{11.32}$$

and substituting these expressions into Eq. (11.30), we obtain

$$\mathbf{P}_t = [k^2 \mathbf{A}_{2t} + ik\mathbf{A}_{1t} + \mathbf{A}_{0t} - \omega^2 \mathbf{M}_t]\mathbf{d}_t \tag{11.33}$$

The vector $\mathbf{P}_t$ is the amplitude vector of external forces acting on the node lines of domain $\mathbf{D}_f$. Here we discuss only the case of $\mathbf{P}_t = 0$. Therefore, the following eigenvalue equation with respect to k can be obtained:

$$0 = [k^2\mathbf{A}_{2t} + ik\mathbf{A}_{1t} + \mathbf{A}_{0t} - \omega^2\mathbf{M}_t]\mathbf{d}_t \qquad (11.34)$$

For a given frequency, Eq. (11.34) can be changed to a standard eigenvalue equation (Dong, 1977):

$$0 = \left(\begin{bmatrix} 0 & \mathbf{I} \\ \omega^2\mathbf{M}_t - \mathbf{A}_{0t} & -i\mathbf{A}_{1t} \end{bmatrix} - k\begin{bmatrix} \mathbf{I} & 0 \\ 0 & \mathbf{A}_{2t} \end{bmatrix}\right)\begin{Bmatrix} \mathbf{d}_t \\ k\mathbf{d}_t \end{Bmatrix} \qquad (11.35)$$

Using a general eigenvalue solver to solve this equation, $2M$ eigenvalues k_j ($j = 1,\ldots,2M$) and their eigenvectors can be obtained. If the upper half of the jth eigenvector corresponding to $\mathbf{d}_t$ is denoted by ϕ_j,

$$\phi_j^T = \{\phi_{1j} \quad \phi_{2j} \quad \cdots \quad \phi_{Mj}\} \qquad (11.36)$$

then the displacement can be written by superposition of the eigenvectors as

$$\mathbf{V}_t = \sum_{j=1}^{2M} C_j\phi_j \exp(ik_jy) = \mathbf{G}(y)\mathbf{C} \qquad (11.37)$$

where C_j are constants to be determined. Equation (11.37) is a fundamental solution for the domain $\mathbf{D}_f$. The constant vector $\mathbf{C}$ can be determined from the boundary conditions on S_3 and S_4 to obtain a special solution for the domain $\mathbf{D}_p$. There are M nodes and $2M$ boundary conditions on S_3 and S_4 to determine $2M$ constants C_j. For half-space problems in the domain of $y \leq 0$, there is no boundary S_4; hence, only M boundary conditions can be used. However, as $y \to -\infty$, the displacement must be bounded. Therefore, only the eigenvalues that satisfy $\text{Im}(k_j) \geq 0$ need to be used, and for eigenvalues with $\text{Im}(k_j) = 0$, a condition $\text{Re}(k_j) \leq 0$ has to be added to ensure that the waves propagate in the negative y-direction. Consequently, only M eigenvalues need to be chosen, which equals exactly the number of the available boundary conditions on the boundary S_3. A process to determine $\mathbf{C}$ and to derive a relationship between the displacement and stress vectors on the boundaries S_3 and S_4 is discussed below.

Using Eq. (11.37), the constant vector $\mathbf{C}$ can be expressed in terms of the displacement vector at the node point on boundaries S_3 and S_4

$$\mathbf{C} = \mathbf{G}_b^{-1}\mathbf{V}_{bt} = \begin{bmatrix} \mathbf{G}_U \\ \mathbf{G}_L \end{bmatrix}^{-1}\mathbf{V}_{bt} \qquad (11.38)$$

where

$$\mathbf{V}_{bt} = \{\mathbf{V}_{bt}^U \quad \mathbf{V}_{bt}^L\} = \{u_1 \quad v_1 \quad u_2 \quad v_2 \quad \cdots \quad u_M \quad v_M\} \qquad (11.39)$$

$$\mathbf{G}_U = \begin{bmatrix} \phi_{11}Y_1^U & \phi_{12}Y_2^U & \cdots & \phi_{1L}Y_L^U \\ \phi_{21}Y_1^U & \phi_{22}Y_2^U & \cdots & \phi_{2L}Y_L^U \\ \vdots & \vdots & \ddots & \vdots \\ \phi_{M1}Y_1^U & \phi_{M2}Y_2^U & \cdots & \phi_{ML}Y_L^U \end{bmatrix} \qquad (11.40)$$

$$Y_j^U = \exp(ik_j y^U) \qquad (11.41)$$

where the superscripts U and L indicate the matrices or variables on the upper boundary S_3 and the lower boundary S_4, respectively. In Eq. (11.39), u_i and v_i are the displacement components in the x and y directions at point i on the boundaries S_3 and S_4. In Eq. (11.41), y^U is the y-coordinate on the upper boundary S_3. The matrix $\mathbf{G}_L$ in Eq. (11.38) is the same as $\mathbf{G}_U$ but Y_j^U is replaced with $Y_j^L = \exp(ik_j y^L)$, where y^L is the y-coordinate on the lower boundary S_4 (see Figure 11.1).

Using Eq. (11.17), the vector of stress on $y =$ constant plane at any point within an element can be given by

$$\mathbf{R}_y = \mathbf{D}'_{yx}\frac{\partial \mathbf{U}}{\partial x} + \mathbf{D}_{yy}\frac{\partial \mathbf{U}}{\partial y} \qquad (11.42)$$

where

$$\mathbf{D}'_{yx} = \begin{bmatrix} c_{13} & c_{33} \\ c_{12} & c_{23} \end{bmatrix} \qquad (11.43)$$

Hence

$$\mathbf{R} = \{\mathbf{R}_y|_{x=0} \quad \mathbf{R}_y|_{x=h/2} \quad \mathbf{R}_y|_{x=h}\}^T = \mathbf{R}_1\mathbf{V} + \mathbf{R}_2\frac{\partial \mathbf{V}}{\partial y} \qquad (11.44)$$

where $\mathbf{R}_1$ and $\mathbf{R}_2$ are given by

$$\mathbf{R}_1 = \frac{1}{h}\begin{bmatrix} -3\mathbf{D}'_{yx} & 4\mathbf{D}'_{yx} & -\mathbf{D}'_{yx} \\ -\mathbf{D}'_{yx} & 0 & \mathbf{D}'_{yx} \\ \mathbf{D}'_{yx} & -4\mathbf{D}'_{yx} & 3\mathbf{D}'_{yx} \end{bmatrix} \qquad (11.45)$$

$$\mathbf{R}_2 = \begin{bmatrix} \mathbf{D}_{yy} & 0 & 0 \\ 0 & \mathbf{D}_{yy} & 0 \\ 0 & 0 & \mathbf{D}_{yy} \end{bmatrix} \qquad (11.46)$$

Assembling the matrices $\mathbf{R}_1$ and $\mathbf{R}_2$ for all the strip elements, we have the stress vector for all the node lines $\mathbf{R}_t$ as follows

$$\mathbf{R}_t = \mathbf{R}_{1t}\mathbf{V}_t + \mathbf{R}_{2t}\frac{\partial \mathbf{V}_t}{\partial y} \tag{11.47}$$

Using Eqs. (11.37), (11.38), and (11.47), we finally have

$$\mathbf{R}_{bt} = \mathbf{K}\mathbf{V}_{bt} \tag{11.48}$$

where $\mathbf{R}_{bt} = \{\mathbf{R}_{bt}^{U}\ \mathbf{R}_{bt}^{L}\}^{T}$ is the external traction vector on boundaries S_3 and S_4, and

$$\mathbf{K} = \begin{bmatrix} \mathbf{R}_{1t} & 0 \\ 0 & \mathbf{R}_{1t} \end{bmatrix} + \begin{bmatrix} \mathbf{R}_{2t}\dfrac{\partial \mathbf{G}_U}{\partial y}\mathbf{G}_b^{-1} \\ \mathbf{R}_{2t}\dfrac{\partial \mathbf{G}_L}{\partial y}\mathbf{G}_b^{-1} \end{bmatrix} \tag{11.49}$$

is the stiffness matrix. Equation (11.48) gives the relationship between the tractions and the displacements on the boundaries S_3 and S_4. For given $\mathbf{R}_{bt}$, $\mathbf{V}_{bt}$ can be solved from Eq. (11.48).

11.3   SEM for Static Problems (Flamant's Problem)

The foregoing formulation can be easily applied to solve static problems. The only modification needed is to set the terms concerning time and frequency to zero. The Flamant's problem, a semi-infinite space loaded by a concentrate force at the origin $(0, 0)$, is considered. To simulate the semi-infinite space by the strip element method, a half-space infinite element is used at $x \geq a$. The infinite strip element is developed using a special interpolation function for infinite elements (see Zienkiewicz et al., 1989). The procedure of developing the formulae for infinite strip element is the same as for regular strip elements, except the use of infinite shape functions. The matrices for the infinite strip element are listed as follows:

$$\mathbf{A}_0 = \frac{1}{30a}\begin{bmatrix} 46\mathbf{D}_{xx} & -52\mathbf{D}_{xx} & 6\mathbf{D}_{xx} \\ -52\mathbf{D}_3 & 64\mathbf{D}_{xx} & -12\mathbf{D}_{xx} \\ 6\mathbf{D}_{xx} & -12\mathbf{D}_{xx} & 6\mathbf{D}_{xx} \end{bmatrix} \tag{11.50}$$

$$\mathbf{A}_2 = \frac{a}{3}\begin{bmatrix} \mathbf{D}_{yy} & -2\mathbf{D}_{yy} & -\mathbf{M}_{\infty}^1 \\ -2\mathbf{D}_{yy} & 16\mathbf{D}_{yy} & \mathbf{M}_{\infty}^1 \\ -\mathbf{M}_{\infty}^1 & \mathbf{M}_{\infty}^1 & \mathbf{M}_{\infty}^2 \end{bmatrix} \tag{11.51}$$

where $\mathbf{M}_\infty^1$ is a matrix of large constant numbers of machine infinity.

$$\mathbf{R}_1 = \frac{1}{2a} \begin{bmatrix} -12\mathbf{D}'_{yx} & 16\mathbf{D}'_{yx} & -4\mathbf{D}'_{yx} \\ -\mathbf{D}'_{yx} & 0 & \mathbf{D}'_{yx} \\ 0 & 0 & 0 \end{bmatrix} \qquad (11.52)$$

Matrices $\mathbf{A}_1$ and $\mathbf{R}_2$ for an infinite strip element are the same as those for the regular strip element matrices.

Using the strip element method and making use of the symmetric conditions of the problem, the semi-infinite space is modeled by a quarter-space with the following boundary conditions:

at $x = 0$

$$\sigma_{xy} = 0, \quad u = 0 \qquad (11.53)$$

and at $y = 0$

$$\sigma_{yy} = \frac{q_0}{h_1}\left(\frac{x}{h_1} - 1\right), \quad \text{for } x \le h_1$$

$$\sigma_{yy} = 0, \qquad\qquad \text{for } x > h_1 \qquad (11.54)$$

$$\sigma_{xy} = 0, \qquad\qquad \text{for } x \ge 0$$

This means that the line load is approximated by a triangularly distributed force. The time-harmonic term has been omitted. The exact solution of Flamant's problem, which can be found in any text on elasticity theory, is

$$\sigma_{xx} = \frac{-2q_0 x^2 y}{\pi(x^2 + y^2)^2} \quad \sigma_{xy} = \frac{-2q_0 xy^2}{\pi(x^2 + y^2)^2} \quad \sigma_{yy} = \frac{-2q_0 y^3}{\pi(x^2 + y^2)^2} \qquad (11.55)$$

Figure 11.2 shows the stress distributions on the line of $y = -0.1$, $x > 0$. The results are obtained by the strip element method using 20 and 30 elements. An infinite element is attached at $x = 0.7$. The convergence is very good. Figure 11.3 shows the results obtained using the strip element method with 30 elements, together with the exact solutions given by Eq. (11.55). Very good agreement is observed.

11.4   SEM for Dynamic Problems

In applying any method of domain discretization for dynamic problems of infinite bodies, so called nonreflecting boundaries have to be used in order to reduce the size of the domain to be discretized. A good review of nonreflecting boundary conditions has been given by Givoli (1991). Three nonreflecting

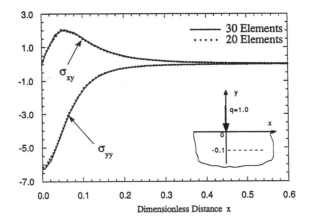

FIGURE 11.2
Convergence of stress distributions on $y = -0.1$ for an isotropic half-space subjected to a static line load, computed by the strip element method with an infinite strip element attached at $x = 0.7$. (From Liu, G. R. and Achenbach, J. D., *ASME Journal of Applied Mechanics*, 61, 270, 1994. With permission.)

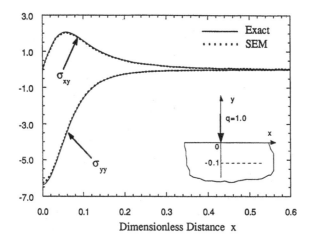

FIGURE 11.3
Stress distributions on $y = -0.1$ for an isotropic half-space subjected to a static line load. Comparison of results by the exact and the strip element method with an infinite strip element attached at $x = 0.7$. (From Liu, G. R. and Achenbach, J. D., *ASME Journal of Applied Mechanics*, 61, 270, 1994. With permission.)

boundary conditions for elastic waves are implemented for the strip element method. A general form of the three nonreflecting boundary conditions can be written as

$$\left.\frac{\partial \mathbf{U}}{\partial x}\right|_{x=a} = \mathbf{BU}|_{x=a} \tag{11.56}$$

The matrix **B** takes different forms, respectively, for the first, second, and third nonreflecting boundary conditions:

$$\mathbf{B} = i\begin{bmatrix} k_l & 0 \\ 0 & k_t \end{bmatrix} \tag{11.57}$$

$$\mathbf{B} = i\mathbf{B}_0 - ik\mathbf{B}_1 + ik^2\mathbf{B}_2 \tag{11.58}$$

$$\mathbf{B} = i\begin{bmatrix} k_r & 0 \\ 0 & k_r \end{bmatrix} \tag{11.59}$$

Equations (11.57) and (11.58) correspond to the first and the second-order nonreflecting boundary conditions given by Clayton et al. (1977). The matrices $\mathbf{B}_i$ ($i = 0, 1, 2$) in Eq. (11.58) have been given by Clayton et al. (1977) and Engquist et al. (1979). The third nonreflecting boundary condition is obtained by replacing c_l and c_t with c_r, where c_r is the velocity of the Rayleigh wave. This modification was proposed by Bamberger et al. (1988) for the absorption of Rayleigh waves on the surface of a semi-infinite space.

Equation (11.56) gives the relationship between the derivative of the displacement with respect to x and the displacement itself on the boundary line of $x = a$. With the help of the constitutive and geometric relationships, the stress vector at $x = a$ can also be expressed in terms of the displacement vector at $x = a$. Substituting the boundary stress vector into Eq. (11.33), a modified eigenvalue equation for the infinite body is obtained, and fundamental solutions can be obtained for the domain $\mathbf{D}_f$ where the nonreflecting boundary condition is used. Finally, the displacement can be obtained from Eq. (11.48) after the imposition of the boundary conditions on the boundaries S_3 and S_4.

The three nonreflecting boundary conditions are implemented in the SEM code. It is found that none of these three nonreflecting boundary conditions works satisfactorily well for the half-space case. Liu and Achenbach (1994) have thus constructed a new viscoelastic nonreflecting boundary by attaching a few artificial viscoelastic elements at $x = a$. By letting the viscosity of the elements increase gradually in the x direction, these viscoelastic elements can very efficiently absorb waves with any incident angle to the boundary. The material constant matrix of the kth viscoelastic element, $\mathbf{c}_k^{ve}$, is assumed to be

$$\mathbf{c}_k^{ve} = \mathbf{c} - i\alpha_0\beta^{k-1}\mathbf{c} \tag{11.60}$$

where $\mathbf{c}$ is the elastic constant matrix. In the following calculations, the factors α_0 and β in Eq. (11.60) are chosen to be 0.01 and 1.624, respectively. In the SEM model, a total of 120 elements are used in the range of $0 \leq x \leq 15\lambda_0$, where 100 elements are elastic, and 20 elements attached to the right surface of the 100th elastic element are viscoelastic.

11.4.1 Harmonic Waves in 2D Space

The elastodynamic response of an unbounded 2D body to a time-harmonic line load is the well-known 2D Green's function that can be found in textbooks, such as that edited by Beskos (1987). To apply the strip element method, only a quarter space with the following boundary conditions has to be considered:

at $x = 0$

$$\sigma_{xy} = 0, \quad u = 0 \tag{11.61}$$

and at $y = 0$

$$\sigma_{yy} = \frac{q_0}{h_1}\left(\frac{x}{h_1} - 1\right), \quad \text{for } x \leq h_1$$

$$\sigma_{yy} = 0, \qquad\qquad \text{for } x > h_1 \tag{11.62}$$

$$u = 0, \qquad\qquad \text{for } x \geq 0$$

Thus, the line load is approximated by a triangularly distributed force. The time-harmonic term has been omitted.

Figure 11.4 shows the displacement v on the line $y = 0$ ($x > 0$). The solid line is obtained by the exact method (Green's function), and the dashed line is obtained by the strip element method with the viscoelastic nonreflecting boundary. Excellent agreement is observed. Figure 11.5 shows the displacements obtained by the strip element method together with the first nonreflecting boundary condition and the SEM with a clamped boundary at $x = 7\lambda_1$. The exact results are also plotted in the same figure. Quite good results are obtained using the first nonreflecting boundary condition. Very poor results are given

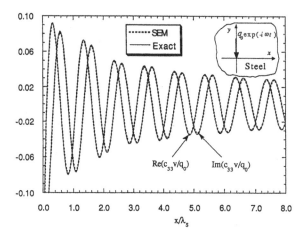

FIGURE 11.4
Comparison of elastodynamic displacements on $y = 0$ for an isotropic full-space subjected to a time-harmonic line load. Comparison of results by the exact and the strip element method with a viscoelastic nonreflecting boundary. (From Liu, G. R. and Achenbach, J. D., *ASME Journal of Applied Mechanics*, 61, 270, 1994. With permission.)

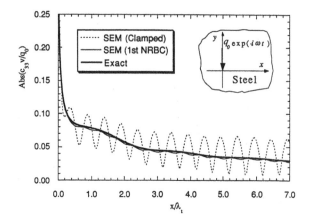

FIGURE 11.5
Elastodynamic displacements on $y = 0$ for an isotropic full-space subjected to a time-harmonic line load. Comparison of results by the exact and the strip element method with different artificial boundaries. (From Liu, G. R. and Achenbach, J. D., *ASME Journal of Applied Mechanics*, 61, 270, 1994. With permission.)

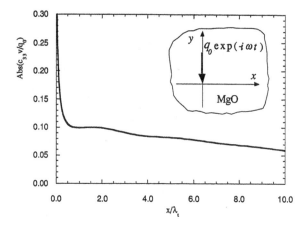

FIGURE 11.6
Elastodynamic displacements on $y = 0$ for an isotropic full-space subjected to a time-harmonic line load, obtained by the strip element method with a viscoelastic nonreflecting boundary. (Form Liu, G. R. and Achenbach, J. D., *ASME Journal of Applied Mechanics*, 61, 270, 1994. With permission.)

using the clamped boundary condition because of the reflection of incident waves from the clamped boundary. Very good results can be obtained using the second nonreflecting boundary condition.

An important feature of the strip element method is its applicability to aniso-tropic solids. For anisotropic solids, there is no simple form of the exact solution available for an artificial boundary like the second nonreflecting boundary con-dition. Figure 11.6 shows the displacement distribution at $x = 0$ for an anisotropic full-space (single crystal MgO) obtained by the strip element method with the use of the viscoelastic nonreflecting boundary. The single crystal MgO is cubic symmetric, and its anisotropic factor is about 1.5. The time-harmonic line periodic load is applied at the origin in the [0 0 1] direction.

11.4.2   Harmonic Waves in Half-Space (Lamb's Problem)

For an isotropic half-space subjected to a normal time-harmonic line load, there is no closed-form of exact elastodynamic solution. However, the following integral form of the solution has been obtained by the application of Fourier transform techniques.

$$\frac{c_{33}}{q_0}u = \frac{1}{2\pi}\int_{-\infty}^{\infty}\frac{i\xi[(k_t^2-2\xi^2)\exp(-\beta y)+2\alpha\beta\exp(-\alpha y)]}{(2\xi^2-k_t^2)^2-4\xi^2\alpha\beta}\exp(-i\xi x)d\xi \quad (11.63)$$

$$\frac{c_{33}}{q_0}v = \frac{1}{2\pi}\int_{-\infty}^{\infty}\frac{i\beta[(k_t^2-2\xi^2)\exp(-\beta y)+2\xi^2\exp(-\alpha y)]}{(2\xi^2-k_t^2)-4\xi^2\alpha\beta}\exp(-i\xi x)d\xi \quad (11.64)$$

where

$$\alpha = \sqrt{\xi^2-k_t^2}, \quad \beta = \sqrt{\xi^2-k_l^2} \quad (11.65)$$

and k_l and k_t are the wavenumbers of the longitudinal and transverse waves. Equations (11.64) and (11.65) are evaluated numerically. Note that near the two poles $\xi = \pm k_r$, where k_r is the wavenumber of the Rayleigh wave. Hence, the technique of complex path described in Chapter 4 needs to be used. Also, because the integrands vary very rapidly near the poles, an adaptive quadrature scheme, which can integrate over rapidly varying integrands, has to be used near these points (see Chapter 4). In addition, the integrand also oscillates at a very high frequency when $\xi \to \pm\infty$, hence again a subroutine for integrations over rapidly oscillating functions has to be used (see Chapter 8).

   In the strip element method, the domain for the numerical calculations is a quarter-space. The boundary conditions are the same as in Eqs. (11.53) and (11.54). Visco-elastic nonreflecting boundary elements are used.

   Figure 11.7 shows the displacement distribution on the surface of a half-space of steel subjected to a time-harmonic normal line load. The dashed lines are obtained by the numerical evaluation of Eqs. (11.63) and (11.64), and the solid lines are obtained by the strip element method with the second nonreflecting boundary condition. Although good agreement is obtained, small oscillations are observed in the curves obtained by the strip element method because the second nonreflecting boundary condition cannot completely

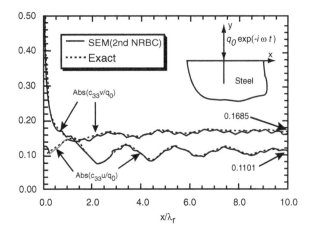

FIGURE 11.7
Elastodynamic displacements on the surface for an isotropic half-space subjected to a time-harmonic line load. Comparison of results by the exact method and the strip element method with the second nonreflecting boundary condition. (From Liu, G. R. and Achenbach, J. D., *ASME Journal of Applied Mechanics, 61, 270, 1994.* With permission.)

absorb waves that are incident on the boundary with large angles of incidence (Clayton et al., 1977). The displacements have also been computed using the first nonreflecting boundary condition, but these results are much worse than the second nonreflecting boundary condition. The third nonreflecting boundary condition is as good as the second because the major wave reflections are for Rayleigh waves, and the third nonreflecting boundary condition absorbs the Rayleigh waves quite well. Figure 11.8 shows the distribution of the absolute value of the displacement obtained by the strip element method with the viscoelastic nonreflecting boundary, together with the exact solutions. Excellent agreement is observed. The curves obtained by the strip element method are a little lower than the exact curves (see Figures 11.7 and 11.8) because the line load is treated as a distributed traction, as given by Eq. (11.54). It has been confirmed that the result of the strip element method approach the exact curves more closely when the width of the first element becomes reduced.

Another checking point is that the absolute values of the two displacement components should approach constant values when $\xi \to \infty$. In this case the constants are $Abs(c_{33}u/q_0) = 0.1101$ and $Abs(c_{33}v/q_0) = 0.1685$ as shown in Figures 11.7 and 11.8.

For anisotropic half-space problems, there is no exact solution. Figure 11.9 shows the displacement distribution on the surface of an anisotropic half-space (single crystal MgO) obtained by the strip element method with the visco-elastic nonreflecting boundary. The normal line load is acting at the origin in the [0 0 1] direction.

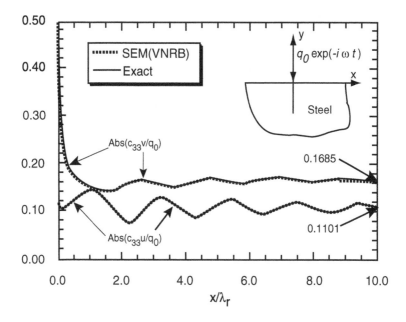

FIGURE 11.8

Elastodynamic displacements on the surface for an isotropic half-space subjected to a time-harmonic line load. Comparison of results by the exact method and the strip element with a viscoelastic nonreflecting boundary condition. (From Liu, G. R. and Achenbach, J. D., *ASME Journal of Applied Mechanics*, 61, 270, 1994. With permission.)

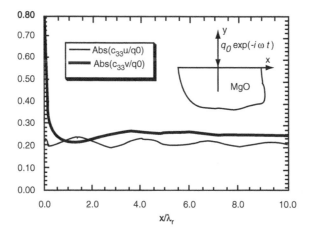

FIGURE 11.9

Elastodynamic displacements on the surface for an anisotropic half-space (MgO) subjected to a time-harmonic line load, obtained by the strip element method with a viscoelastic nonreflecting boundary. (From Liu, G. R. and Achenbach, J. D., *ASME Journal of Applied Mechanics*, 61, 270, 1994. With permission.)

11.5   Remarks

Since the finite element discretization produces a large number of node variables, the matrix size in the finite element method is usually very large. In the boundary element method, a fundamental problem for a full-space is first solved exactly, and then the fundamental solutions (Green's functions) are used to derive boundary integral equations that are solved by discretizing the boundary. In the strip element method, the problem domain is first discretized into strip elements to obtain approximate governing equations that are then solved analytically. Because of the lower levels of discretization, the matrix size in both the boundary element method and the strip element method is very small compared to the finite element method. Table 11.1 gives a simple example to show how much memory can be saved using the strip element method. In Table 11.1, a square domain is considered and linear interpolation functions are used in the strip element method, the boundary element method, and the finite element method. Table 11.1 shows that the matrix size in the boundary element method is four times larger than that in the strip element method. If 100 elements in the horizontal direction are needed to solve the problem, the matrix size in the finite element method is 50 times larger than that in the strip element method, even if the bandwidth storage technique is used in the finite element method.

TABLE 11.1

Comparison of Matrix Sizes for the Strip Element Method (SEM), Boundary
Element Method (BEM), and Finite Element Method (FEM)

SEM	BEM	FEM

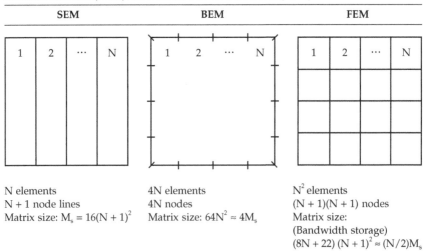

SEM	BEM	FEM
N elements	$4N$ elements	N^2 elements
$N + 1$ node lines	$4N$ nodes	$(N + 1)(N + 1)$ nodes
Matrix size: $M_s = 16(N + 1)^2$	Matrix size: $64N^2 \approx 4M_s$	Matrix size:
		(Bandwidth storage)
		$(8N + 22)(N + 1)^2 \approx (N/2)M_s$

Source: Liu, G. R. and Achenbach, J. D., *ASME Journal of Applied Mechanics*, 61, 270, 1994.
With permission.

To use the strip element method efficiently, some special numerical techniques should be used, based on the following considerations:

- Since the matrices given before Eq. (11.35) are narrowly banded symmetric or asymmetric matrices, it is recommended to use bandwidth storage techniques to reduce the storage and the computation time.
- Since the matrices after Eq. (11.35) are usually full and complex-valued matrices, and some of them contain exponential functions, it is strongly recommended to use subroutines for solving linear algebraic equations instead of subroutines for solving inverse matrices, to save computation time and maintain accuracy.
- If the matrices contain exponential functions, which are often very large or very small, an efficient way to avoid loss of precision is to normalize these matrices before calling equation solvers. Alternatively, one can use the technique described in Chapter 4 by dividing the eigenvalues into two groups (up-going wave modes and down-going wave modes) before constructing a displacement vector using Eq. (11.37).
- Since the computation time in a strip element method program is mainly consumed by solving the eigenvalue equation, Eq. (11.35), to obtain eigenvalues and eigenvectors, an efficient eigenvalue solver is required. The most efficient method is first to calculate Eq. (11.35) for eigenvalues using an eigenvalue subroutine, and then solve Eq. (11.34) for the corresponding eigenvectors using a subroutine for solving linear algebraic equations because the number of equations in Eq. (11.34) is half that of Eq. (11.35).

This chapter focuses on the development of the formulation for the SEM and the validation of the SEM. It also proposes techniques for dealing with non-reflecting boundary conditions. The strength of the SEM is demonstrated in the following chapters through a number of applications to wave scattering problems in anisotropic laminates.

12

Wave Scattering by Cracks in Composite Laminates

12.1 Introduction

In the preceding chapters, the composite laminates or plates were assumed to be perfect. In practice, defects, such as cracks and other flaws could be generated in these structures during the process of manufacturing, operation, and/or service. The growth of such defects may eventually cause catastrophic failure of the entire structures. Techniques using elastic waves, such as ultrasonic testing, play a very important role in detecting cracks in materials. Cracks in composite laminated plates can also be detected and characterized by conventional ultrasonic techniques, although reliable results are much more difficult to obtain than for traditional materials, such as metals, that are homogeneous to a much smaller scale. In fiber-reinforced composites, the individual fibers are micrometers in diameter, and the thicknesses of plies are of the order of one tenth of a millimeter. It would not be productive to attempt detection of cracks with signals whose dominant wavelength is smaller than the ply thickness. Such signals would generate waves that have too much interference with the inhomogeneity of the composite material, and they would be rapidly attenuated. Lower frequency signals whose dominant wavelength is of the order of the thickness of laminated plates, say millimeters, will propagate over longer distances since the waves will have less interference with the inhomogeneity of the material at such lower frequency signals. On the other hand, the lower the frequency of the input signals, the lower the resolution is. Hence, it would be difficult to use very low frequency signals whose dominant wavelength is much longer than the plate thickness to detect cracks whose sizes are of the order of the plate thickness, even though they propagate far. Therefore, signals with intermediate frequency whose dominant wavelength is about the thickness of the laminate are most preferred. Waves at such an intermediate frequency are Lamb waves of dispersive nature. Traditional ultrasonic techniques using bulk waves will not be able to produce clear images. Analysis of such waves requires also proper methodology. Theories based on bulk waves (nondispersion waves) are not applicable, and classic theories of bending plates will not work either because the wavelength is not long enough and the displacement distribution across the cross-section of the plates is much more complex than the plane deformation assumption (Kirchhoff or Mindlin assumption). Methods described in Chapters 5 and 8 have proven effective. The task of this chapter is to extend these methods for anisotropic laminates with cracks.

The most important thing in using techniques of elastic waves successfully is to find out the relationships between the parameters of a crack and the scattered elastic wave fields. However, the finding of the relationships is usually very difficult for anisotropic laminates. The problem for finding the relationship is basically an inverse problem, which needs a lot of computational efforts. To solve an inverse problem, many corresponding forward problems have to be solved. Hence, a very efficient method and suitable modeling technique for solving the forward problem are a prerequisite.

Investigation into the scattering of waves in elastic materials with cracks have been carried out by a number of researchers, and many methods and models have been proposed for the problems of scattering of waves in elastic materials. One of these methods is the technique combining an analytical procedure with the finite element method (FEM) (Al-Nassar et al., 1991; Karunasena et al., 1991c; Karim et al., 1992). The boundary element method (BEM) is an alternative to investigate wave propagation problems (Ahn et al., 1991). A combination of the BEM and FEM can also be very efficient to analyze scattering of waves in plate structures (S. W. Liu et al., 1991, 1993; Datta et al, 1992; Liu et al., 1992).

This chapter introduces a strip element method (SEM) for analyzing waves scattered by cracks in composite laminated plates. The approach of using SEM was originally proposed by Liu and Achenbach (1995). It was later extended for the development of techniques for crack detection (Liu and Lam, 1994; Lam et al., 1995, 1997). First, the SEM developed previously is extended to obtain the solution in the frequency domain for uncracked anisotropic laminated plates subjected to an external loading. Then, a general SEM solution for a loaded laminated plate with boundaries perpendicular to the plane of the plate (vertical boundaries) is given by combining the particular solution for the loaded laminate and the fundamental SEM solution (the complementary solution of the associated homogeneous form of the governing differential equations). A set of general SEM equations that gives the relationship between the traction and displacement vectors at nodal points on the vertical boundaries is obtained from the general SEM solution. For a plate with a crack, the plate is divided into several domains. A general SEM equation is formulated for each domain. By assembling all the equations, a set of algebraic equations for the entire cracked laminate is obtained using the conditions on the vertical boundaries between the domains. To obtain the response of a plate in the time domain, a Fourier transformation technique is used, and an exponential window method (EWM) is introduced to avoid singularities of the integrand in the Fourier integration. Furthermore, based on the SEM, techniques are introduced for the characterization of horizontal and vertical cracks in anisotropic laminated plates.

12.2 Governing Differential Equations

Consider a composite laminated plate of thickness H with a horizontal or vertical crack, as shown in Figure 12.1. The plate is made of an arbitrary number of linearly elastic laminae. The bonding between plies is assumed to be perfect except in the region of the crack. Deformations of the plate are assumed small under a harmonic excitation. The Cartesian coordinate system is used. The plate is infinitely long in the x direction. To simplify the problem, we assume that the crack is throughout in the y-direction. Therefore, the problem is two-dimensional. The upper surface of the plate is subjected to a transverse line load of

$$q(x, t) = q_0 \delta(x) \exp(-i\omega t) \tag{12.1}$$

where q_0 is a constant, and $\delta(x)$ is the Dirac delta function of x. The load is uniformly distributed along the y-axis.

Boundary conditions are separated into two categories. There are boundary conditions on horizontal planes (HBCs) for boundaries that are parallel to the plane of the plate, and on vertical planes (VBCs) for boundaries that are perpendicular to the plate plane.

The SEM is a combination of the finite element method, modal analysis method, and the Fourier transformation technique. It first uses layer elements to discretize the plate through the thick direction and derives a reduced governing equation, and then it uses the modal analysis method and the Fourier transformation technique to obtain the solution of the resulting equation. We employ three-nodal-line elements to model the plate along the thick direction. The degrees of freedom u_l, w_l, u_m, w_m, u_u, and w_u are, respectively, unknown horizontal and vertical displacement amplitudes at the lower,

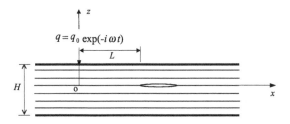

FIGURE 12.1
Composite laminate containing a crack.

middle, and upper nodal-lines l, m, and u, of the element. Following the procedure given in Chapter 11, and noticing the change of the coordinate system, the displacements within the strip element are related to the unknown displacement amplitudes $\mathbf{V}$ at the nodal lines as

$$\mathbf{u}(x, z, t) = \mathbf{N}(z)\mathbf{V}(x)\exp(-i\omega t) \tag{12.2}$$

where $\mathbf{N}(z)$ is the matrix of quadratic shape functions of z as given in Chapter 11. Applying the principle of virtual work to each element, a set of approximate differential equations can be obtained for each element. Assembling all the elements and using the HBCs, a system of approximate governing differential equations for the total plate can finally be obtained

$$-\mathbf{A}_{2t}\frac{d^2\mathbf{V}}{dx^2} + \mathbf{A}_{1t}\frac{d\mathbf{V}}{dx} + (\mathbf{A}_{0t} - \omega^2\mathbf{M}_t)\mathbf{V} = \mathbf{q}_0 \tag{12.3}$$

The vector $\mathbf{q}_0$ is related to the traction vector $\mathbf{q}$ by

$$\mathbf{q} = \mathbf{q}_0\exp(-i\omega t) \tag{12.4}$$

For a plate with vertical boundaries, the general solution of Eq. (12.3) consists of two parts, namely, a particular solution which satisfies Eq. (12.3), and the complementary solution which satisfies the associated homogeneous equation of Eq. (12.3). In the next section, we derive a particular solution for general anisotropic laminated plates.

12.3   Particular Solution

A particular solution for a transversely isotropic half-space has been given by Seale and Kausel (1989), which is an extension of the solution for isotropic layered media given by Waas (1972) and Kausel (1981). The method is termed a seminumerical method (SNM). The formulation of SNM for a particular solution for general anisotropic laminates is as follows. We introduce first the Fourier transformation with respect to the horizontal coordinates x as follows

$$\tilde{\mathbf{V}}_p(k) = \int_{-\infty}^{\infty} \mathbf{V}_p(x)e^{ikx}dx \tag{12.5}$$

where k is the wavenumber and the subscript p indicates the particular solution. The application of the Fourier transform to Eq. (12.3) leads to the following characteristic equations for the plate in the transform domain:

$$[k^2\mathbf{A}_{2t} - ik\mathbf{A}_{1t} + \mathbf{A}_{0t} - \omega^2\mathbf{M}_t]\tilde{\mathbf{V}}_p = \tilde{\mathbf{q}}_0 \tag{12.6}$$

This equation can be rewritten in a more compact form:

$$[A - kB]d = p \tag{12.7}$$

where

$$A = \begin{bmatrix} 0 & I \\ \omega^2 M - A_0 & iA_1 \end{bmatrix} \tag{12.8}$$

$$B = \begin{bmatrix} I & 0 \\ 0 & A_2 \end{bmatrix} \tag{12.9}$$

$$p = \begin{Bmatrix} 0 \\ -\tilde{q}_0 \end{Bmatrix}, \quad d = \begin{Bmatrix} \tilde{V}_p \\ k\tilde{V}_p \end{Bmatrix} \tag{12.10}$$

Applying the analysis of mode technique to Eq. (12.10), we have

$$d = \sum_{m=1}^{2M} \frac{\varphi_m^L P \varphi_m^R}{(k - k_m)B_m} \tag{12.11}$$

where $M = 3(2N + 1)$ is the number of rows of the square matrices A_{it} and M_t, and

$$B_m = \varphi_m^L B \varphi_m^R \tag{12.12}$$

In Eqs. (12.11) and (12.12), k_m, φ_m^L, and φ_m^R are, respectively, the eigenvalues and the left and right eigenvectors obtained from the following characteristic equations:

$$0 = [A - k_m B]\varphi_m^R, \quad 0 = \varphi_m^L[A - k_m B] \tag{12.13}$$

The eigenvectors φ_m^L and φ_m^R obtained from Eq. (12.13) can be written as

$$\varphi_m^L = [\varphi_{m1}^L \quad \varphi_{m2}^L], \quad \varphi_m^R = \begin{Bmatrix} \varphi_{m1}^R \\ \varphi_{m2}^R \end{Bmatrix} \tag{12.14}$$

where φ_{m1}^L, φ_{m2}^L, φ_{m1}^R, and φ_{m2}^R have the dimension of M. Hence we have

$$\tilde{V}_p = -\sum_{m=1}^{2M} \frac{\varphi_{m2}^L \tilde{q} \varphi_{m1}^R}{(k - k_m)B_m} \tag{12.15}$$

If we specialize the spatial variation of the load to that of a line load acting at $x = x_0$, we may write

$$\mathbf{q} = \mathbf{q}_0 \delta(x - x_0) \tag{12.16}$$

where δ is the Dirac delta function. In this case, the Fourier transform of the external force becomes

$$\tilde{\mathbf{q}} = \mathbf{q}_0 e^{ikx_0} \tag{12.17}$$

Hence, Eq. (12.15) reduces to

$$\tilde{\mathbf{V}}_p = -\sum_{m=1}^{2M} \frac{\varphi_{m2}^L \mathbf{q}_0 e^{ikx_0} \varphi_{m1}^R}{(k - k_m) B_m} \tag{12.18}$$

Applying the inverse Fourier transform to Eq. (12.18), the displacement in the spatial domain can be expressed by

$$\mathbf{V}_p = -\frac{1}{2\pi} \int_{-\infty}^{\infty} \sum_{m=1}^{2M} \frac{\varphi_{m2}^L \mathbf{q}_0 e^{ikx_0} \varphi_{m1}^R}{(k - k_m) B_m} e^{-ikx} \, dk \tag{12.19}$$

Notice that Eq. (12.19) is very similar to Eq. (1.43) in Chapter 1, in which the integral can be carried out analytically. A close examination of Eq. (12.19) reveals that φ_{m2}^L, φ_{m1}^R, $\mathbf{q}_0$, and B_m are independent of k. In view of complex k, the integral in Eq. (12.19) is an analytic function of k, except at $2M$ poles ($k = k_m$) in the complex k-plane. An example of distribution of the poles is given in Figure 12.2. The integrations in Eq. (12.19) can be carried out analytically using Cauchy's theorem as done in Chapter 1. To that end we found

$$\mathbf{V}_p = \begin{cases} -i \sum_{m=1}^{M} \dfrac{\varphi_{m2}^{+L} \mathbf{q}_0 \varphi_{m1}^{+R}}{B_m^+} e^{-ik_m^+(x - x_0)}, & \text{for } x \geq x_0 \\[3mm] i \sum_{m=1}^{M} \dfrac{\varphi_{m2}^{-L} \mathbf{q}_0 \varphi_{m1}^{-R}}{B_m^-} e^{-ik_m^-(x - x_0)}, & \text{for } x < x_0 \end{cases} \tag{12.20}$$

where the superscript + denotes variables evaluated at the poles corresponding to waves propagating in the positive x direction (leftward) included by the lower semicircular loop (dashed line in Figure 12.3). These poles satisfy

$$\text{Im}(k_m) < 0$$
$$\text{Re}(k_m) < 0 \quad \text{when } \text{Im}(k_m) = 0 \tag{12.21}$$

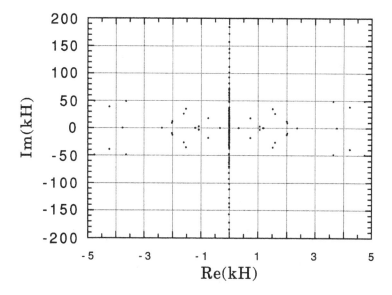

FIGURE 12.2
Poles in the complex wavenumber plane for laminate [C0/G45/G–45]$_s$.

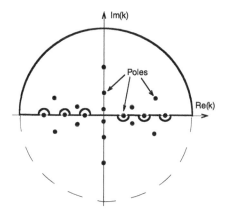

FIGURE 12.3
Integral contours. (From Liu, G. R. and Achenbach, J. D., *ASME Journal of Applied Mechanics*, 62, 607, 1995. With permission.)

The superscript – denotes variables evaluated at poles corresponding to waves propagating in the negative x direction (rightward) included by the upper semicircular loop (solid line in Figure 12.3). These poles satisfy

$$\mathrm{Im}(k_m) > 0$$
$$\mathrm{Re}(k_m) > 0 \quad \text{when } \mathrm{Im}(k_m) = 0 \tag{12.22}$$

Equation (12.20) gives a 2D Green's function for the displacement for anisotropic laminated plates. The Green's function for stresses can be obtained using Eq. (12.20). This Green's function can be employed in the boundary element method (BEM) for solving boundary value problems. If the external force is distributed in the x direction, the solution can be obtained in the form of a superposition integral over $\mathbf{V}_p$.

12.4 General Solution

Consider now a rectangular block of a laminate bounded by its upper and lower surfaces of the laminate and left and right vertical boundaries. The length of the block is denoted by L. The complementary solution of the associated homogeneous equation of Eq. (12.3) can be expressed by superposition of the right eigenvectors φ_m^R, i.e.,

$$\mathbf{V}_c = \sum_{m=1}^{M} C_m \varphi_{m1}^{+R} \exp(ik_m^+(x-L)) + \sum_{m=1}^{M} C_{(M+m)} \varphi_{m1}^{-R} \exp(ik_m^- x) = \mathbf{G}(x)\mathbf{C} \quad (12.23)$$

where the subscript c denotes the complementary solution and the coefficient vector $\mathbf{C}$ can be determined using VBCs. This form of summation of eigenvalues and eigenvectors in Eq. (12.23) ensures stability in numerical calculation, as there will be no exponentially growing terms in matrix $\mathbf{G}$ regardless the value of L. For a semi-infinite laminate defined in $x \geq 0$, we have $L \to \infty$, and the 2nd summation in Eq. (12.23) will vanish since k_m^- satisfies Eq. (12.22). For a semi-infinite laminate defined $x \leq 0$, we have $L \to \infty$, and the 1st summation in Eq. (12.23) will vanish since k_m^- satisfies Eq. (12.21). In these cases, the number of unknown constants in $\mathbf{C}$ is halved.

The sum of the particular and complementary solutions yields the general solution of Eq. (12.3) in the form of

$$\mathbf{V}_g = \mathbf{V}_c + \mathbf{V}_p = \mathbf{G}(z)\mathbf{C} + \mathbf{V}_p \quad (12.24)$$

where the subscript g denotes the general solution. Thus, the coefficient vector $\mathbf{C}$ can be expressed in terms of particular and general solutions at vertical boundaries.

$$\mathbf{C} = \mathbf{G}_b^{-1}(\mathbf{V}_b - \mathbf{V}_{pb}) \quad (12.25)$$

where the subscript b denotes boundaries. Substitution of Eq. (12.25) into Eq. (12.24) gives

$$\mathbf{V}_g = \mathbf{G}(z)\mathbf{G}_b^{-1}(\mathbf{V}_b - \mathbf{V}_{pb}) + \mathbf{V}_p \quad (12.26)$$

Following the formulation given in Section 11.2, stresses acting on a plane of x = constant can be written as

$$\mathbf{R}_x = \mathbf{L}_x^T \mathbf{c} \mathbf{L} \mathbf{U} = \mathbf{L}_x^T \mathbf{c} \mathbf{L}_x \frac{\partial \mathbf{U}}{\partial x} + \mathbf{L}_x^T \mathbf{c} \mathbf{L}_z \frac{\partial \mathbf{U}}{\partial z} = \mathbf{D}_{xx} \frac{\partial \mathbf{U}}{\partial x} + \mathbf{D}_{xz}' \frac{\partial \mathbf{U}}{\partial z} \qquad (12.27)$$

where $\mathbf{D}_{xx}$ has been given in Section 11.2, and $\mathbf{D}_{xz}'$ is found to be

$$\mathbf{D}_{xz}' = \mathbf{L}_x^T \mathbf{c} \mathbf{L}_z = \begin{bmatrix} c_{13} & c_{12} \\ c_{23} & c_{22} \end{bmatrix} \qquad (12.28)$$

The vector of stresses on the x = constant plane at the nodal lines can be obtained as

$$\mathbf{R} = \{ \mathbf{R}_x|_{y=0} \quad \mathbf{R}_x|_{y=h/2} \quad \mathbf{R}_x|_{y=h} \}^T = \mathbf{R}_1 \mathbf{V} + \mathbf{R}_2 \frac{\partial \mathbf{V}}{\partial x} \qquad (12.29)$$

where $\mathbf{R}_1$ and $\mathbf{R}_2$ are given by

$$\mathbf{R}_1 = \frac{1}{h} \begin{bmatrix} -3\mathbf{D}_{xz}' & 4\mathbf{D}_{xz}' & -3\mathbf{D}_{xz}' \\ -\mathbf{D}_{xz}' & 0 & \mathbf{D}_{xz}' \\ \mathbf{D}_{xz}' & -4\mathbf{D}_{xz}' & 3\mathbf{D}_{xz}' \end{bmatrix} \qquad (12.30)$$

$$\mathbf{R}_2 = \begin{bmatrix} \mathbf{D}_{xx} & 0 & 0 \\ 0 & \mathbf{D}_{xx} & 0 \\ 0 & 0 & \mathbf{D}_{xx} \end{bmatrix} \qquad (12.31)$$

Assembling the matrices $\mathbf{R}_1$ and $\mathbf{R}_2$ for all the strip elements, we have the stress vector for all the node lines $\mathbf{R}_t$ as follows:

$$\mathbf{R}_t = \mathbf{R}_{1t} \mathbf{V}_g + \mathbf{R}_{2t} \frac{\partial \mathbf{V}_g}{\partial y} \qquad (12.32)$$

Using Eqs. (12.26) and (12.32), we finally have the stress at the two vertical boundaries (x = constant) of the block:

$$\mathbf{R}_b = \mathbf{K} \mathbf{V}_b + \mathbf{S}_p \qquad (12.33)$$

where

$$\mathbf{R}_b = \begin{Bmatrix} \mathbf{R}_b^L \\ \mathbf{R}_b^R \end{Bmatrix}, \quad \mathbf{V}_b = \begin{Bmatrix} \mathbf{V}_b^L \\ \mathbf{V}_b^R \end{Bmatrix} \qquad (12.34)$$

are the external traction and displacement vectors on the vertical boundaries. The superscripts L and R indicate that the matrices or vectors are evaluated on the left and right vertical boundaries. The stiffness matrix in Eq. (12.33) is given by

$$
\mathbf{K} = \begin{bmatrix} \mathbf{R}_{1t} & 0 \\ 0 & \mathbf{R}_{1t} \end{bmatrix} + \begin{bmatrix} \mathbf{R}_{2t}\dfrac{\partial \mathbf{G}^{L}}{\partial x}\mathbf{G}_{b}^{-1} \\ \mathbf{R}_{2t}\dfrac{\partial \mathbf{G}^{R}}{\partial x}\mathbf{G}_{b}^{-1} \end{bmatrix}
\tag{12.35}
$$

The vector $\mathbf{S}_p$ in Eq. (12.33) is found to be

$$
\mathbf{S}_{p} = \begin{bmatrix} \mathbf{R}_{1t} & 0 \\ 0 & \mathbf{R}_{1t} \end{bmatrix} \begin{Bmatrix} \dfrac{\partial \mathbf{V}_{p}^{L}}{\partial x} \\ \dfrac{\partial \mathbf{V}_{p}^{R}}{\partial x} \end{Bmatrix} - \begin{bmatrix} \mathbf{R}_{2t}\dfrac{\partial \mathbf{G}^{L}}{\partial x}\mathbf{G}_{b}^{-1} \\ \mathbf{R}_{2t}\dfrac{\partial \mathbf{G}^{R}}{\partial x}\mathbf{G}_{b}^{-1} \end{bmatrix} \begin{Bmatrix} \mathbf{V}_{p}^{L} \\ \mathbf{V}_{p}^{R} \end{Bmatrix}
\tag{12.36}
$$

which is called the equivalent external force acting on the vertical boundaries. If the plate has only one boundary on the left or right, the condition at infinity on the x-axis has to be used, and the size of the system given by Eq. (12.33) is reduced by half.

12.5   Application of the SEM to Cracked Laminates

To apply the SEM to a plate that contains cracks, the plate has to be divided into domains in which the SEM is applicable. To illustrate the procedure we consider first an infinite uncracked plate, which is divided into left and right domains by a vertical junction. The SEM is applied to the two domains. For each domain, Eq. (12.33) that gives a relationship between the stresses and displacements on the junction, can be obtained. By assembling the two sets of equations for the two domains, a relationship between the displacement and the traction acting on the junction can be obtained. The boundary conditions on the junction subsequently yield the system equation of the entire laminate, which can be solved using standard routines for the displacement. The constant vector $\mathbf{C}$ and the complete displacement field at any point in the lami-nate can be obtained using Eqs. (12.25) and (12.24). For an uncracked plate, it is not necessary to use the SEM. However, this is a good approach to verify the SEM formulation because the results obtained by the SEM for uncracked plates can be compared with the particular solution. For infinite

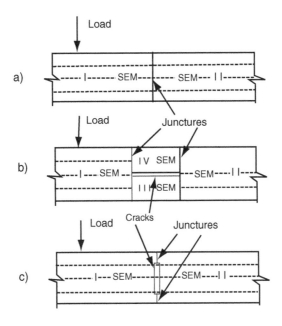

FIGURE 12.4
Division of laminates into subdomains in which the SEM can be applied. (a) An uncracked plate is divide d into left and right domains; (b) a plate with a horizontal crack is divided into four domains; (c) a plate with a vertical crack is divided into two domains. (From Liu, G. R. and Achenbach, J. D., *ASME Journal of Applied Mechanics*, 62, 607, 1995. With permission.)

plates with horizontal or vertical cracks, the plate needs to be divided into sub-domains, to which the SEM can be applied, as shown in Figure 12.4. Detailed formulation and the process of assembling equations for the sub-domains will be given in later sections, when the SEM is applied for the characterization of cracks.

12.6 Solution in the Time Domain

To obtain the solution in the time domain, Fourier superposition will be applied:

$$u_t(t) = \frac{1}{2\pi} \int_{-\infty}^{\infty} u(\omega)F(\omega)e^{i\omega t}\, d\omega \tag{12.37}$$

where $u(\omega)$ is a solution in the frequency domain, while $F(\omega)$ is the Fourier transform of the time dependence of the external force, $f(t)$, acting on the plate. The subscript t indicates that the variable is in the time domain.

The integration of Eq. (12.37) usually has to be carried out numerically. Difficulties with the integration come from singularities of $u(\omega)$ at $\omega = 0$ and at the cutoff frequencies (frequencies at $k = 0$), as discussed by Vasudevan and Mal (1985). The situation here is very similar to that discussed in Section 1.6 for the 1D problem. To overcome these difficulties, an exponential window method (EWM) is employed here. The EWM has been used by Vasudevan and Mal (1985) to calculate the transient response of an elastic plate and by Kausel et al. (1992) to obtain the transient response for damped and undamped vibrating structures. The EWM is also detailed in Section 4.6. Using the EWM, the integration path in Eq. (12.37) can be shifted downward in the complex ω-plane by an arbitrary amount η without changing the results. Hence, Eq. (12.37) can equivalently be written as

$$u_t(t) = \frac{e^{\eta t}}{2\pi} \int_{-\infty}^{\infty} u(\omega - i\eta) F(\omega - i\eta) e^{i\omega t} d\omega \qquad (12.38)$$

where

$$F(\omega - i\eta) = \int_0^{t_d} e^{-\eta t} f(t) e^{-i\omega t} dt \qquad (12.39)$$

in which t_d is the duration of the external force. Using Eq. (12.38) instead of Eq. (12.37), the singularity of $u(\omega)$ at $\omega = 0$ is avoided since $u(\omega - i\eta)$ is well behaved for $-\infty < \omega < \infty$. The displacement in the time domain can therefore be obtained.

12.7 Examples of Scattered Wave Fields

In this section, several examples are used to demonstrate the solution procedure of the SEM for analyzing wave scattering by cracks. Numerical results are presented in the form of normal displacements on the upper surface of the plate. A steel plate, $(C90/G0)_s$, $(C0/G90)_s$, and $(C90/G45/G-45)_s$ laminates are studied. The material properties of the plates are listed in Table 3.1.

In all calculations, the following dimensionless parameters are used:

$$\bar{x} = x/H, \quad \bar{c} = c/c(4, 4), \quad \bar{w} = c(4, 4)w/q_0, \quad \bar{u} = c(4, 4)u/H$$
$$\bar{\omega} = \omega H/c_s, \quad \bar{c} = c/c_s, \quad c_s = \sqrt{c(4, 4)/\rho} \qquad (12.40)$$

where c and ρ are the matrix of material constant and mass density of a reference material. The reference material for a steel plate is the steel plate itself, while the reference material for a hybrid composite plate is C0. In subsequent calculations, $\bar{\omega} = 3.14$ and $L = 4.0H$. The plate is divided into 16 elements vertically in the subsequent calculations.

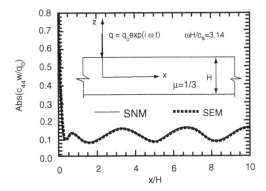

FIGURE 12.5
Displacements on the upper surface of an isotropic plate subjected to a time-harmonic line load. Comparison of the results computed using the seminumerical method (SNM) and the strip element method (SEM). (From Liu, G. R. and Achenbach, J. D., *ASME Journal of Applied Mechanics*, 62, 607, 1995. With permission.)

12.7.1 Response in Frequency Domain

Displacement responses of an isotropic plate with a Poisson's ratio of one-third subjected to a time-harmonic line load have been computed by the SNM and SEM programs. The results are shown in Figure 12.5 for comparison. Sixteen elements were used in both programs. The junction for the SEM is at $x = 4H$. Excellent agreement is observed in Figure 12.5. The convergence of method is also confirmed using different numbers of elements, and it is found that both programs give no significant difference between the results obtained using 4 elements and 16 elements at dimensionless frequency of 3.14, as shown in Figure 12.6. This example has also been solved using the original SEM (Chapter 11), in which the plate is divided into elements in the vertical direction and a viscoelastic nonreflecting boundary is used at the end of the regular elements. Very good agreement has been observed as shown in Figure 12.7.

Displacement responses of a cracked isotropic plate have also been calculated, and the results are shown in Figure 12.8 for a plate with a horizontal crack over $4H \le x \le 5H$ in the midplane of the plate. Additional peaks in the spectrum are observed as compared with that for the uncracked plate. These peaks have also been observed and discussed by Datta et al. (1992) and Karim et al. (1992). We consider the absolute value of the displacement on the plate surfaces. Oscillations appear between the load and the crack because reflections of waves from the crack interact with the incident field.

For a plate $[C90/G0]_s$, the effect of a vertical surface-breaking crack in the upper C90 layer has been calculated. For a time-harmonic load that acts at a distance of $4H$ from the crack, the displacement w on the upper surface of the plate is shown in Figure 12.9, together with the results for the uncracked

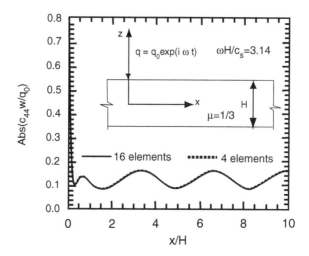

FIGURE 12.6
Convergence of displacement on the upper surface of an isotropic plate subjected to a time-harmonic line load computed using the strip element method (SEM).

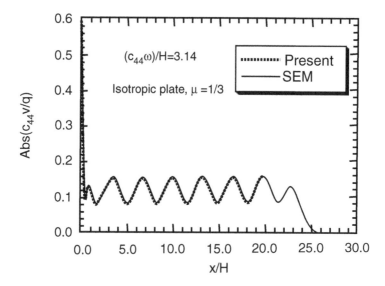

FIGURE 12.7
Displacements on the upper surface of an isotropic plate subjected to a time-harmonic line load. Comparison of the results computed using the present and the original SEM.

plate. The displacement discontinues at $x = 4H$, although this is not visible in the figure. The presence of the crack only slightly affects the solution in the far field. As expected, an oscillation of the absolute value of the displacement between the load point and the crack is observed. From Figure 12.9,

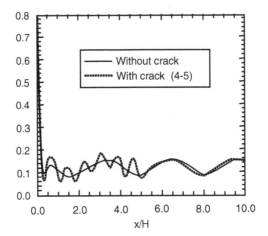

FIGURE 12.8
Dimensionless vertical displacements on the upper surface of an isotropic plate subjected to a time-harmonic line load. Comparison of the results for a perfect plate and plate with a horizontal in the mid plane of the plate from $x = 4H$ to $5H$.

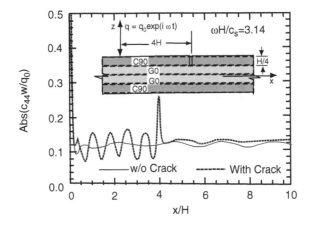

FIGURE 12.9
Displacements on the upper surface of a laminated plate subjected to a time-harmonic line load. Comparison of the results obtained by the SEM for an uncracked and for a plate with a vertical surface-breaking crack. (From Liu, G. R. and Achenbach, J. D., *ASME Journal of Applied Mechanics*, 62, 607, 1995. With permission.)

the position of the crack is obvious. It is also found that the amplitude of the additional oscillation is related to the depth of the crack.

For the plate $[C90/G0]_s$, the effect of an interior vertical crack assumed in the G90 layers has been investigated. Figure 12.10 shows the displacement responses for the plate with and without a crack. An oscillation of the absolute value of the displacement between the load point and the crack is

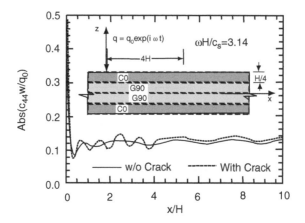

FIGURE 12.10

Displacements on the upper surface of a laminated plate subjected to a time-harmonic line load. Comparison of the results obtained by the SEM for an uncracked and for a plate with an interior vertical crack. (From Liu, G. R. and Achenbach, J. D., *ASME Journal of Applied Mechanics*, 62, 607, 1995. With permission.)

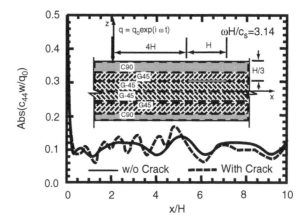

FIGURE 12.11

Displacements on the upper surface of a laminated plate subjected to a time-harmonic line load. Comparison of the results obtained by the SEM for an uncracked and for a plate with a horizontal crack. (From Liu, G. R. and Achenbach, J. D., *ASME Journal of Applied Mechanics*, 62, 607, 1995. With permission.)

observed. The effects of the cracks are smaller than for a surface-breaking crack in the plate. For the subsurface internal crack, the displacement is continuous, but the position of the crack can still be detected.

For a plate [C90/G45/G–45]$_s$, the effect of a delamination between the G45 and G–45 layers has been studied. Figure 12.11 shows the scattered displacement field in comparison with the uncracked plate. Again, a superimposed oscillation is observed between the load and the right crack tip. It is noted that the superimposed oscillation ends at the right crack tip. This is because a crack tip functions as a wave source that emits waves in all directions. Hence, at any observation point on the left of the right crack tip, there are waves moving to the right generated by the load and waves moving to the left generated by the right crack tip. The interference of these waves generates phase changes of the displacement responses, which appear as the additional oscillations in the absolute values of the displacement. On the other hand, at any observation point on the right of the right crack tip, there are only waves moving to the right. This is because all sources in the plate (the load, the left and right crack tips) are located to the left of the observation points. More detailed examination of this phenomenon is given in the next chapter.

12.7.2 Response in Time Domain

A Gaussian modulated sinusoid is used as the time history of the applied load, i.e.,

$$f(t) = \frac{2}{\sigma\sqrt{2\pi}} e^{-(t-t_0)^2/(2\sigma^2)} \sin(\omega_c t), \quad 0 \leq t < t_d \tag{12.41}$$

where t is the time, σ is a parameter that controls the duration of the pulse, t_0 determines the time delay of the pulse, and ω_c is the center angular frequency of the pulse. The time history of the applied load used in this chapter is shown in Figure 12.12. The Fourier transform of the applied load is

$$F(\omega) = \int_0^{t_d} f(t) e^{-i\omega t} dt \tag{12.42}$$

A numerical evaluation of $F(\omega)$ shows that the absolute value of $F(\omega)$ has peak amplitudes at two values of ω and approaches zero rapidly as $|\omega|$ increases.

Responses of a plate excited by a pulsed line load given by Eq. (12.41) can be obtained from Eq. (12.38), which is obtained by introducing the exponential window method (EWM). To check the validity of the results, displacement responses of a perfect isotropic plate with a Poisson's ratio of one-third have been computed by the SEM and a layer element method and the results are shown in Figure 12.13. Sixteen elements are used in both methods. Excellent agreement is observed.

It should be noted that the shifting constant η in Eqs. (12.38) and (12.39) has to be chosen carefully (see Kausel et al., 1992). Various η are tested for our cases, and it is found that a larger η will introduce severe numerical precision loss in the long-time responses. A small η (say, $\eta = 0.1H/c_s$) is

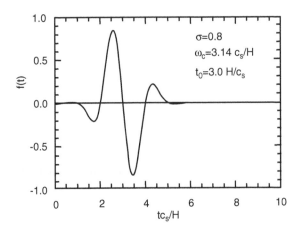

FIGURE 12.12
Time history of the applied load. (From Liu, G. R. and Achenbach, J. D., *ASME Journal of Applied Mechanics*, 62, 607, 1995. With permission.)

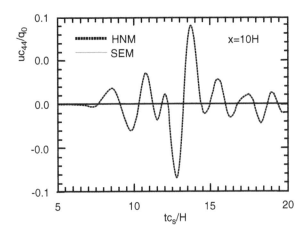

FIGURE 12.13
Transient displacement response on the upper surface of an isotropic plate. Comparison of the hybrid numerical method (HNM) and SEM. (From Liu, G. R. and Achenbach, J. D., *ASME Journal of Applied Mechanics*, 62, 607, 1995. With permission.)

enough to avoid the singularity of $u(\omega)$ at $\omega = 0$ and the rapid changes at the cutoff frequencies.

Figure 12.14 shows the scattered displacement field on the surface of the plate $[C90/G45/G{-}45]_s$ with and without a horizontal crack. The observation point is at $x = 10H$. Significant differences are observed when $tc_s/H > 12$.

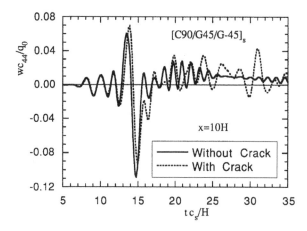

FIGURE 12.14
Transient displacement response on the upper surface of a laminated plate. Comparison of the results obtained by the SEM for an uncracked plate and for a plate with a horizontal crack. (From Liu, G. R. and Achenbach, J. D., *ASME Journal of Applied Mechanics*, 62, 607, 1995. With permission.)

12.8 Characterization of Horizontal Cracks

12.8.1 SEM Formulation

Consider an infinite laminate containing a horizontal crack, as shown in Figure 12.15. The material of the laminate can be isotropic or anisotropic, and inhomogeneous in the thickness direction such as in composite laminates. The length of the horizontal crack is denoted by a_c, and the depth of the crack from the upper surface of the laminate is denoted by d_c. This section provides a possible technique to detect the crack and determine both the length and depth of the crack by exciting the laminate with harmonic loads and analyzing the scattered wave field by cracks in laminate.

The laminate is first excited by a harmonic loading that does not vary with the y-direction. Hence, the problem is two dimensional and is in the frequency domain. The load acting on the surface of the laminate is given by

$$F(x) = q_0 f(x) \exp(-i\omega t) \qquad (12.43)$$

where q_0 is a constant, and $f(x)$ represents the distribution of the load along the x-direction as shown in Figure 12.15. In this chapter, $f(x)$ is assumed as

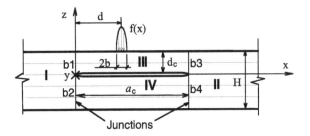

FIGURE 12.15
A distributed load scanning on the surface of the plate over the crack.

follows:

$$f(x) = \begin{cases} \exp[-\{(x+d)/\alpha\}^2], & |x-d| \leq b \\ 0, & |x-d| \geq b \end{cases} \qquad (12.44)$$

Equation (12.44) is a simulation of a load generated by a laser beam, where d is the offset of the center of the load from the origin (left crack tip), and b is the half-width of the applied load, while α is a parameter that controls the distribution of the load. In this chapter, b and α are chosen to be $0.1H$ and $0.05H$, respectively.

The cracked laminate is divided into four subdomains as denoted by Roman numerals in Figure 12.15. Domain **I** is bounded by boundaries $b1$, $b2$, and the upper and lower surfaces of the laminate. Domain **II** is bounded by boundaries $b3$, $b4$, and the upper and lower surfaces of the laminate. Domain **III** is bounded by boundary $b1$, $b3$, upper surface of the laminate, and the upper surface of the crack. Domain **IV** is bounded by boundaries $b2$ and $b4$, lower surfaces of the crack, and lower surface of the laminate. In each domain, the SEM is used to obtain sets of equations that give the relationship between the displacements and stresses at the nodal points on the vertical boundaries of the subdomains. For domain **I**, we can obtain (see Eq. (12.30))

$$\mathbf{R}_b^I = \mathbf{K}^I \mathbf{V}_b^I + \mathbf{S}_b^I \qquad (12.45)$$

where the superscript of the Roman numeral indicates the domain that the equation governs, while **K** is the stiffness matrix of the domain. Vectors **R**, **V**, and **S** are, respectively, the external force vector acting on the vertical boundary, displacement vector on the vertical boundary, and the equivalent stress vector caused by the external force acting on the horizontal boundaries of the domain (see, Liu and Achenbach, 1994). Equation (12.45) can be further divided into submatrices as follows:

$$
\begin{Bmatrix} \mathbf{R}_{b1}^{I} \\ \mathbf{R}_{bL}^{I} \\ \mathbf{R}_{b2}^{I} \end{Bmatrix} = \begin{bmatrix} \mathbf{K}_{11}^{I} & \mathbf{K}_{12}^{I} & \mathbf{K}_{13}^{I} \\ \mathbf{K}_{21}^{I} & \mathbf{K}_{22}^{I} & \mathbf{K}_{23}^{I} \\ \mathbf{K}_{31}^{I} & \mathbf{K}_{32}^{I} & \mathbf{K}_{33}^{I} \end{bmatrix} \begin{bmatrix} \mathbf{V}_{b1}^{I} \\ \mathbf{V}_{bL}^{I} \\ \mathbf{V}_{b2}^{I} \end{bmatrix} + \begin{Bmatrix} \mathbf{S}_{b1}^{I} \\ \mathbf{S}_{bL}^{I} \\ \mathbf{S}_{b2}^{I} \end{Bmatrix}
\tag{12.46}
$$

where the subscripts $b1$ and $b2$ stand, respectively, for the vectors on the upper and lower right vertical boundary of the domain excluding the left crack tip. The subscript bL stands for the vectors at the left crack tip.

Similarly, for domain **II** we obtain

$$
\begin{Bmatrix} \mathbf{R}_{b3}^{II} \\ \mathbf{R}_{bR}^{II} \\ \mathbf{R}_{b4}^{II} \end{Bmatrix} = \begin{bmatrix} \mathbf{K}_{11}^{II} & \mathbf{K}_{12}^{II} & \mathbf{K}_{13}^{II} \\ \mathbf{K}_{21}^{II} & \mathbf{K}_{22}^{II} & \mathbf{K}_{23}^{II} \\ \mathbf{K}_{31}^{II} & \mathbf{K}_{32}^{II} & \mathbf{K}_{33}^{II} \end{bmatrix} \begin{bmatrix} \mathbf{V}_{b3}^{II} \\ \mathbf{V}_{bR}^{II} \\ \mathbf{V}_{b4}^{II} \end{bmatrix} + \begin{Bmatrix} \mathbf{S}_{b3}^{II} \\ \mathbf{S}_{bR}^{II} \\ \mathbf{S}_{b4}^{II} \end{Bmatrix}
\tag{12.47}
$$

where the subscripts $b3$ and $b4$ stand, respectively, for the vectors on the upper and lower left vertical boundary of domain **II** excluding the right crack tip. The subscript bR stands for the vectors at the right crack tip. For domains **III** and **IV**, we have

$$
\begin{Bmatrix} \mathbf{R}_{b1}^{III} \\ \mathbf{R}_{bL}^{III} \\ \mathbf{R}_{b3}^{III} \\ \mathbf{R}_{bR}^{III} \end{Bmatrix} = \begin{bmatrix} \mathbf{K}_{11}^{III} & \mathbf{K}_{12}^{III} & \mathbf{K}_{13}^{III} & \mathbf{K}_{14}^{III} \\ \mathbf{K}_{21}^{III} & \mathbf{K}_{22}^{III} & \mathbf{K}_{23}^{III} & \mathbf{K}_{24}^{III} \\ \mathbf{K}_{31}^{III} & \mathbf{K}_{32}^{III} & \mathbf{K}_{33}^{III} & \mathbf{K}_{34}^{III} \\ \mathbf{K}_{41}^{III} & \mathbf{K}_{42}^{III} & \mathbf{K}_{43}^{III} & \mathbf{K}_{44}^{III} \end{bmatrix} \begin{bmatrix} \mathbf{V}_{b1}^{III} \\ \mathbf{V}_{bL}^{III} \\ \mathbf{V}_{b3}^{III} \\ \mathbf{V}_{bR}^{III} \end{bmatrix} + \begin{Bmatrix} \mathbf{S}_{b1}^{III} \\ \mathbf{S}_{bL}^{III} \\ \mathbf{S}_{b3}^{III} \\ \mathbf{S}_{bR}^{III} \end{Bmatrix}
\tag{12.48}
$$

and

$$
\begin{Bmatrix} \mathbf{R}_{bL}^{IV} \\ \mathbf{R}_{b2}^{IV} \\ \mathbf{R}_{bR}^{IV} \\ \mathbf{R}_{b4}^{IV} \end{Bmatrix} = \begin{bmatrix} \mathbf{K}_{11}^{IV} & \mathbf{K}_{12}^{IV} & \mathbf{K}_{13}^{IV} & \mathbf{K}_{14}^{IV} \\ \mathbf{K}_{21}^{IV} & \mathbf{K}_{22}^{IV} & \mathbf{K}_{23}^{IV} & \mathbf{K}_{24}^{IV} \\ \mathbf{K}_{31}^{IV} & \mathbf{K}_{32}^{IV} & \mathbf{K}_{33}^{IV} & \mathbf{K}_{34}^{IV} \\ \mathbf{K}_{41}^{IV} & \mathbf{K}_{42}^{IV} & \mathbf{K}_{43}^{IV} & \mathbf{K}_{44}^{IV} \end{bmatrix} \begin{bmatrix} \mathbf{V}_{bL}^{IV} \\ \mathbf{V}_{b2}^{IV} \\ \mathbf{V}_{bR}^{IV} \\ \mathbf{V}_{b4}^{IV} \end{bmatrix} + \begin{Bmatrix} \mathbf{S}_{bL}^{IV} \\ \mathbf{S}_{b2}^{IV} \\ \mathbf{S}_{bR}^{IV} \\ \mathbf{S}_{b4}^{IV} \end{Bmatrix}
\tag{12.49}
$$

Assemble all sets of the equations for all the domains by using the following continuity conditions at the junctions that divide the laminate into subdomains:

$$\mathbf{R}_{b1} = \mathbf{R}_{b1}^{I} - \mathbf{R}_{b1}^{III}, \quad \mathbf{V}_{b1} = \mathbf{V}_{b1}^{I} = \mathbf{V}_{b1}^{III} \tag{12.50}$$

$$\mathbf{R}_{bL} = \mathbf{R}_{bL}^{I} - \mathbf{R}_{bL}^{III} - \mathbf{R}_{bL}^{IV}, \quad \mathbf{V}_{bL} = \mathbf{V}_{bL}^{I} = \mathbf{V}_{bL}^{III} = \mathbf{V}_{bL}^{IV} \tag{12.51}$$

$$\mathbf{R}_{b2} = \mathbf{R}_{b2}^{I} - \mathbf{R}_{b2}^{IV}, \quad \mathbf{V}_{b2} = \mathbf{V}_{b2}^{I} = \mathbf{V}_{b2}^{IV} \tag{12.52}$$

$$\mathbf{R}_{b3} = \mathbf{R}_{b3}^{III} - \mathbf{R}_{b3}^{II}, \quad \mathbf{V}_{b3} = \mathbf{V}_{b3}^{II} = \mathbf{V}_{b3}^{III} \tag{12.53}$$

$$\mathbf{R}_{bR} = \mathbf{R}_{bR}^{III} + \mathbf{R}_{bR}^{IV} - \mathbf{R}_{bR}^{II}, \quad \mathbf{V}_{bR} = \mathbf{V}_{bR}^{II} = \mathbf{V}_{bR}^{III} = \mathbf{V}_{bR}^{IV} \tag{12.54}$$

$$\mathbf{R}_{b4} = \mathbf{R}_{b4}^{IV} - \mathbf{R}_{b4}^{II}, \quad \mathbf{V}_{b4} = \mathbf{V}_{b4}^{II} = \mathbf{V}_{b4}^{IV} \tag{12.55}$$

A set of equations that gives the relationship between the displacements and stresses at all nodal points on the junctions can then be obtained. The equations are

$$\mathbf{R}_{J} = \mathbf{K}\mathbf{V}_{J} + \mathbf{S}_{J} \tag{12.56}$$

where $\mathbf{R}_{J}$ is the external force vector acting on the junctions given by

$$\mathbf{R}_{J} = \{\mathbf{R}_{b1} \quad \mathbf{R}_{bL} \quad \mathbf{R}_{b2} \quad \mathbf{R}_{b3} \quad \mathbf{R}_{bR} \quad \mathbf{R}_{b4}\}^{T} \tag{12.57}$$

and $\mathbf{V}_{J}$ is the displacement vector acting on the junctions as follows:

$$\mathbf{V}_{J} = \{\mathbf{V}_{b1} \quad \mathbf{V}_{bL} \quad \mathbf{V}_{b2} \quad \mathbf{V}_{b3} \quad \mathbf{V}_{bR} \quad \mathbf{V}_{b4}\}^{T} \tag{12.58}$$

In Eq. (12.56), $\mathbf{S}_{J}$ is given by

$$\mathbf{S}_{J} = \begin{Bmatrix} \mathbf{S}_{b1}^{I} - \mathbf{S}_{b1}^{III} \\ \mathbf{S}_{bL}^{I} - \mathbf{S}_{bL}^{III} - \mathbf{S}_{bL}^{IV} \\ \mathbf{S}_{b2}^{I} - \mathbf{S}_{b2}^{IV} \\ -\mathbf{S}_{b3}^{II} + \mathbf{S}_{b3}^{III} \\ -\mathbf{S}_{bR}^{II} + \mathbf{S}_{bR}^{III} + \mathbf{S}_{bR}^{IV} \\ -\mathbf{S}_{b4}^{II} + \mathbf{S}_{b4}^{IV} \end{Bmatrix} \tag{12.59}$$

In Eq. (12.56), $\mathbf{K}$ is the stiffness matrix for the cracked laminate given by

$$
\mathbf{K} =
\begin{bmatrix}
\mathbf{K}_{11}^{I} - \mathbf{K}_{11}^{III} & \mathbf{K}_{12}^{I} - \mathbf{K}_{12}^{III} & \mathbf{K}_{13}^{I} & -\mathbf{K}_{13}^{III} & -\mathbf{K}_{14}^{III} & 0 \\
\mathbf{K}_{21}^{I} - \mathbf{K}_{21}^{III} & \mathbf{K}_{22}^{I} - \mathbf{K}_{22}^{III} - \mathbf{K}_{11}^{IV} & \mathbf{K}_{23}^{I} - \mathbf{K}_{12}^{IV} & -\mathbf{K}_{23}^{III} & -\mathbf{K}_{24}^{III} - \mathbf{K}_{13}^{IV} & -\mathbf{K}_{14}^{IV} \\
\mathbf{K}_{31}^{I} & \mathbf{K}_{32}^{I} - \mathbf{K}_{21}^{IV} & \mathbf{K}_{33}^{I} - \mathbf{K}_{22}^{IV} & 0 & -\mathbf{K}_{23}^{IV} & -\mathbf{K}_{24}^{IV} \\
\mathbf{K}_{31}^{III} & \mathbf{K}_{32}^{III} & 0 & \mathbf{K}_{33}^{III} - \mathbf{K}_{11}^{II} & \mathbf{K}_{34}^{III} - \mathbf{K}_{12}^{II} & -\mathbf{K}_{13}^{II} \\
\mathbf{K}_{41}^{III} & \mathbf{K}_{42}^{III} + \mathbf{K}_{31}^{IV} & \mathbf{K}_{32}^{IV} & \mathbf{K}_{43}^{III} - \mathbf{K}_{21}^{II} & \mathbf{K}_{44}^{III} - \mathbf{K}_{22}^{II} + \mathbf{K}_{33}^{IV} & \mathbf{K}_{34}^{IV} - \mathbf{K}_{23}^{II} \\
0 & \mathbf{K}_{41}^{IV} & \mathbf{K}_{42}^{IV} & -\mathbf{K}_{31}^{II} & \mathbf{K}_{43}^{IV} - \mathbf{K}_{32}^{II} & \mathbf{K}_{44}^{IV} - \mathbf{K}_{33}^{II}
\end{bmatrix}
$$

$$(12.60)$$

Solving Eq. (12.56), the displacements at the junctions can be obtained, and then the whole displacement field can finally be obtained.

12.8.2 Technique for Crack Detection

The ultimate goal of the analytical modeling is to find the relationship between the characteristics of cracks in the laminate and the responses of the laminate to the imposed loads. A technique to find that relationship is to move the loading along the surface of the laminate (change the value of d in Eq. (12.44)), and, in the meantime, pick up the resulting displacement in the z-direction at the central point of the distributed load ($x = d$); then plot the responses of the laminate while the load scans over the laminate. For a laminate without cracks, the scanning result is a horizontal line because the responses of the uncracked laminate at the central point of load are independent of the location of the load, as infinitely long laminates are considered in this chapter. However, for a cracked laminate, the scanning result will no longer be a constant, and the responses can be expected to have a significant change when the load passes over the crack tips. This provides a way to determine the length of the crack in the laminate. Moreover, the responses in the region between the two crack tips can be expected to contain information related to the thickness of the domain **III** (see Figure 12.15) and the crack depth. By investigating these responses, the depth of the crack can also be found. The following section will show how this technique works by numerical experiments.

12.8.3 Detection of a Crack in an Isotropic Plate

Numerical experiments are made for a homogeneous isotropic plate with a Poisson's ratio of $\mu = 1/3$ and a hybrid laminate denoted by $[C90/G45/G\text{--}45]_s$ with horizontal cracks. Material constants of carbon/epoxy and glass/epoxy can be found in Liu and Tani et al. (1991).

In this section, the following additional dimensionless parameters are used:

$$\bar{a}_c = a_c/H, \quad \bar{d}_c = d_c/H \tag{12.61}$$

For dimensionless parameters defined in (12.40), C90 is chosen as the reference material [C90/G45/G−45]$_s$ laminate.

Crack Length

The responses of the plate subjected to the harmonic load are dependent on the frequencies of the load. The frequency dependency of the scanning result of a cracked plate is first investigated to find better frequency of crack characterization. Figures 12.16–12.18 show the scanning results of the absolute

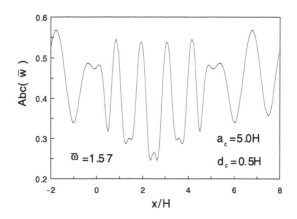

FIGURE 12.16
Scanning results of the absolute value of the vertical displacement amplitude for an isotropic plate ($\mu = 1/3$) with a horizontal crack. The left crack tip of the crack is at the origin.

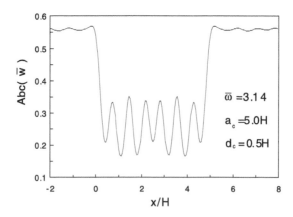

FIGURE 12.17
Scanning results of the absolute value of the vertical displacement amplitude for an isotropic plate ($\mu = 1/3$) with a horizontal crack. The left crack tip of the crack is at the origin.

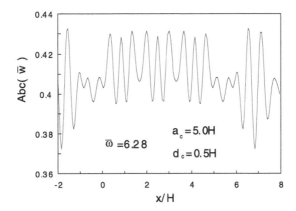

FIGURE 12.18
Scanning results of the absolute value of the vertical displacement amplitude for an isotropic plate ($\mu = 1/3$) with a horizontal crack. The left crack tip of the crack is at the origin.

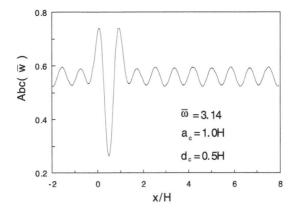

FIGURE 12.19
Scanning results of the absolute value of the vertical displacement amplitude for an isotropic plate ($\mu = 1/3$) with a horizontal crack. The left crack tip of the crack is at the origin.

value of displacement amplitude of the plate excited by the load as given by Eq. (12.1). The results are obtained by using the SEM, and the frequencies of the load are $1.57\bar{\omega}$ (Figure 12.16), $3.14\bar{\omega}$ (Figure 12.17), and $6.28\bar{\omega}$ (Figure 12.18), whose corresponding wavelengths of shear waves are $4.0H$, $2.0H$, and $1.0H$, respectively. From those figures note that the crack length can be approximately determined from the pattern change of the results when the load scans over the crack. Frequency of $3.14\bar{\omega}$ is the best to determine the crack length because the results between the two crack tips are significantly different from the results in the uncracked region. Hence, the following results are obtained by using a harmonic load with a frequency of $3.14\bar{\omega}$.

Scanning results for other crack lengths, $a_c = 1.0H$, $2.0H$, and $3.0H$, are shown in Figures 12.19–12.21. In all the cases, the positions of the cracks can

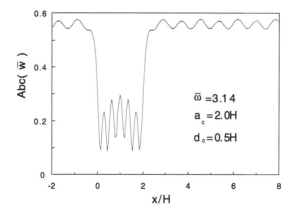

FIGURE 12.20
Scanning results of the absolute value of the vertical displacement amplitude for an isotropic plate ($\mu = 1/3$) with a horizontal crack. The left crack tip of the crack is at the origin.

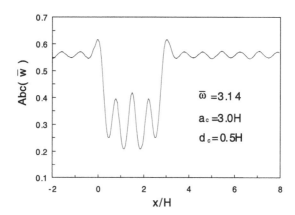

FIGURE 12.21
Scanning results of the absolute value of the vertical displacement amplitude for an isotropic plate ($\mu = 1/3$) with a horizontal crack. The left crack tip of the crack is at the origin.

be determined approximately from the scanning result. Although the exact crack length is difficult to determine from the scanning result, clear images of the cracks can be obtained.

Depth of Crack
In Figure 12.17, an important fact can be observed. The spacing of the peaks of the oscillations in the region of $0 < x < a_c$ are about the same. Comparing Figures 12.17 and 12.21, it is found that the spacings of the peaks of the oscillations in $0 < x < a_c$ for $a_c = 3.0H$ are almost the same as for $a_c = 5.0H$. This fact indicates that the peak spacing of the oscillation is less dependent on the crack length.

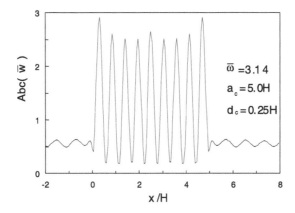

FIGURE 12.22
Scanning results of the absolute value of the vertical displacement amplitude for an isotropic plate ($\mu = 1/3$) with a horizontal crack. The left crack tip of the crack is at the origin.

Figure 12.22 shows the results of changing the depth of the crack to $0.25H$ for the case of $a_c = 5.0H$: Again the peak spacings of the oscillation in the region of $0 < x < a_c$ are almost the same. However, the peak spacing for the $d_c = 0.25H$ (Figure 12.22) is much smaller than for $d_c = 0.5H$ (Figure 12.17). This is a very important fact because it indicates that the depth of the crack is related to the peak spacing of the oscillations in the region of $0 < x < a_c$.

To find the relationship between the crack depth and the peak spacing of the oscillations, we investigated the scanning results of the plate using the following procedure.

1. Calculate the scanning results for plates with various crack depths and lengths, and output the results over the region of $0 < x < a_c$.

2. Transfer the results to wavenumber domain:

$$\tilde{w}(k) = \int_0^{a_c} \overline{w}(x) \exp(-ikx) dx \qquad (12.62)$$

where the k is the wavenumber. The spectra of the results are shown in Figures 12.23–12.26. Note that the wavenumber, at which the response has the maximum value, is related to the peak spacings of the oscillation in $0 < x < a_c$. In this chapter, we call this wavenumber the *central wavenumber*, and denoted it by k_{max}. Comparing Figure 12.23 to Figure 12.26, it can be found that the central wavenumbers are almost the same for the same crack depths (Figures 12.23 and 12.24; Figures 12.25 and 12.26) and are significantly different for different crack depths (Figures 12.23 and 12.25; Figures 12.24 and 12.26).

3. Select the central wavenumbers, k_{max}, and plot them against the crack depths, d_c; the final results are shown in Figure 12.27.

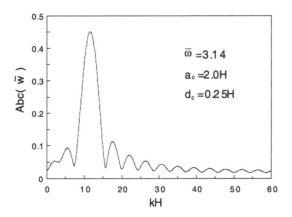

FIGURE 12.23
Spectrum of scanning results for an isotropic plate ($\mu = 1/3$) with a horizontal crack.

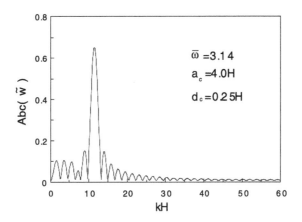

FIGURE 12.24
Spectrum of scanning results for an isotropic plate ($\mu = 1/3$) with a horizontal crack.

From Figure 12.27, the crack depth is easily found from the central wave-number of the scanning results. It is also noted that the $k_{max} - d_c$ curves are less dependent on the crack length for the case of $a_c \geq 2.0H$. Hence the crack depth can be obtained quite accurately even though the crack length obtained is an approximate value. Using polynomial regression of degree 3, the relationship between the crack depth and the central wavenumber can be obtained as follows:

$$\bar{d}_c = 5.0279 - 0.97988p + 0.067041p^2 - 0.0015653p^3 \qquad (12.63)$$

where $p = k_{max}H$. Equation (12.63) is valid for $8.86 \leq p \leq 15.3$, $0.125H \leq d_c \leq 0.5H$, $2.0H \leq a_c \leq 5.0H$. The reason to limit the usage of Eq. (12.63) is that the

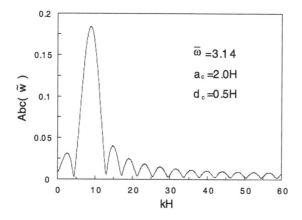

FIGURE 12.25
Spectrum of scanning results for an isotropic plate ($\mu = 1/3$) with a horizontal crack.

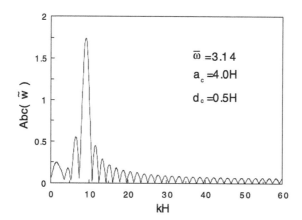

FIGURE 12.26
Spectrum of scanning results for an isotropic plate ($\mu = 1/3$) with a horizontal crack.

samples for obtaining it have been taken in these regions. To expand the usage of Eq. (12.63), further investigation has to be made. We have used Eq. (12.63) to predict the crack depth of a cracked plate with crack length of $10.0H$ and crack depth of $0.25H$ by the following procedure:

1. The scanning results in the region of $0.0 < x < a_c$ are computed for the cracked plate with crack length of $10.0H$ and crack depth of $0.25H$.

2. The results in the wavenumber domain are then calculated using Eq. (12.62), and the central wavenumber obtained to be $11.3343/H$.

3. The predicted depth from Eq. (12.63) is $0.2478H$, compared to $0.25H$ as given. Obviously, this is a very good prediction.

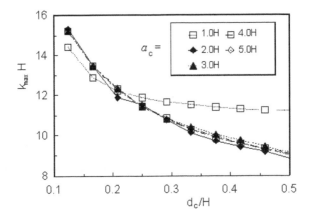

FIGURE 12.27
Depth dependency of the central wavenumber on the depth of the crack for an isotropic plate ($\mu = 1/3$).

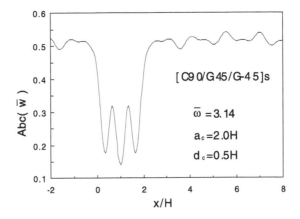

FIGURE 12.28
Scanning results of the absolute value of the vertical displacement amplitude for an anisotropic laminated plate with a horizontal crack. The left crack tip of the crack is at the origin.

12.8.4   Detection of a Crack in an Anisotropic Laminate

Crack Length
Figures 12.28 and 12.29 show the scanning results of a hybrid laminate [C90/G+45/G–45]$_s$ for crack lengths of 2.0H and 4.0H. Clear images of crack length in the laminates can be obtained.

Depth of the Crack
Following the same procedure described previously, the $k_{max} - d_c$ curves for the [C90/G + 45/G–45]$_s$ laminate can be obtained and is shown in Figure 12.30.

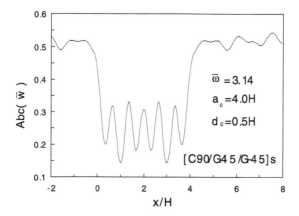

FIGURE 12.29
Scanning results of the absolute value of the vertical displacement amplitude for an anisotropic laminated plate with a horizontal crack. The left crack tip of the crack is at the origin.

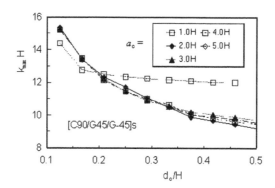

FIGURE 12.30
Depth dependency of the central wavenumber on the depth of the crack for a hybrid laminated plate.

Using polynomial regression of degree 3, a formulation of the relationship between the crack depth and the central wavenumber can be obtained as follows:

$$\bar{d}_c = 6.1914 - 1.2144p + 0.082547p^2 - 0.0019016p^3 \qquad (12.64)$$

where $p = k_{max}H$. Equation (12.64) is valid for $9.19 < p < 15.4$, $0.125H < d_c < 0.5H$, $2.0H < a_c < 5.0H$.

12.9 Characterization of a Vertical Surface-Breaking Crack

12.9.1 SEM Formulation

An infinite laminate with a finite thickness H and a vertical crack is considered in this section. It is assumed that the laminate occupies the region $-\infty \leq (x, z) \leq \infty$, $-H/2 \leq y \leq H/2$. The material of the laminate can be either isotropic or anisotropic. The elastic constants and density of the material are denoted by c_{ij} ($i, j = 1, 2, 6$) and ρ. To simplify the problem, it is assumed that the crack is throughout in the z-direction. We consider a vertical surface-breaking crack, as shown in Figure 12.31. The length of the crack is denoted by a_c, and the distance from the crack to the load is denoted by d_c.

The harmonic line load applied to the laminate is given by Eq. (12.1). It is assumed that the load does not vary in the z-direction. The SEM is first used to investigate the scattering of wave fields in laminate by a vertical crack; a numerical procedure is then presented for detecting the crack. In the numerical examples, we define the following dimensionless parameters:

$$\bar{x} = x/H, \quad \bar{w} = c_{66}w, \quad \bar{\omega} = \omega H/v_s,$$

$$v_s = \sqrt{c_{66}/\rho}, \quad c_{66} = \mathbf{c}(3, 3), \quad \bar{a}_c = a_c/H \tag{12.65}$$

where w is the displacement in y direction, and $\mathbf{c}$, ρ, and v_s are the material constants matrix, density, and shear wave velocity of the material, and a_c is the length of the vertical crack.

Using the SEM, the laminate with a vertical surface-breaking crack is divided into two subdomains as denoted in Figure 12.31. The laminate is separated into two subdomains by the crack and the vertical junction b. In each subdomain, the SEM is used to obtain sets of equations that give the relationship between the displacements and stresses at the nodal points on the vertical boundaries of the subdomains. For domain **I**, we can obtain

$$\mathbf{R}_b^I = \mathbf{K}^I \mathbf{V}_b^I + \mathbf{S}_b^I \tag{12.66}$$

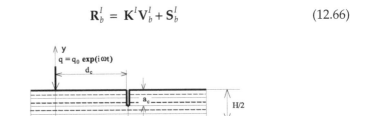

FIGURE 12.31
Division of a plate with a vertical surface-breaking crack into domains in which the SEM can be applied.

where the superscript of the Roman numeral indicates the domain that the equation governs, while **K** is the stiffness matrix of the domain. Vectors **R**, **V**, and **S** are, respectively, the stress vector, displacement vector, and the equivalent external force vector on the vertical boundary of the domain. Equation (12.66) can be further divided into submatrices as follows:

$$
\begin{Bmatrix} \mathbf{R}_{cL}^{I} \\ \mathbf{R}_{b}^{I} \end{Bmatrix} = \begin{bmatrix} \mathbf{K}_{11}^{I} & \mathbf{K}_{12}^{I} \\ \mathbf{K}_{21}^{I} & \mathbf{K}_{22}^{I} \end{bmatrix} \begin{Bmatrix} \mathbf{V}_{cL}^{I} \\ \mathbf{V}_{b}^{I} \end{Bmatrix} + \begin{Bmatrix} \mathbf{S}_{cL}^{I} \\ \mathbf{S}_{b}^{I} \end{Bmatrix}
\tag{12.67}
$$

where the subscript cL stands for the vectors on the left surface of the crack excluding the crack tip. The subscript b stands for the vectors on the vertical boundary, the junction b, including the crack tip.

Similarly, for domain **II** we obtain

$$
\begin{Bmatrix} \mathbf{R}_{cR}^{II} \\ \mathbf{R}_{b}^{II} \end{Bmatrix} = \begin{bmatrix} \mathbf{K}_{11}^{II} & \mathbf{K}_{12}^{II} \\ \mathbf{K}_{21}^{II} & \mathbf{K}_{22}^{II} \end{bmatrix} \begin{Bmatrix} \mathbf{V}_{cR}^{II} \\ \mathbf{V}_{b}^{II} \end{Bmatrix} + \begin{Bmatrix} \mathbf{S}_{cR}^{II} \\ \mathbf{S}_{b}^{II} \end{Bmatrix}
\tag{12.68}
$$

where the subscript cR stands for the vectors on the right surface of the crack. Because the surfaces of the crack are free, we have

$$
\mathbf{R}_{cL}^{I} = 0
\tag{12.69}
$$

$$
\mathbf{R}_{cR}^{II} = 0
\tag{12.70}
$$

Assembling all the sets of the equations for all the domains by using the boundary conditions and continuity conditions at the junction, b, we obtain the following constraint equations:

$$
\mathbf{R}_{c} = \mathbf{R}_{cL}^{I} = \mathbf{R}_{cR}^{II} = 0
\tag{12.71}
$$

$$
\mathbf{R}_{b} = \mathbf{R}_{b}^{I} - \mathbf{R}_{b}^{II}, \quad \mathbf{V}_{b} = \mathbf{V}_{b}^{I} = \mathbf{V}_{b}^{II}
\tag{12.72}
$$

A set of equations that gives the relationship between the displacements and stresses at whole nodal points on the junction can be obtained from

$$
\mathbf{R}_{J} = \mathbf{K}\mathbf{V}_{J} + \mathbf{S}_{J}
\tag{12.73}
$$

where $\mathbf{R}_{J}$ is the external force vector acting on the junction and the crack surfaces given by

$$
\mathbf{R}_{J} = \{ \mathbf{R}_{cL}^{I} \quad \mathbf{R}_{cR}^{II} \quad \mathbf{R}_{b} \}^{T}
\tag{12.74}
$$

and $\mathbf{V}_J$ is the displacement vector on the junction and the crack surfaces as follows:

$$\mathbf{V}_J = \{\mathbf{V}_{cL}^I \quad \mathbf{V}_{cR}^{II} \quad \mathbf{V}_b\}^T \tag{12.75}$$

The vector $\mathbf{S}_J$ is given by

$$\mathbf{S}_J = \left\{ \begin{array}{c} \mathbf{S}_{cL}^I \\ \mathbf{S}_{cR}^{II} \\ \mathbf{S}_b^I - \mathbf{S}_b^{II} \end{array} \right\} \tag{12.76}$$

and $\mathbf{K}$ is the stiffness matrix for the cracked laminate given by

$$\mathbf{K} = \begin{bmatrix} \mathbf{K}_{11}^I & 0 & \mathbf{K}_{12}^I \\ 0 & \mathbf{K}_{11}^{II} & \mathbf{K}_{12}^{II} \\ \mathbf{K}_{21}^I & -\mathbf{K}_{21}^{II} & \mathbf{K}_{22}^I - \mathbf{K}_{22}^{II} \end{bmatrix} \tag{12.77}$$

Solving Eq. (12.72), the displacements at the junction and the surface of the crack can be obtained, and then the whole displacement field can finally be computed.

12.9.2 Technique for Crack Detection

There are two parameters, the location and crack length, to be characterized for a vertical surface-breaking crack. The ultimate goal is to find the relationship between these two parameters and the responses of the laminate to the imposed loads. To achieve this, the displacement field in an isotropic plate scattered by a vertical surface-breaking crack is investigated using the SEM.

Figure 12.32 shows a comparison of the absolute values of the displacement distributions for an isotropic plate with and without a crack, where the crack length $a_c = 0.5H$. The plate is loaded by a time-harmonic load at a distance of $4H$ away from the crack. It is found that the displacement is discontinuous at the location of the crack. Thus, it can be concluded that the crack position can be determined from the distribution of the displacement on the surface of the plate; this conclusion can be drawn from Figures 12.33 and 12.34. However, determining the crack length requires further detailed investigation, discussed below.

From Figure 12.32, it is found that there are additional oscillations on the curve of the absolute values of the displacement in the region between the load and the crack. It is also found that the longer the crack, the larger the amplitudes of the additional oscillations. This is because in the region between the load and the crack there are waves in two directions. One is coming from

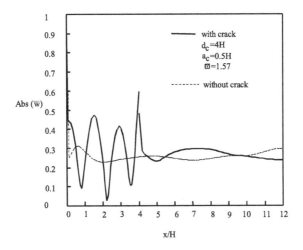

FIGURE 12.32
Displacement responses on the upper surface of an isotropic plate to a time-harmonic line load. Comparison of the result obtained by the SEM for an uncracked plate and for a plate with a vertical surface-breaking crack.

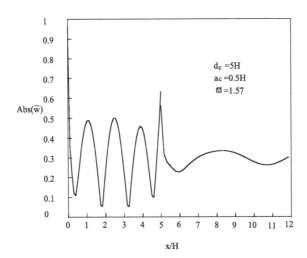

FIGURE 12.33
Displacement responses on the upper surface of an isotropic plate($\mu = 1/3$) with a vertical surface-breaking crack.

the source (load point) and propagating in the positive x direction, and the other is reflected from the crack surface and propagating in the negative x direction. These waves at two different directions interfere, resulting in additional oscillations. Because the reflected waves are related to the crack length, there is the possibility of determining the crack length from the amplitude of the additional oscillations. A technique to determine the crack depth is

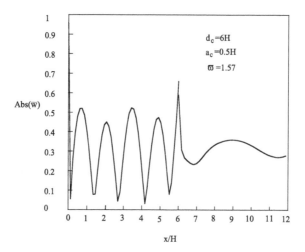

FIGURE 12.34

Displacement responses on the upper surface of an isotropic plate ($\mu = 1/3$) with a vertical surface-breaking crack.

by first exciting the plate with a time-harmonic load. The displacement field is computed, and the crack length is then determined from the distribution pattern of the absolute values of the displacement between the excitation point and the crack. The following detailed investigation using SEM leads to the procedure of crack characterization.

12.9.3 Frequency Dependency

A homogeneous isotropic plate with a Poison's ratio of $\mu = 1/3$ is considered in the numerical investigation. Wave motion in the plate is affected by three parameters: the frequency of the load, ω, distance between the load and the crack, d_c, and the crack length, a_c. We first investigate the relationship between the load frequency and the responses of the cracked plate. By using the SEM, we obtain the displacement distribution on the upper surface of the plate excited by loads with frequencies of 0.5π, $3\pi/4$, π, 1.5π, 2π. It is found that the displacement fields depend strongly on the frequency of the load, and the frequency of $1.57\bar{\omega}$ is the best value to use to determine the crack position. Hence, the results shown in the following subsections are obtained using a harmonic load with a fixed frequency of $1.57\bar{\omega}$.

12.9.4 Dependency of Loading Position

From Figures 12.32–12.34, it is observed that the displacement responses are similar in pattern and different in the number of peaks for different distances between the load and the crack. It is also observed that the spacing of the

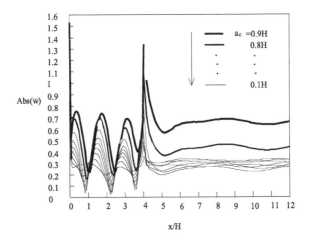

FIGURE 12.35
Displacement responses on the upper surface of an isotropic plate. Comparison of the results for different crack lengths ($d_c = 4.0H$).

peaks of the oscillations in the region of $0 < x < d_c$ are about the same, and that the corresponding peak values change insignificantly. This indicates strongly that the peak spacing and the peak values of the oscillation are less dependent upon the loading position, d_c.

12.9.5 Determination of Crack Length

The results obtained from varying the length of the crack from $0.1H$ to $0.9H$ by a step of $0.1H$ and computing the absolute values of the vertical displacement on the upper surface of the plate are presented in Figure 12.35. The peak spacing and peak positions of the oscillation are almost the same, and the peak values of the oscillations increase following the increase of the crack length. This indicates that the crack length is related to the peak values of the oscillations in the region of $0 < x < d_c$.

Note that the displacement responses in the region between the load and the crack on the plate surface are contributed from the reflections of waves from the crack and the incident source. The additional oscillations appearing in the absolute values of the displacement are caused by the interference of the incident waves and the reflected waves. Since the reflections of waves are related to the crack length, it is a natural to expect a relationship between the crack length and the peak values of the oscillations. The procedure to obtain such a relationship is outlined below:

1. Calculate the displacement field of the cracked plate for various a_c and d_c.

2. Subtract the displacement of the uncracked plate, $w_u(x)$, from the displacement of the cracked plate, $w_c(x)$,

$$w_a(x) = w_c(x) - w_u(x) \tag{12.78}$$

where $w_a(x)$ is related to the displacement of the additional oscillation.

3. Calculate the mean value $\bar{w}_a$ of the displacement $w_a(x)$ in the region of $x_1 < x < x_2$, in which x_1 and x_2 are positions where $w_a(x)$ has the first and last extreme values in the region of $0 < x < d_c$ (see Figure 12.36).

$$\bar{w}_a = \frac{1}{m} \sum w_a(x_i) \tag{12.79}$$

Subtract the mean value $\bar{w}_a$ from the displacement $w_a(x)$ to obtain

$$\tilde{w}_a(x) = w_a(x) - \bar{w} \tag{12.80}$$

Calculate the mean value $\bar{w}_m$ by

$$\bar{w}_m = \frac{1}{n} \sum_{i=1}^{n} Abs(w_i) \tag{12.81}$$

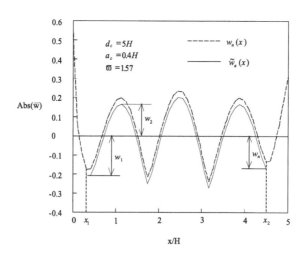

FIGURE 12.36
Displacements related to the additional oscillations.

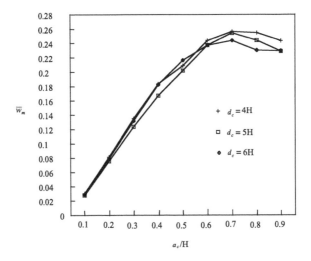

FIGURE 12.37
Relationship between the crack depths and mean values of the absolute extreme value of the additional oscillations.

where w_i $(i = 1, 2, ..., n)$ are the extreme values of $\tilde{w}_a(x)$ in the region of $x_1 < x < x_2$

4. Plot $\overline{w}_m$ against the crack length, a_c, as shown in Figure 12.37. From Figure 12.37, the crack length can be determined from the mean value $\overline{w}_m$ of the absolute extreme values. It is also noted that the $\overline{w}_m \sim a_c$ curves for the different distances between the load and crack are very close. This implies that the $\overline{w}_m \sim a_c$ curves are less dependent on the distance d_c. This is a very important fact because one does not need to pay much attention to the loading position. Using polynomial regression of degree 3, the relationship between the crack length and the mean value $\overline{w}_m$ can be obtained as follows:

$$\bar{a}_c = 0.0147 + 3.2985\overline{w}_m - 15.1438\overline{w}_m^2 + 49.7855\overline{w}_m^3 \qquad (12.82)$$

Equation (12.82) is obtained for the region of $0.1 \leq \bar{a}_c \leq 0.7$. Equation (12.82) is used to predict the crack length for a plate with a vertical surface-breaking crack with an actual crack length of $0.25H$. The procedure is to first compute the displacement responses for the cracked plate subject to a load applied at a distance of $d_c = 7H$. Following the above-mentioned steps, the mean value $\overline{w}_m$ has been obtained to be 0.09906. Substituting this value into Eq. (12.82), the predicted crack length is found to be $0.2412H$. Compared with the actual value of $0.25H$, it is a very good prediction.

12.10   Characterization of Middle Interior Vertical Cracks

The 2D configuration of an interior middle vertical crack in an infinite laminate is presented in Figure 12.38. To simplify the problem, we assume again that the crack is located at the middle of the plate.

12.10.1   SEM Formulation

The procedure for obtaining the SEM equations is the same as that given in the previous section. The laminate is divided into two subdomains, domain **I** and domain **II**, as denoted in Figure 12.38. In each subdomain, the relationship between the displacements and stresses at the nodal points on the vertical boundaries can be obtained as a set of equations. For domain **I**, we have

$$
\begin{Bmatrix} \mathbf{R}^I_{b1} \\ \mathbf{R}^I_{cL} \\ \mathbf{R}^I_{b2} \end{Bmatrix} =
\begin{bmatrix} \mathbf{K}^I_{11} & \mathbf{K}^I_{12} & \mathbf{K}^I_{13} \\ \mathbf{K}^I_{21} & \mathbf{K}^I_{22} & \mathbf{K}^I_{23} \\ \mathbf{K}^I_{31} & \mathbf{K}^I_{32} & \mathbf{K}^I_{33} \end{bmatrix}
\begin{Bmatrix} \mathbf{V}^I_{b1} \\ \mathbf{V}^I_{cL} \\ \mathbf{V}^I_{b2} \end{Bmatrix} +
\begin{Bmatrix} \mathbf{S}^I_{b1} \\ \mathbf{S}^I_{cL} \\ \mathbf{S}^I_{b2} \end{Bmatrix}
\tag{12.83}
$$

where the subscripts $b1$ and $b2$ stand for the vectors on the vertical junctions $b1$ and $b2$, respectively, including the crack tip. The subscript cL stands for the vector on the left surface of the crack excluding the crack tip. Similarly, for domain **II** we obtain

$$
\begin{Bmatrix} \mathbf{R}^{II}_{b1} \\ \mathbf{R}^{II}_{cR} \\ \mathbf{R}^{II}_{b2} \end{Bmatrix} =
\begin{bmatrix} \mathbf{K}^{II}_{11} & \mathbf{K}^{II}_{12} & \mathbf{K}^{II}_{13} \\ \mathbf{K}^{II}_{21} & \mathbf{K}^{II}_{22} & \mathbf{K}^{II}_{23} \\ \mathbf{K}^{II}_{31} & \mathbf{K}^{II}_{32} & \mathbf{K}^{II}_{33} \end{bmatrix}
\begin{Bmatrix} \mathbf{V}^{II}_{b1} \\ \mathbf{V}^{II}_{cR} \\ \mathbf{V}^{II}_{b2} \end{Bmatrix} +
\begin{Bmatrix} \mathbf{S}^{II}_{b1} \\ \mathbf{S}^{II}_{cR} \\ \mathbf{S}^{II}_{b2} \end{Bmatrix}
\tag{12.84}
$$

where the subscript cR stands for the vectors on the right surface of the crack.

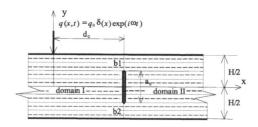

FIGURE 12.38
Plate with a vertical interior crack.

Considering the boundary conditions of stress-free surfaces of the crack, we have

$$\mathbf{R}_{cL}^{I} = \mathbf{0} \tag{12.85}$$

$$\mathbf{R}_{cL}^{II} = \mathbf{0} \tag{12.86}$$

Furthermore, the following boundary conditions and continuity conditions on the junctions should be satisfied:

$$\mathbf{R}_{c} = \mathbf{R}_{cL}^{I} = \mathbf{R}_{cR}^{II} = \mathbf{0} \tag{12.87}$$

$$\mathbf{R}_{b1} = \mathbf{R}_{b1}^{I} - \mathbf{R}_{b1}^{II}, \quad \mathbf{V}_{b1} = \mathbf{V}_{b1}^{I} = \mathbf{V}_{b1}^{II} \tag{12.88}$$

$$\mathbf{R}_{b2} = \mathbf{R}_{b2}^{I} - \mathbf{R}_{b2}^{II}, \quad \mathbf{V}_{b2} = \mathbf{V}_{b2}^{I} = \mathbf{V}_{b2}^{II} \tag{12.89}$$

Assembling all the sets of the equations for the two domains using the boundary conditions and continuity conditions, a set of equations that gives the relationship between the displacements and stresses at all the nodal points on the junctions is obtained as

$$\mathbf{R}_{J} = \mathbf{K}\mathbf{V}_{J} + \mathbf{S}_{J} \tag{12.90}$$

where

$$\mathbf{R}_{J} = \{\mathbf{R}_{b1} \quad \mathbf{R}_{cL}^{I} \quad \mathbf{R}_{cR}^{II} \quad \mathbf{R}_{b2}\}^{T} \tag{12.91}$$

is the external force vector acting on the junctions and the crack surfaces and $\mathbf{V}_{J}$ is the displacement vector on the junctions and the crack surfaces as follows:

$$\mathbf{V}_{J} = \{\mathbf{V}_{b1} \quad \mathbf{V}_{cL}^{I} \quad \mathbf{V}_{cR}^{II} \quad \mathbf{V}_{b2}\}^{T} \tag{12.92}$$

The vector $\mathbf{S}_{J}$ and matrix $\mathbf{K}$ for the cracked plate are given as follows:

$$\mathbf{S}_{J} = \left\{ \begin{array}{c} \mathbf{S}_{b1}^{I} - \mathbf{S}_{b1}^{II} \\ \mathbf{S}_{cL}^{I} \\ \mathbf{S}_{cR}^{II} \\ \mathbf{S}_{b2}^{I} - \mathbf{S}_{b2}^{II} \end{array} \right\} \tag{12.93}$$

$$\mathbf{K} = \begin{bmatrix} \mathbf{K}_{11}^{I} - \mathbf{K}_{11}^{II} & \mathbf{K}_{12}^{I} & -\mathbf{K}_{12}^{II} & \mathbf{K}_{13}^{I} - \mathbf{K}_{13}^{II} \\ \mathbf{K}_{21}^{I} & \mathbf{K}_{22}^{I} & 0 & \mathbf{K}_{23}^{I} \\ \mathbf{K}_{21}^{II} & 0 & \mathbf{K}_{22}^{II} & \mathbf{K}_{23}^{II} \\ \mathbf{K}_{31}^{I} - \mathbf{K}_{31}^{II} & \mathbf{K}_{32}^{I} & -\mathbf{K}_{32}^{II} & \mathbf{K}_{33}^{I} - \mathbf{K}_{33}^{II} \end{bmatrix} \tag{12.94}$$

Equation (12.90) can be solved to obtain the displacements at the junctions and the surface of the crack. Then the entire displacement field in the cracked plate can finally be obtained.

12.10.2 Technique for Crack Detection

We use an isotropic plate to study issues related to crack detection. Since it is assumed the crack is located in the middle of the plate, there are two parameters that need to be characterized: crack length and location (the distance between the load and the crack). The aim of the numerical modeling is to reveal the relationship between the parameters of the crack and the scattered wave field in the plate. The scattered displacement field by the interior vertical crack are computed in detail using the SEM. The results are shown in Figures 12.39–12.44. Comparison of the absolute values of the displacement distributions on the upper surface of the plates with and without a crack are shown in Figure 12.39. The crack length is $a_c = 0.6H$, and the crack is applied on the plate at a distance of $6H$ away from the load. The displacement

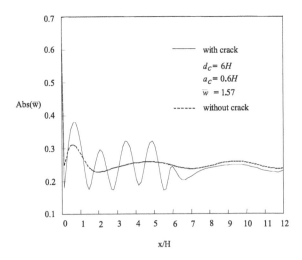

FIGURE 12.39
Displacement distribution on the upper surface of an isotropic plate excited by a time-harmonic line load. Comparison of the result for an uncracked plate and for a plate with a vertical interior crack, $d_c = 6H$.

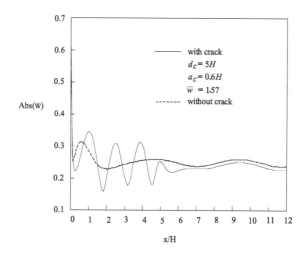

FIGURE 12.40
Displacement distribution on the upper surface of an isotropic plate excited by a time-harmonic line load. Comparison of the result for an uncracked plate and for a plate with a vertical interior crack, $d_c = 5H$.

distributions for a cracked plate are found to be quite different from the uncracked plate between the load and the crack. Additional superimposed oscillations are found on the curve of the absolute values of the displacement in the region between the load and the crack (see Figure 12.39). These oscillations end at the position of the crack. The same behavior is also observed in Figures 12.40 and 12.41. This behavior can be used for detecting the position of the crack. However, it is more difficult to determine the crack length, and it requires further investigation as follows.

12.10.3 Dependency of Load Frequency

An isotropic plate of Poison's ratio of $\mu = 1/3$ is used in this numerical investigation. The loading frequency effects on the displacement of the plate are investigated. The displacement field excited by loads with frequencies of 0.5π, $3\pi/4$, π, 1.5π, and 2π are computed. It is found that the frequency of $1.57\overline{\omega}$ is most suited for determining the crack location because the results between the load and crack are significantly different from those elsewhere. Thus, all subsequent numerical work will be performed using a harmonic load with a fixed frequency of $1.57\overline{\omega}$.

12.10.4 Dependency of Loading Position

The spacing between the peaks of the oscillations in the region of $0 < x < d_c$ is about the same and is clearly seen in Figure 12.39. Comparing Figures 12.39 and 12.41, it is also observed that the displacement responses are

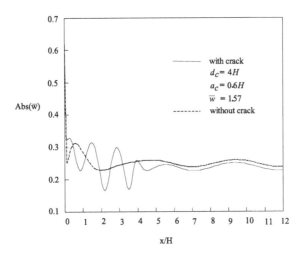

FIGURE 12.41
Displacement distribution on the upper surface of an isotropic plate excited by a time-harmonic line load. Comparison of the result for an uncracked plate and for a plate with a vertical interior crack, $d_c = 4H$.

similar in the pattern but different in the number of peaks for different distance between load and crack. Furthermore, for different distances d_c, the spacing of the peaks of the oscillations between the load and the crack are about the same, and the corresponding peak values change insignificantly. Hence, it can be concluded that the peak spacing and the peak value of the oscillation are less dependent on the loading position. This conclusion is similar to that for the case of surface-breaking cracks.

12.10.5 Determination of Crack Length

The displacement fields on the upper surface of the plate have been computed for different crack lengths. Figure 12.42 shows the computed results of absolute values of displacements in which the crack length changes from $0.1H$ to $0.9H$ by a step increment of $0.1H$. It is observed again that the peak spacing and peak positions of the oscillation are almost the same, and that the peak values of the oscillations increase with increasing crack length. This indicates that the crack length is closely related to the peak value of the oscillations in the region of $0 < x < d_c$.

We use now the numerical procedure given in section 12.9.5 to obtain the relationship between the crack length and the peak values of the oscillations. The relationship between the mean value $\bar{w}_m$ and the crack depth, a_c, is shown in Figure 12.43.

The crack depth can be determined using Figure 12.43 if the mean value $\bar{w}_m$ of the absolute extreme values is obtained. From Figure 12.43, an important fact is observed: the $\bar{w}_m \sim a_c$ curves are very close for different values of

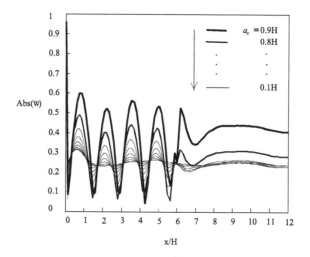

FIGURE 12.42
Displacement on the upper surface of an isotropic plate. Comparison of the results for different crack length ($d_c = 6.0H$).

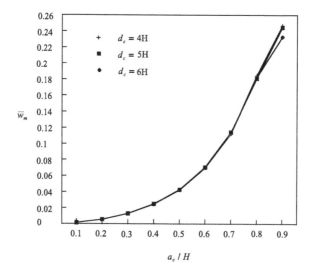

FIGURE 12.43
Relationship between the crack depths and mean value of the absolute extreme value of the additional oscillations.

distance between the load and crack. Especially when $a_c < 0.8H$, it is hard to distinguish them. This proves that the $\overline{w}_m \sim a_c$ curves are less dependent on the distance d_c. Therefore, one does not need to pay too much attention to the loading position. Using polynomial regression of degree four, the relationship between the crack depth and the mean value $\bar{w}_m$ can be obtained as

$$\bar{a}_c = 0.1112 + 14.6548\bar{w}_m - 151.1776\bar{w}_m^2 + 724.7152\bar{w}_m^3 - 1217.1198\bar{w}_m^4 \quad (12.95)$$

Equation (12.95) is obtained for the region of $0.1 \le \bar{a}_c \le 0.9$.

To validate the accuracy of Eq. (12.95), it is used to predict the crack length for a plate with an interior vertical crack with an actual length $0.45H$. The distance between the load and the crack is $6.5H$. The displacement responses are first computed for the cracked plate, and the mean value $\bar{w}_m$ has been obtained to be 0.032703. Substituting this value to Eq. (12.95), the predicted length of crack is found to be $0.4527H$. Compared with the actual $0.45H$, it is indeed a very good prediction.

12.11 Characterization of Arbitrary Interior Vertical Cracks

For vertical surface-breaking and interior middle vertical cracks, only two parameters need to be determined: the location and the length of the crack. Thus far, the crack location has been determined from the distribution of the displacement field. Additional oscillations have been found in the curves of the absolute values of the displacement between the load and the crack. The crack length has been determined from the mean value of the peaks of the additional superimposed oscillations. It is found that the mean value of the peaks of the additional oscillations is predominantly dependent on the crack length and far less dependent on the distance between the load and the crack.

For the characterization of the interior vertical cracks of arbitrary position, it is, however, more challenging. The reason is that there are three parameters to be determined: the location, length, and depth of the crack. This section presents a new technique to characterize an interior vertical crack of arbitrary position in a plate.

12.11.1 Technique for Crack Detection

Consider an infinite plate of finite thickness H, as shown in Figure 12.44. The plate is loaded by two time-harmonic line loads. The crack length is denoted by a_c, and the center of the crack is located at a distance e_c (eccentric distance) from the middle plane of the plate. The eccentric distance e_c can be positive or negative. The distance from the crack to the load is denoted by l_c.

For an interior vertical crack, there are three parameters to be characterized: the location (the distance between the load and the crack) l_c, the crack length a_c, and the eccentric distance e_c or the crack depth (the distance between the crack tip and the plate surface) d_t. In the case of the surface-breaking crack

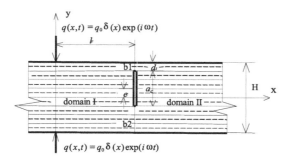

FIGURE 12.44

Plate with a vertical interior crack. The plate is divided into two domains in which the SEM can be applied.

and interior middle crack, the scattered waves affected only by crack length and the crack can be characterized by the displacement field on the upper surface of the plate. For eccentric cracks, it is not possible to determine the length and the depth of the crack from only the upper surface displacement field, as both the length and the depth of the crack will affect the scattered waves. As more than one displacement response is required to determine the two parameters, the most convenient method is to use the displacements on both the upper and lower surfaces of a plate. We apply two symmetric loads on both the upper and lower surface of the plate. For cracks located at the center of the plate, the vertical displacements are symmetrical because the geometry and the load are symmetrical to the x-axis. For eccentric cracks, the displacements fields should be different on the two surfaces. The SEM is used to compute the scattered displacement fields. The plate is symmetrically excited at $x = 0$ on the upper and lower surfaces of the plate by two time-harmonic line loads given by Eq. (12.1).

12.11.2 Wave Scattering by Arbitrary Interior Vertical Cracks

The SEM program is used to analyze wave scattering in a plate with an interior vertical crack. Computations of responses in the frequency domain were carried out for an isotropic plate with Poison's ratio $\mu = 1/3$. This section uses the following additional dimensionless parameters:

$$\bar{e}_c = e_c/H, \quad \bar{d}_t = d_t/H \tag{12.96}$$

The responses of the plate to the time-harmonic loads depend on the frequencies of the loads, and the harmonic loads with a fixed frequency of $\bar{\omega} = 1.57$ have been found most preferable to determine the crack position; it is also used in this section.

Comparisons of the absolute values of the displacement distributions on the upper and lower surfaces of the plates with and without a crack are shown in Figures 12.45–12.47. The crack length a_c and crack depth d_t are, respectively, $0.6H$ and $0.05H$. The crack location l_c is $7H$, $5H$, and $4H$ for Figures 12.45 to 12.47, respectively. Comparing the two curves for the plate

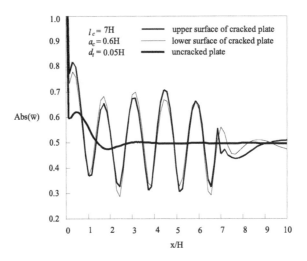

FIGURE 12.45
Displacement on the upper and lower surface of an isotropic plate subjected to two time-harmonic line loads on the two surfaces of the plate. Comparison of the results for an uncracked plate and for a plate with a vertical interior crack ($l_c = 6H$).

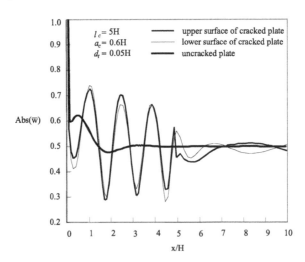

FIGURE 12.46
Displacement on the upper and lower surfaces of an isotropic plate subjected to two time-harmonic line loads on the two surfaces of the plate. Comparison of the results for an uncracked plate and for a plate with a vertical interior crack ($l_c = 5H$).

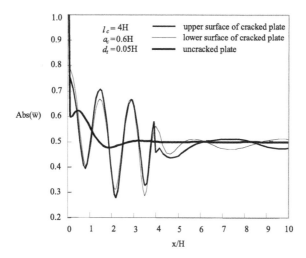

FIGURE 12.47
Displacement on the upper and lower surfaces of an isotropic plate subjected to two time-harmonic line loads on the two surfaces of the plate. Comparison of the results for an uncracked plate and for a plate with a vertical interior crack ($l_c = 4H$).

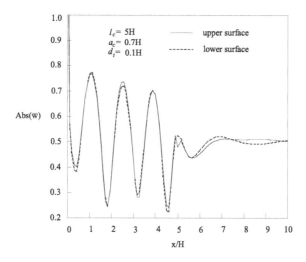

FIGURE 12.48
Displacement response in the upper and lower surfaces of an isotropic plate. The crack length is $0.7H$.

with and without a crack, it is found again that there are additional superimposed oscillations on the curves of the absolute values of the displacement for the cracked plate in the region between the load and the crack. The displacement response on the upper surface is different from that of the lower

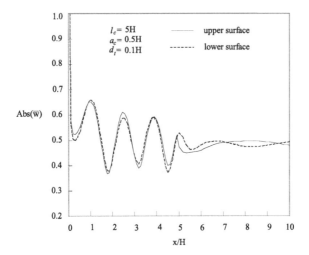

FIGURE 49
Displacement response in the upper and lower surface of an isotropic plate. The crack length is 0.5*H*.

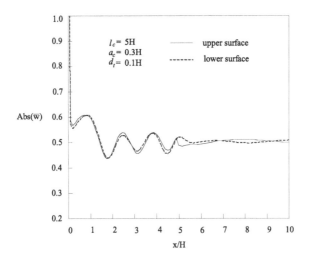

FIGURE 50
Displacement response in the upper and lower surface of an isotropic plate. The crack length is 0.3*H*.

surface. More examples for various crack lengths and depths are shown in Figures 12.48–12.55. It is found that peak values of the oscillations are related to both the crack length and crack depth. These features of the oscillation can be used to determine the crack parameters.

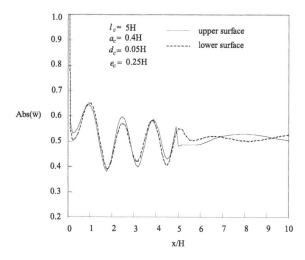

FIGURE 51

Displacement response in the upper and lower surface of an isotropic plate. The crack length is 0.25*H*.

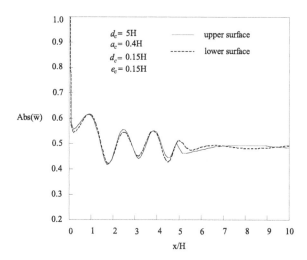

FIGURE 52

Displacement response in the upper and lower surface of an isotropic plate. The crack length is 0.15*H*.

12.11.3 Dependency of Loading Position

From Figures 12.45–12.47 some important facts can be observed. First, the oscillations begin at the load point and end at the position of the crack. Second, for different distances between load and crack, the displacement

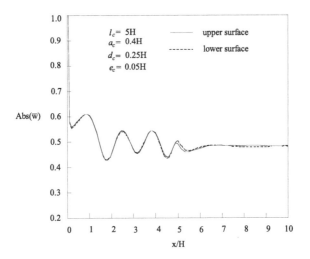

FIGURE 53

Displacement response in the upper and lower surface of an isotropic plate. The crack length is $0.05H$.

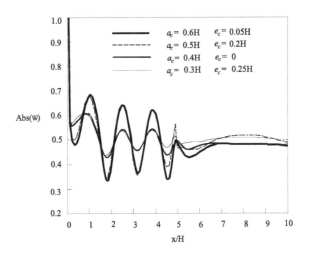

FIGURE 54

Comparison of displacement response on the upper surface for different crack length and different eccentric distance, $l_c = 5H$.

distributions are similar in pattern but different in peak numbers. Therefore, the crack position can be approximately determined from the distribution pattern of the displacements on the plate surfaces. However, determining the crack length and depth is much more difficult, and this is addressed in detail in the following section.

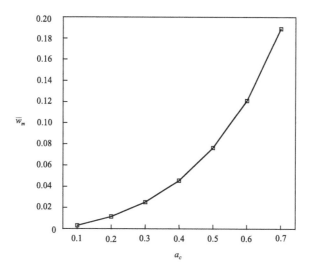

FIGURE 55

Relationship between crack length a_c and the value $\bar{w}_m$.

12.11.4 Determination of Crack Length

Figures 12.48–12.53 show the displacement fields on the upper (thick line) and lower (thin line) surfaces of the plate for various crack lengths and depths. For a crack depth of $d_t = 0.1H$, the absolute values of the displacement are shown in Figures 12.48–12.50 for $a_c = 0.7H$, $0.5H$ and $0.3H$, respectively. It is observed that the peak values of the oscillations decrease as the crack length decreases. Figures 12.51–12.53 show the absolute values of displacement for $a_c = 0.4$ and $e_c = 0.25H$, $0.15H$, and $0.05H$, respectively. It is found that the peak values of the oscillations increase as the crack depth e_c increases. On the other hand, the displacement fields between the load and the crack can be very close for different combinations of eccentric distance and crack length (see Figure 12.54). This indicates that the peak values of the oscillations in the region of $0 < x < l_c$ are influenced by both the length and depth of the crack. Comparing Figures 12.51 and 12.53, another important fact surfaces: the larger the eccentric distance e_c is, the larger is the difference between the peak values of the superimposed oscillations in the upper and lower surfaces.

From the above observations, it is correct to say that the crack length and eccentric distance affect both the peak values of the displacement fields and the difference between superimposed oscillations on the upper and lower surfaces. These two parameters (a_c and e_c) cannot be determined by the simple method used in the case of surface-breaking crack and interior middle crack. However, we have found that the difference between the peak values of the superimposed oscillations on the upper and lower surfaces increases as the eccentric distance e_c increases for a given crack length a_c. This means that the eccentric distance e_c can be determined after the crack length a_c is determined. The technique described below can be used to determine the relationship between the crack length and the peak values of the oscillations.

1. Calculate the displacement fields, $w_{uu}(x)$, on the surface of uncracked plate.
2. Calculate the displacement fields on both the upper and lower surfaces for cracked plate with various a_c and e_c.
3. For a cracked plate with crack length a_c and eccentric distance e_c, subtract the surface displacement of the uncracked plate, $w_{uu}(x)$, from the upper surface displacement of the cracked plate, $w_{uc}(x)$. We obtain

$$w_{au}(x) = w_{uc}(x) - w_{uu}(x) \tag{12.97}$$

where $w_{au}(x)$ is related to the displacement of the additional oscillation on the upper surface. Then subtract the lower surface displacement $w_{lc}(x)$ from the upper surface displacement $w_{uc}(x)$. We get the displacement difference between the upper and lower surface, i.e.,

$$w_b(x) = w_{uc}(x) - w_{lc}(x) \tag{12.98}$$

4. Calculate the mean value $\overline{w}_{au}$ of the displacement $w_{au}(x)$ in the region of $x_1 < x < x_2$ in which x_1 and x_2 are positions where $w_{au}(x)$ has the first and last maximum values in the region of $0 < x < l_c$.

$$\overline{w}_{au} = \frac{1}{m}\sum_{i=1}^{m} w_{au}(x_i) \tag{12.99}$$

where m is the number of observation points in the region of $x_1 < x < x_2$.

5. Subtract the mean value $\overline{w}_{au}$ from the displacement $w_{au}(x)$ and the result can be denoted as $\tilde{w}_{au}(x)$ and is given by

$$\tilde{w}_{au}(x) = w_{au}(x) - \overline{w}_{au} \tag{12.100}$$

6. Calculating the mean value $\overline{w}_{aum}$ and $\overline{w}_{bm}$ by

$$\overline{w}_{aum} = \frac{1}{n}\sum_{i=1}^{n} Abs(w_{aui}) \tag{12.101}$$

$$\overline{w}_{bm} = \frac{1}{N}\sum_{i=1}^{N} Abs(w_{bi}) \tag{12.102}$$

where w_{aui} ($i = 1, 2,...,n$) and w_{bi} ($i = 1, 2,...,N$) are, respectively, the extreme values of $\tilde{w}_{au}(x)$ and $w_b(x)$ in the region of $x_1 < x < x_2$.

7. Calculate w_m by subtracting the mean value $\overline{w}_{bm}$ from $\overline{w}_{aum}$

$$w_m = \overline{w}_{aum} - \overline{w}_{bm} \tag{12.103}$$

It is found that w_m is dependent on the crack length, but almost independent of eccentric distance when the same crack length has little change.

8. Repeat steps from 3–7, and calculate the mean value $\overline{w}_m$ of w_m for various eccentric distances at certain crack length.

9. Repeat steps 3–8; calculate the value $\overline{w}_m$ for different crack lengths. Plotting $\overline{w}_m$ against the crack lengths, $\bar{a}_c$, the final results are shown in Figure 12.55.

The crack length can be determined from the value $\overline{w}_m$ using Figure 12.55. Using polynomial regression of degree four, the relationship between the crack length and the value $\overline{w}_m$ can be obtained as follows:

$$\bar{a}_c = 0.0680 + 12.4458\overline{w}_m - 149.5156\overline{w}_m^2 + 953.6342\overline{w}_m^3 - 2208.2745\overline{w}_m^4$$

$$\tag{12.104}$$

Equation (12.104) is obtained for the region of $0.1 \leq \bar{a}_c \leq 0.7$.

12.11.5 Determination of the Crack Depth

It has been found that the difference between the displacements on the upper and lower surfaces increase as the eccentric distance e_c increases. This fact can also be observed from Figure 12.56. For a given e_c, it is observed from Figure 12.57 that a larger a_c leads to a larger difference between the displacements on the upper and the lower surfaces. That is, the difference of the upper surface displacement and the lower surface displacement is affected by both the eccentric distance e_c and crack length a_c. An intensive study has been carried out, and it is also found that $\bar{a}_c\sqrt{|\bar{e}_c|}$ is a combined quantity of e_c and a_c related to the displacement difference.

When the crack length has been determined, the eccentric distance can be determined from the difference between the displacement on the upper and lower surfaces. The procedure to obtain the e_c is outlined below:

1. Calculate $w_b(x)$ using Eq. (12.98).
2. Calculate $\overline{w}_{bm}$ using Eq. (12.102).
3. Plot $\overline{w}_{bm}$ against the value $\bar{a}_c\sqrt{|\bar{e}_c|}$ as shown in Figure 12.58.

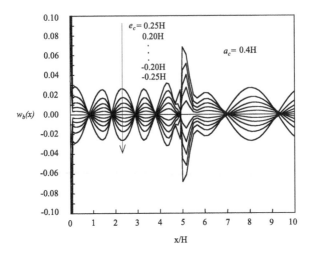

FIGURE 12.56
Comparison of the displacement difference $w_b(x)$ between the upper and lower surfaces for the case of same crack length with different eccentric distance.

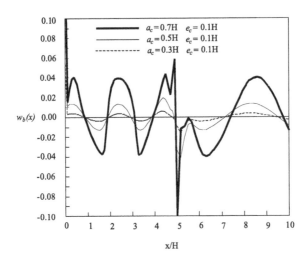

FIGURE 12.57
Comparison of the displacement difference $w_b(x)$ between the upper and lower surfaces for the case of different crack length with same eccentric distance.

4. Calculate $\overline{w}_{alm}$ of lower surface displacement according to the computing method of $\overline{w}_{aum}$. It is found that the eccentric distance e_c is positive if $\overline{w}_{aum} > \overline{w}_{alm}$, and negative otherwise.

The value $\bar{a}_c \sqrt{|\bar{e}_c|}$ can be determined from the value $\overline{w}_{bm}$ using Figure 12.58. The eccentric distance e_c can then be obtained after determining the crack

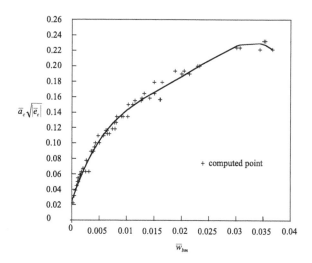

FIGURE 12.58

Relationship between the geometry parameter $\bar{a}_c\sqrt{|\bar{e}_c|}$ and the mean value $\bar{w}_{bm}$.

length and comparing the value of $\bar{w}_{aum}$ and $\bar{w}_{alm}$. Using polynomial regression of degree four, the relationship between the value $\bar{a}_c\sqrt{|\bar{e}_c|}$ and the value $\bar{w}_{bm}$ can be obtained as follows:

$$\bar{a}_c\sqrt{|\bar{e}_c|} = 0.0249 + 21.1657\,\bar{w}_{bm} - 1348.0561\,\bar{w}_{bm}^2 + 46230.6478\,\bar{w}_{bm}^3$$
$$- 578669.4297\,\bar{w}_{bm}^4 \tag{12.105}$$

Equation (12.105) is obtained for the region of $0.1 \le \bar{a}_c \le 0.7$, $0.05 \le |\bar{e}_c| \le 0.4$. Once the eccentric distance is obtained, the crack depth can be obtained using the following simple equation:

$$\bar{d}_t = 0.5 - \bar{a}_c/2 - \bar{e}_c \tag{12.106}$$

To verify Eqs. (12.104) and (12.105), a plate with an interior vertical crack with an actual crack length of 0.35H is used. The crack depth is 0.25H, the eccentric distance is $e_c = 0.075$, and the distance between the load and the crack is 6.5H. The displacement responses are first computed for the cracked plate. By using the above-mentioned steps, the response results are investigated. The mean values $\bar{w}_{aum}$, $\bar{w}_{alm}$, and $\bar{w}_{bm}$ have been obtained to be 0.03601, 0.03549, and 0.00438, respectively. The predicted length of crack is found from Eq. (12.104) to be 0.3400H. The predicted eccentric distance is then found from Eqs. (12.105) to be 0.0787H, and crack depth is calculated using Eq. (12.106) to be 0.2513H. Compared with the true values of 0.35H, 0.075H, and 0.25H, respectively, these values are indeed a very good prediction.

12.12   Remarks

The SEM has been applied to study wave-scattering problems in laminates. A number of formulae have been obtained to calculate the parameters of cracks in laminates. The formulae are expected to be used in practice for the characterization of cracks.

In detecting defects in composite specimens by means of ultrasound, fluid is generally used as a coupling medium. Because of fluid-structural coupling effects, the fluid presence could affect the accuracy of evaluation. Based on the present method, Liu and Xi et al. (1999) have developed a SEM for analyzing waves scattered by cracks in immersed composite laminated plates.

13

Wave Scattering by Flaws in Composite Laminates

13.1 Introduction

The failure mechanism of composites is much more complex than for metals because of the anisotropic, inhomogeneous, and layered nature of the composites. In many cases, composites such as sandwich panels are damaged in a certain region with multiple fine cracks or voids rather than with a single crack. Quantitative characterization of this damaged zone, called *flaw* here, in composites has become of increasing interest. Sandwich panels or plates are widely used in spacecraft, aircraft, and ships (Robson, 1989; Hellbratt et al., 1989). The sandwich plate is a special class of laminates that consists of two thin, stiff face (outer) layers and one thick flexible inner core layer. Structural efficiency is achieved by separating the stiff facing material with a thick lightweight core material. The structural reliability depends highly on the support of the core material. The stiffness and strength of the structure would be heavily affected by the failure of the weaker core material. Usually, the core materials are more easily damaged during the manufacturing process and practical services. Therefore, the quantitative characterization of this damaged zone or flaw in sandwich plates has become one of increasing interest. However, conventional nondestructive testing (NDT) methods are not sufficient for sandwich plates because of the material's inhomogeneity, and some special techniques, such as the thermographic NDT methods, have been used for locating flaws in sandwich plates with foam core materials (Vikström, 1989). Elastic waves can be used to detect such flaws in sandwich plates. Incident waves with a certain single frequency can be used to excite the sandwich plates, and the responses of the plate to the excitation can be obtained to determine the parameters of the flaw. This technique can possibly be used for different sandwich plates with different types of core materials because an incident wave with a different frequency can be used for a different material. To obtain the responses of a sandwich plate with a flaw, a very efficient method together with suitable modeling techniques is a necessary prerequisite.

This chapter focuses on waves scattered by flaws in composite laminates and the detection of flaws in sandwich plates. The methods used in this chapter were discussed by Liu, Lam, and Shang (1996) and Liu and Lam (1996). When investigating the scattering of waves by flaws in anisotropic laminates, the plate with a rectangular flaw is divided appropriately into several domains to which the SEM is applicable, and a SEM equation for each domain can then be obtained. By assembling all the equations of these domains, a set of algebraic equations is obtained. These equations are then solved using the boundary conditions on the vertical boundaries of the domain. Scattered wave fields in the frequency domain for hybrid composite plates and sandwich panels with rectangular flaws are computed. The results are discussed in comparison with results for laminates without a flaw.

Sandwich plates often consist of three layers of transversely isotropic materials. Therefore, pure SH waves can exist in this type of sandwich plate and uncoupled with Lamb wave modes. SEM formulae for SH waves in sandwich plates with and without flaws are also derived. Scattered SH wave fields in the frequency domain for a sandwich plate with rectangular flaws are computed. The results are also discussed in comparison with results for laminates without a flaw. A technique for characterizing such a flaw is also presented. The present procedure has a wide range of applications in the shipbuilding and aerospace industries in that it can systematically characterize the loss of structural integrity in sandwich structures.

13.2   Application of the SEM to Plates Containing Flaws

To apply the SEM to a plate with a rectangular flaw, the plate has to be divided into three subdomains, as denoted by Roman numerals in Figure 13.1. Domain **I** is bounded by boundaries b1 and the upper and lower surfaces of the plate. Domain **II** is bounded by boundaries b2 and the upper and lower surfaces of the plate. Domain **III** is bounded by boundaries b1, b2, and the upper and

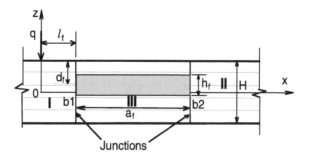

FIGURE 13.1
Division of a laminate with a horizontal flaw into domains in which the SEM can be applied.
(From Liu, G. R., Lam K. Y., and Shang H. M., *Composites Part B*, 27B, 431, 1996. With permission.)

lower surfaces of the plate. In each domain the SEM is used to obtain sets of equations that give the relationship between the displacements and stresses at the nodal points on the vertical boundaries of the subdomains. For domain **I**, we can obtain

$$\mathbf{R}^I_{b1} = \mathbf{K}^I \mathbf{V}^I_{b1} + \mathbf{S}^I_{b1} \tag{13.1}$$

where the superscript of the Roman numeral indicates the domain that the equation governs. Similarly, for domains **II** and **III** we obtain

$$\mathbf{R}^{II}_{b2} = \mathbf{K}^{II} \mathbf{V}^{II}_{b2} + \mathbf{S}^{II}_{b2} \tag{13.2}$$

$$\mathbf{R}^{III}_{b} = \mathbf{K}^{III} \mathbf{V}^{III}_{b} + \mathbf{S}^{III}_{b} \tag{13.3}$$

where in Eq. (13.3) the subscript b stands for the two vertical boundaries of domain **III**. Equation (13.3) can be further divided into submatrices as follows.

$$\begin{Bmatrix} \mathbf{R}^{III}_{b1} \\ \mathbf{R}^{III}_{b2} \end{Bmatrix} = \begin{bmatrix} \mathbf{K}^{III}_{11} & \mathbf{K}^{III}_{12} \\ \mathbf{K}^{III}_{21} & \mathbf{K}^{III}_{22} \end{bmatrix} \begin{Bmatrix} \mathbf{V}^{III}_{b1} \\ \mathbf{V}^{III}_{b2} \end{Bmatrix} + \begin{Bmatrix} \mathbf{S}^{III}_{b1} \\ \mathbf{S}^{III}_{b2} \end{Bmatrix} \tag{13.4}$$

In Eqs.(13.1), (13.2), and (13.4), the subscripts $b1$ and $b2$ stand, respectively, for the vectors on the vertical boundaries of the corresponding domains.

Assembling all the sets of the equations for all the subdomains using the following continuity conditions at the two junctions that divide the plate into subdomains,

$$\mathbf{R}_{b1} = \mathbf{R}^I_{b1} - \mathbf{R}^{III}_{b1}, \quad \mathbf{V}_{b1} = \mathbf{V}^I_{b1} = \mathbf{V}^{III}_{b1} \tag{13.5}$$

$$\mathbf{R}_{b2} = \mathbf{R}^{III}_{b2} - \mathbf{R}^{II}_{b2}, \quad \mathbf{V}_{b2} = \mathbf{V}^{II}_{b2} = \mathbf{V}^{III}_{b2} \tag{13.6}$$

a set of equations that gives the relationship between the displacements and stresses at whole nodal points on the junctions can be obtained. The equations are

$$\mathbf{R}_J = \mathbf{K}_J \mathbf{V}_J + \mathbf{S}_J \tag{13.7}$$

where $\mathbf{R}_J$ is the external force vector acting at the junctions given by

$$\mathbf{R}_J = \{\mathbf{R}_{b1} \quad \mathbf{R}_{b2}\}^T \tag{13.8}$$

and $\mathbf{V}_J$ is the displacement vector at the junctions as follows:

$$\mathbf{V}_J = \{\mathbf{V}_{b1} \quad \mathbf{V}_{b2}\}^T \tag{13.9}$$

In Eq. (13.7), $\mathbf{S}_j$ is given by

$$\mathbf{S}_j = \left\{ \begin{array}{c} \mathbf{S}_{b1}^I - \mathbf{S}_{b1}^{III} \\ \mathbf{S}_{b2}^{III} - \mathbf{S}_{b2}^{II} \end{array} \right\} \tag{13.10}$$

and $\mathbf{K}_j$ is the stiffness matrix for the plate with a rectangular flaw, and is given by

$$\mathbf{K}_j = \begin{bmatrix} \mathbf{K}^I - \mathbf{K}_{11}^{III} & -\mathbf{K}_{12}^{III} \\ \mathbf{K}_{21}^{III} & -\mathbf{K}^{II} + \mathbf{K}_{22}^{III} \end{bmatrix} \tag{13.11}$$

The boundary conditions at the junctions subsequently yield the displacement. The complete displacement field can then be obtained by

$$\mathbf{V}_g(x) = \mathbf{G}(x)\mathbf{G}_d^{-1}(\mathbf{V}_J - \mathbf{V}_P) + \mathbf{V}_p(x) \tag{13.12}$$

where

$$\mathbf{V}_P = \{\mathbf{V}_p^I \quad \mathbf{V}_p^r\}^T \tag{13.13}$$

13.3 Examples for Wave Scattering in Laminates

Computations of responses in the frequency domain using SEM code were carried out for a hybrid laminate denoted by $[C90/G45/G-45]_s$. The same has also been carried out for an isotropic plate with Poisson's ratio of one-third for the purpose of comparison. The material constants of carbon/epoxy and glass/epoxy used for the computations are given in Table 13.1.

The following dimensionless parameters in this section are used:

$$\bar{x} = x/H, \quad \bar{w} = c_{44}w/q_0, \quad \bar{\omega} = \omega H/c_s$$
$$c_s = \sqrt{c_{44}/\rho}, \quad c_{44} = \mathbf{c}(4,4), \quad \bar{d}_f = d_f/H \tag{13.14}$$

where w is the displacement in the z direction, and $\mathbf{c}$, ρ, and c_s are the material constants matrix, density, and shear wave velocity of the reference material.

TABLE 13.1

Dimensions and Material Parameters

	Thickness (mm)	G_{yz} (GPa)	G_{xy} (GPa)	ρ (g/cm^3)
Face layer	5	6.42	6.42	1760
Core	50	0.05	0.05	130

For the homogeneous isotropic plate, the reference material is the material of the plate. For the $[C90/G45/G{-}45]_s$ plate, C90 is chosen as the reference material.

The flaw considered is assumed to be a rectangular damaged zone in which the equivalent elastic constants are modified by a damage factor β_f as follows:

$$\mathbf{c}_f = (1 - \beta_f)\mathbf{c} \tag{13.15}$$

There is no change in the density in the damaged zone since the material does not disappear even if it is damaged. The height, length, and depth of the flaw are denoted by h_f, a_f, and d_f, respectively (see Figure 13.1). The depth of the flaw is defined as the distance between the upper surface of the plate and horizontal center line of the flaw. The distance between the load and the left boundary of the rectangular flaw is denoted by l_f.

First, the displacement responses of a hybrid laminate $[C90/G45/G{-}45]_s$ without a flaw subjected to a time-harmonic line load were computed using the SEM program, and a program of the exact method described in Chapter 4. The load is acting on the upper surface of the plate. The dimensionless frequency used in the examples is 3.14. At this frequency, the wavelength of the slowest shear wave for materials in G45 is about $2.0H$. Hence, the thickness of the elements should be less than $0.5H$ or more than three elements have to be used (see discussion in Chapter 8). Since the hybrid laminate is under consideration, 24 elements were used in the SEM program to obtain the present results. Figure 13.2 gives the responses of the displacement in the vertical direction on the upper surface of the laminate. In the SEM program, d_f, a_f, l_f, and h_f are set, respectively, to be $0.5H$, $4.0H$, $4.0H$, and $0.5H$.

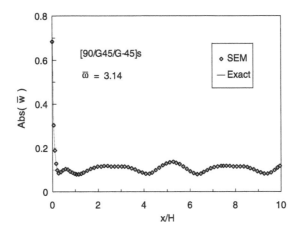

FIGURE 13.2
Displacement on the upper surface of a hybrid laminate subjected to a time-harmonic line load acting on the upper surface of the plate. Comparison of the results obtained by the strip element method (SEM), and an exact method given in Chapter 4. (From Liu, G. R., Lam K. Y., and Shang H. M., *Composites Part B*, 27B, 431, 1996. With permission.)

However, the damage factor β_f is set to be 1.0, namely no flaws in the plate. Excellent agreement is observed. The convergence of the results was also confirmed using a different number of elements. No significant difference was found between the results obtained using 12 elements and 24 elements used in the SEM program. An example for a perfect isotropic plate has also been solved by the present SEM program made for plates with flaws and using the original SEM formulated in Chapter 11. In the original SEM program, the plate is divided into elements in the vertical direction and a viscoelastic nonreflection boundary is used at the end of the regular elements. Although not shown in a figure, the results obtained by these two SEM programs agree very well. This kind of cross-checking is very important to validate the codes before moving into investigation studies.

The responses of an isotropic plate with Poisson's ratio of one-third to harmonic loads of various frequencies acting on the surface of the plate are first investigated. Figure 13.3 shows the responses of the displacement in the vertical direction on the surface of the plate. It is found that there are super-imposed oscillations between the load point and the right boundary of the flaw. Also, excitation with higher frequency results in a more frequent super-imposed oscillation of smaller amplitudes. These superimposed oscillations end at the right boundary of the flaw. This phenomena is also seen in the previous chapter but not explained in detail. The following is a more detailed argument for the reason for this phenomenon.

In the frequency domain, responses of the displacement are complex-valued and have real and imaginary parts. For a given frequency, interferences of waves propagating in opposite directions produce additional oscillations in the absolute values of the displacement. On the other hand, interferences of waves propagating in same directions do not produce this kind of additional oscillation.

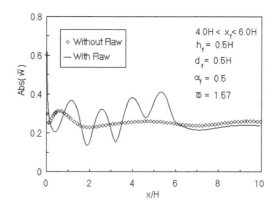

FIGURE 13.3

Displacement on the upper surface of an isotropic plate subjected to a time-harmonic line load. Comparison of the results for plates with and without flaw. ($l_f = 4H$, $a_f = 2H$, $h_f = H/2$, $d_f = H/2$, $\beta_f = 0.5$, $\bar{\omega} = 1.57$.) (From Liu, G. R., Lam K. Y., and Shang H. M., *Composites Part B*, 27B, 431, 1996. With permission.)

To give a qualitative explanation for this argument, we consider a simple case of an interference of a single propagating wave mode with a real wavenumber k_0. First, we consider a wave with a frequency ω, say, wave A, propagating in the positive direction of x. At an observation point x, the displacement contributed from this rightward wave A should have the form of

$$d_A = A[\cos(k_0 x) - i\sin(k_0 x)] \tag{13.16}$$

where A is a constant related to the amplitude of wave A. Note that we omitted the harmonic time term. Next, we consider a wave B with the same frequency ω but propagating in the negative direction of x. At the observation point x, the displacement contributed from the leftward wave, wave B, can be written as

$$d_B = B[\cos(-k_0 x) - i\sin(-k_0 x)] \tag{13.17}$$

where B is a constant related to the amplitude of wave B, and B could be different from A since wave B may come from a different source. It is observed that the absolute values of w_A and w_B are the constants A and B, respectively. However, the absolute value of the summation of waves A and B should be

$$|d_A + d_B| = \sqrt{(A^2 + B^2) + 2AB\cos(2k_0 x)} \tag{13.18}$$

which is an oscillatory function of x. This proves that the interference of two opposite waves could result in an additional oscillation.

Next, we consider an interference of two waves propagating in the same direction, say, rightward. Assuming there is a wave C with the same frequency ω coming from a source at the left side of the observation point x, but, rather than the source for wave A, the displacement contributed from wave C can be written as

$$d_C = C[\cos(k_0 x) - i\sin(k_0 x)] \tag{13.19}$$

where C is a constant related to the amplitude of wave C. The absolute value of the summation of wave A and C is simply the constant $(A + C)$. It is proven that the interference of two waves propagating in the same direction does not result in any additional oscillation.

For waves in a plate, there are many wave modes. The real situation can be much more complex. The modes can be propagating and evanescent, and the displacement at a frequency is a summation of all the wave modes. Hence the absolute value of displacement is no longer a constant but a function of x, even if there are only waves propagating in one direction, as shown in Figure 13.2. However, if there are waves propagating in two opposite directions, the interferences could produce additional oscillations over the curve of the absolute value of the displacement since the argument given in the above paragraphs is applicable for each propagating mode.

It can be seen from Figure 13.1 that in the region between the load and the right boundary of the flaw, there are rightward waves generated by the load

and leftward waves reflected from the right boundary of the flaw. The interferences of the leftward and rightward waves produce the additional oscillations in the absolute values of the displacement. On the other hand, in the
region immediately to the right boundary of the flaw, there are only rightward
waves because all sources in the plate (the load, the left and right boundaries
of the flaw) are located to the left of the observation points. There are no
interferences of two opposite moving waves and, therefore, no additional oscillations over the curve of absolute values of the displacement (see Figure 13.3).

This difference in the distribution of the displacement responses between
plates with and without a flaw makes it possible to determine the length
of the flaw. A technique to determine the length of the rectangular flaw needs
the following two steps. First, by obtaining Figure 13.3, we can determine the
approximate position of the right boundary of the flaw. Then, by applying the
load on the right side of the flaw, we can also determine the position of the left
boundary of the flaw.

Figure 13.4 shows the comparison of responses of the displacement on the
upper surface of the plates with and without a flaw. The dimensionless
frequency is 1.57. In flaw parameters, d_f, a_f, l_f, h_f, and β_f are, respectively, $0.5H$,
$4.0H$, $4.0H$, $0.5H$ and $0.5H$. In this case, both the right and left boundaries of
the flaw can be approximately determined. Considering responses at a point
between the load point and the left boundary of the flaw, the responses consist
of three parts. The first is generated by the load, the second comes from the
reflection from the left boundary, and the third comes from the refraction from
the right boundary. Hence, there is a possibility that the reflected and the
refracted waves interfere with each other, so that the phases of responses in
the region between the load point and the left boundary of the flaw might
not have significant change, and, therefore, there are no superimposed oscillations in the responses. In these cases like Figure 13.4, the image of the flaw

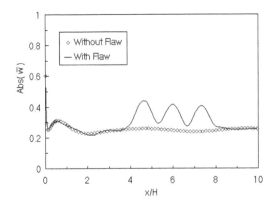

FIGURE 13.4
Displacement on the upper surface of an isotropic plate subjected to a time-harmonic line load.
Comparison of the results for plates with and without flaw. ($l_f = 4H$, $af = 4H$, $h_f = H/2$, $d_f = H/2$,
$\beta_f = 0.5$, $\bar{\omega} = 1.57$.) (From Liu, G. R., Lam K. Y., and Shang H. M., *Composites Part B*, 27B, 431,
1996. With permission.)

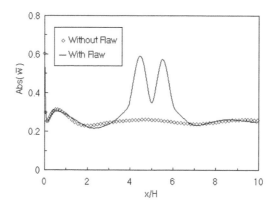

FIGURE 13.5

Displacement on the upper surface of an isotropic plate subjected to a time-harmonic line load. Comparison of the results for plates with and without flaw. ($l_f = 4H$, $a_f = 2H$, $h_f = H/2$, $d_f = H/2$, $\beta_f = 0.9$, $\bar{\omega} = 1.57$.) (From Liu, G. R., Lam K. Y., and Shang H. M., *Composites Part B*, 27B, 431, 1996. With permission.)

can be observed directly by only one step of loading. However, such cases are special and might not always happen. For general cases, the technique stated at the end of the pervious paragraph has to be used.

Figure 13.5 shows the displacement responses on the upper surface of the isotropic plate with a flaw of $\beta_f = 0.9$. The length of the flaw is $2.0H$. Because α_f is much smaller than the previous cases, images of the flaws can be quite clearly observed.

Figures 13.6–13.8 show the displacement responses on the upper surface of the hybrid laminate $[C90/G45/G-45]_s$ with a flaw length of $2.0H$. The upper G45 and G−45 layers are assumed damaged, and the factor β_f is assumed to be 0.1 (Figure 13.6), 0.5 (Figure 13.7), and 0.9 (Figure 13.8). The dimensionless frequency of the load is 1.57. For $\beta_f = 0.1$ and 0.5, the right boundary of the flaw can be approximately determined from the distribution of the displacement responses of the plate, and the left boundary of the flaw can be determined from the responses of the plate by applying the load on the right side of the right boundary of the flaw. For $\beta_f = 0.9$, a quite clear image of the flaw can be observed; hence, both the left and right boundaries of the flaw can be approximately determined.

13.4 SH Waves in Sandwich Plates

Consider a 2D problem of SH wave (horizontally polarized transverse share wave) in sandwich plates, whose layers can be assumed orthotropic. Assume that SH waves are propagating in the x direction that is also on the material principal axis. The external excitation and the field dependencies are independent of the y-axis. For antiplane (x-z plane) motion (SH wave) in an

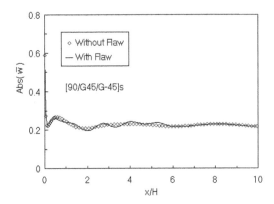

FIGURE 13.6
Displacement on the upper surface of a hybrid laminate subjected to a time-harmonic line load. Comparison of the results for plates with and without flaw. ($l_f = 4H$, $a_f = 2H$, $h_f = H/3$, $d_f = H/3$, $\beta_f = 0.1$, $\bar{\omega} = 1.57$.) (From Liu, G. R., Lam K. Y., and Shang H. M., *Composites Part B*, 27B, 431, 1996. With permission.)

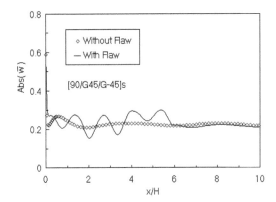

FIGURE 13.7
Displacement on the upper surface of a hybrid laminate subjected to a time-harmonic line load. Comparison of the results for plates with and without flaw. ($l_f = 4H$, $a_f = 2H$, $h_f = H/3$, $d_f = H/3$, $\beta_f = 0.5$, $\bar{\omega} = 1.57$.) (From Liu, G. R., Lam K. Y., and Shang H. M., *Composites Part B*, 27B, 431, 1996. With permission.)

orthotropic material, the system of governing differential equations (in the case free of external body force) is expressed as

$$\frac{\partial \sigma_{xy}}{\partial x} + \frac{\partial \sigma_{yz}}{\partial z} = \rho \ddot{v} \tag{13.20}$$

where the dot indicates differentiation with respect to time, ρ is the mass density, and v is the displacement components in the y directions, while σ_{xy} and σ_{yz} are the shear stresses. For an orthotropic material with x-axis being

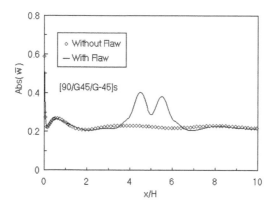

FIGURE 13.8

Displacement on the upper surface of a hybrid laminate subjected to a time-harmonic line load. Comparison of the results for plates with and without flaw. ($l_f = 4H$, $a_f = 2H$, $h_f = H/3$, $d_f = H/3$, $\beta_f = 0.9$, $\bar{\omega} = 1.57$.) (From Liu, G. R., Lam K. Y., and Shang H. M., *Composites Part B, 27B*, 431, 1996. With permission.)

one of the material principal axis, the stress-strain relations can be given as

$$\sigma_{yz} = G_{yz}\varepsilon_{yz} \tag{13.21}$$

$$\sigma_{xy} = G_{xy}\varepsilon_{xy} \tag{13.22}$$

where ε_{xy}, ε_{xy} are the shear strains, and G_{yz} and G_{xy} are shear moduli in the y-z and x-y planes, respectively. The relationship between the strains and the displacement is given by

$$\varepsilon_{xy} = \frac{\partial v}{\partial x} \tag{13.23}$$

$$\varepsilon_{yz} = \frac{\partial v}{\partial z} \tag{13.24}$$

Using Eqs. (13.21) to (13.24), the equation of motion (13.20) can be rewritten in terms of displacement v as

$$G_{yz}\frac{\partial^2 v}{\partial z^2} + G_{xy}\frac{\partial^2 v}{\partial x^2} = \rho\ddot{v} \tag{13.25}$$

13.5   Strip Element Equation for SH Waves

Consider a sandwich plate with three orthotropic layers. We follow exactly the procedure of SEM developed in Chapters 11 and 12 for anisotropic laminates. The thickness of the plate is denoted by H, as shown in Figure 13.9. The plate is divided into N strip elements with more than one element in

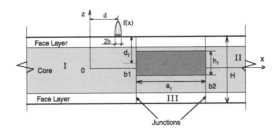

FIGURE 13.9

Division of a sandwich plate with a horizontal flaw into domains in which the SEM can be applied. (From Liu G. R. and Lam K. Y., *Composites Structure*, 34, 1996. With permission.)

each layer. The thickness, two shear moduli, and the density of an element are defined by h, G_{yz}, G_{xy}, and ρ. These parameters could be different from one element to another.

In each element, the displacement $v(x, z, t)$ is assumed to be of the form

$$v(x, z, t) = \mathbf{N}(z)\mathbf{V}(x)\exp(-i\omega t) \tag{13.26}$$

By applying the principle of virtual work to each element, a set of approximate differential equations can be obtained for each element. The procedure is the same as that given in Chapter 8. Assembling all the elements and using the boundary conditions on the horizontal node lines, a system of approximate governing differential equations for the total plate can finally be obtained as follows:

$$\mathbf{q}_t = \left[-\mathbf{A}_{1t}\frac{\partial^2 \mathbf{V}_t}{\partial x^2} + \mathbf{A}_{0t}\mathbf{V}_t - \omega^2 \mathbf{M}_t \mathbf{V}_t \right] \tag{13.27}$$

where the matrices $\mathbf{A}_{1t}$, $\mathbf{A}_{0t}$, $\mathbf{M}_t$, and the vector $\mathbf{q}_t$ can be obtained by assembling the corresponding matrices $\mathbf{A}_1$, $\mathbf{A}_0$, $\mathbf{M}$, and vector $\mathbf{V}_e$ of adjacent elements just as is done in the FEM. The matrices $\mathbf{A}_1$, $\mathbf{A}_0$, and $\mathbf{M}$ for an element are given by

$$\mathbf{A}_0 = \frac{G_{yz}}{3h}\begin{bmatrix} 7 & -8 & 1 \\ & 16 & -8 \\ \text{sym.} & & 7 \end{bmatrix} \tag{13.28}$$

$$\mathbf{A}_1 = \frac{hG_{xy}}{30}\begin{bmatrix} 4 & 2 & -1 \\ & 16 & 2 \\ \text{sym.} & & 4 \end{bmatrix} \tag{13.29}$$

$$\mathbf{M} = \frac{\rho h}{30}\begin{bmatrix} 4 & 2 & -1 \\ & 16 & 2 \\ \text{sym.} & & 4 \end{bmatrix} \tag{13.30}$$

The vector $\mathbf{q}_t$ is the amplitude vector of the shear traction vector applied to the horizontal nodal lines of the plate.

Note that the dimensions of matrices $\mathbf{A}_{1t}$, $\mathbf{A}_{0t}$, $\mathbf{M}_t$ in Eq. (13.27) is $M \times M$ where $M = 2N + 1$ because there is only one degree of freedom at a node. The dimension of vectors $\mathbf{q}_t$ and $\mathbf{V}_t$ is $M \times 1$.

For a plate with vertical boundaries, the general solution of Eq. (13.27) consists of two parts. One is a particular solution that satisfies Eq. (13.27), the other is the complementary solution that satisfies the associated homogeneous equation of Eq. (13.27).

13.6   Particular Solution

A particular solution to Eq. (13.27) can be obtained using the seminumerical method (SNM) described in Chapter 12. The method was originally developed by Waas (1972) for Lame waves in layered isotropic structures. The application of the Fourier transform to Eq. (13.27) leads to the following equation in the wave number domain:

$$\tilde{\mathbf{q}}_t = [k^2 \mathbf{A}_{1t} + \mathbf{A}_{0t} - \omega^2 \mathbf{M}_t] \tilde{\mathbf{V}}_p \tag{13.31}$$

Applying the modal analysis technique to Eq. (13.31), we find

$$\tilde{\mathbf{V}}_p = -\sum_{m=1}^{M} \frac{\varphi_m^T \tilde{\mathbf{q}}_t \varphi_m}{(k^2 - k_m^2) \varphi_m^T \mathbf{A}_{1t} \varphi_m} \tag{13.32}$$

where M is the number of rows of square matrices $\mathbf{A}_{0t}$ (or $\mathbf{A}_{1t}$, or $\mathbf{M}_t$), and k_m and φ_m are the mth eigenvalue and eigenvector obtained for a given ω from the following eigenvalue equation:

$$0 = [(\omega^2 \mathbf{M}_t - \mathbf{A}_{0t}) - k_m^2 \mathbf{A}_{1t}] \varphi_m \tag{13.33}$$

Finally,

$$\mathbf{V}_p = \begin{cases} i \sum_{m=1}^{M} \dfrac{\varphi_m^T \mathbf{q}_0 \varphi_m}{\varphi_m^T \mathbf{A}_{1t} \varphi_m} \dfrac{\exp[-i k_m^+ (x - x_0)]}{2 k_m^+}, & \text{for } x \geq x_0 \\[4mm] -i \sum_{m=1}^{M} \dfrac{\varphi_m^T \mathbf{q}_0 \varphi_m}{\varphi_m^T \mathbf{A}_{1t} \varphi_m} \dfrac{\exp[-i k_m^- (x - x_0)]}{2 k_m^-}, & \text{for } x < x_0 \end{cases} \tag{13.34}$$

It may be noted that eigenvalues k_m^2 are real due to matrix $\mathbf{A}_{0t}$, $\mathbf{A}_{1t}$, and $\mathbf{M}_t$ are real symmetrical matrices. However, k_m^2 could be negative. Hence, k_m can be real and pure imaginary, and the 2M poles are located on either the real or the imaginary axis in the complex plane of k. In Eq. (13.34), k_m^+ denotes a pole locating at negative parts of real and imaginary axes. These poles correspond to waves propagating in the positive x direction (leftward). And k_m^- denotes a pole locating at positive parts of real and imaginary axes. These poles correspond to waves propagating in the negative x direction (rightward).

From Eq. (13.34), it can be found that the SH wave field in an infinite transversely isotropic plate consists of two groups of wave modes. One is the group of propagating wave modes corresponding to the poles on the real wavenumber axis; the other is the group of nonpropagating wave modes corresponding to the poles on the imaginary wavenumber axis. For a wave field in a general anisotropic plate, there is one more group of wave modes, evanescent-propagating wave modes, corresponding to the poles with both non-zero real and non-zero imaginary parts.

Equation (13.34) gives a 2D Green's function of the displacement for an antiplane problem for transversely isotropic laminates. The Green's function of stresses can easily be obtained using Eq. (13.34). If the external shear force is distributed in the x direction, the solution can be obtained in the form of a superposition integral over $\mathbf{V}_p$.

13.7 Complementary Solution

The complementary solution of the associated homogeneous equation of Eq. (13.27) can be expressed by applying superposition of the eigenvectors obtained from Eq. (13.33):

$$\mathbf{V}_c = \sum_{m=1}^{M} C_m \varphi_m e^{ik_m^+(x-L)} + \sum_{m=1}^{M} C_{M+m} \varphi_m e^{ik_m^- x} = \mathbf{B}(x)\mathbf{C} \qquad (13.35)$$

where the subscript c denotes the complementary solution. The unknown constant vector $\mathbf{C}$ can be determined using boundary conditions on boundaries perpendicular to the plate plane. The value of L can be arbitrary, but it is usually the distance between the two boundaries.

13.8 General Solution

The general solution of Eq. (13.27) can be given by adding the particular solution (13.34) and the complementary solution (13.35):

$$\mathbf{V}_g = \mathbf{V}_c + \mathbf{V}_p = \mathbf{B}(x)\mathbf{C} + \mathbf{V}_p \qquad (13.36)$$

where the subscript g denotes the general solution. Using the boundary conditions on the vertical boundaries of the plate, a relationship between the tractions and the displacements on the vertical boundaries is obtained as

$$\mathbf{R}_b = \mathbf{K}\mathbf{V}_b + \mathbf{S}_b \tag{13.37}$$

where

$$\mathbf{R}_b = \{\mathbf{R}_b^L \quad \mathbf{R}_b^R\}^T, \quad \mathbf{V}_b = \{\mathbf{V}_b^L \quad \mathbf{V}_b^R\}^T \tag{13.38}$$

are the external traction and displacement vectors on the vertical boundaries. The matrix $\mathbf{K}$ is the stiffness matrix given by

$$\mathbf{K} = \begin{bmatrix} \mathbf{R}_t \, \mathbf{B}'_L \, \mathbf{B}_d^{-1} \\ \mathbf{R}_t \mathbf{B}'_R \, \mathbf{B}_d^{-1} \end{bmatrix} \tag{13.39}$$

and the vector $\mathbf{S}_b$ is the equivalent external force acting on the vertical boundaries.

$$\mathbf{S}_p = -\begin{bmatrix} \mathbf{R}_t \, \mathbf{B}'_L \, \mathbf{B}_d^{-1} \\ \mathbf{R}_t \mathbf{B}'_R \, \mathbf{B}_d^{-1} \end{bmatrix} \begin{Bmatrix} \mathbf{V}_p^L \\ \mathbf{V}_p^R \end{Bmatrix} \tag{13.40}$$

In the above equations, the prime indicates differentiation with respect to x, and the matrix $\mathbf{B}_d$ is given by

$$\mathbf{B}_d = \begin{bmatrix} \mathbf{B}_L \\ \mathbf{B}_R \end{bmatrix} \tag{13.41}$$

The matrix $\mathbf{R}_t$ is obtained by combining matrices $\mathbf{R}$ for all the elements. The matrix $\mathbf{R}$ for an element is given by

$$\mathbf{R} = G_{xy} \begin{bmatrix} 1 & 0 & 0 \\ 0 & 1 & 0 \\ 0 & 0 & 1 \end{bmatrix} \tag{13.42}$$

The superscripts or subscripts L and R indicate that the matrix or vector has been evaluated on the left and right vertical boundaries.

13.9 SH Waves Scattered by Flaws

In this section, computations of SH wave responses in the frequency domain were carried out for a sandwich plate using the SEM code developed above. The dimensions and material constants for the sandwich plate used for the computations are given in Table 13.1.

The following dimensionless parameters are used:

$$\bar{x} = x/H, \quad \bar{v} = G_0 v/q_0, \quad \bar{\omega} = \omega H/c_s$$
$$c_s = \sqrt{G_0/\rho_0}, \quad \bar{d}_f = d_f/H \tag{13.43}$$

where G_0, ρ_0, and c_s are the shear module G_{xy}, density, and shear wave velocity of the reference material, the face layer.

The flaw considered is assumed a rectangular damaged zone in which the density of the material remains the same, but the equivalent elastic constants are reduced to be

$$G_{xy}^f = (1 - \beta_f)G_{xy}, \quad G_{yz}^f = (1 - \beta_f)G_{yz} \tag{13.44}$$

where β_f is the damage factor ($0 \le \beta_f \le 1$). The height, length, and depth of the flaw are denoted, respectively, by h_f, a_f, and d_f (see Figure 13.9). The depth of the flaw is defined as the distance between the upper surface of the plate and horizontal center line of the flaw.

Displacement responses of a sandwich plate without a flaw subjected to a time-harmonic line shear load were computed by the SNM and SEM programs. The load is assumed to act at $x = 0$ and on the upper surface of the plate. Figure 13.10 gives the responses of the displacement in the y direction on the upper surface of the plate. Sixteen elements were used in both programs. In the SEM program, d_f, a_f, and h_f are set, respectively, to be $0.5H$, $4.0H$, and $5H/6$. However, the damage factor β_f is set to be 0.0, namely no flaws in the plate. Excellent agreement is observed. The convergence of the results was also confirmed using a different number of elements, and no significant difference between the results obtained using 8 elements and 16 elements for both programs.

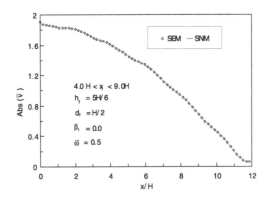

FIGURE 13.10
Displacement on the upper surface of a sandwich plate subjected to a time-harmonic line shear load acting at $x = 0$ and on the upper surface of the plate. Comparison of the results according to the seminumerical method (SNM) and the strip element method (SEM). (From Liu, G. R. and Lam K. Y., *Composites Structure*, 34, 1996. With permission.)

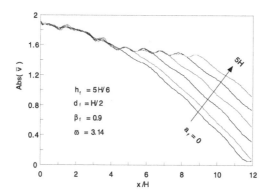

FIGURE 13.11

Displacement on the upper surface of a sandwich plate subjected to a time-harmonic line shear load acting at $x = 0$ and on the upper surface of the plate. Comparison of the results for plates with and without flaw. The left boundary of the flaw is at $x = 4.0H$. (From Liu, G. R. and Lam K. Y., *Composites Structure*, 34, 1996. With permission.)

Figure 13.11 shows the comparison of responses of the displacement v on the upper surface of the plates with and without flaw. The plate is loaded by a time-harmonic line shear load with a dimensionless frequency of 3.14. For the plate with a flaw, d_f, β_f, and h_f are, respectively, $0.5H$, 0.9, and $5H/6$, and the crack length varies from $1.0H$ to $5.0H$ with an increment of $1.0H$. From Figure 13.11, it is found that there are superimposed oscillations between the load point and the right boundary of the flaw. These superimposed oscillations end at the right boundary of the flaw. The reason for this phenomenon is the same as that explained in Section 13.3. The presence of superimposed displacement makes it possible to determine the length of the flaw. A technique to determine the length of the flaw needs the following two steps. First, by obtaining Figure 13.11, we can determine the approximate position of the right boundary of the flaw. Then, by applying the load on the right side of the flaw, we can also determine the position of the left boundary of the flaw.

Figure 13.12 shows the comparison of responses of the displacement v on the upper surface of the plates with flaws of various damage factors. It can be found that the smaller the damage factor, the less scattered are the wave fields. Therefore, the damage factor can be qualitatively determined by comparing the scattered wave fields.

Another technique to determine the length of the flaw is to move a distributing load along the surface of the plate (change the value of d in Figure 13.9), and in the meantime obtain the resulting displacement in the y-direction at the central point of the distributed load ($x = d$). The displacement response can be plotted while the load scans over the flaw. For a plate without flaw, the scanning result is a horizontal line because the responses of the plate without a flaw at the central point of load are independent of the location of the load. However, for a plate with a flaw, the scanning result will no longer be a constant, and the responses can be expected to have a significant

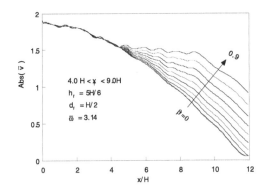

FIGURE 13.12
Displacement on the upper surface of a sandwich plate subjected to a time-harmonic line shear load acting at $x = 0$ and on the upper surface of the plate. Comparison of the results for plates with flaws of various damage factors. (From Liu, G. R. and Lam K. Y., *Composites Structure*, 34, 1996. With permission.)

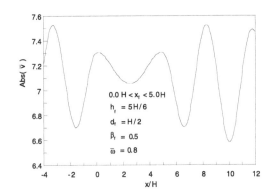

FIGURE 13.13
Scanning results of a sandwich plate with a rectangular flaw. The left boundary of the flaw is on the origin. The dimensionless frequency is 0.8. (From Liu, G. R. and Lam K. Y., *Composites Structure*, 34, 1996. With permission.)

change when the load passes over the boundaries of the flaw. This provides a way to evaluate the length of the flaw in the plate.

The load acting on the surface of the plate is given by Eqs. (12.43) and (12.44). The distribution of the load along the x-direction is shown in Figure 13.9. Parameters b and α in Eq. (12.44) are chosen to be $0.1H$ and $0.05H$, respectively.

The scanning results for the displacement in the thickness direction on the surface of a sandwich plate due to harmonic loads with variations of frequencies are first investigated and are shown in Figures 13.13 to 13.15. The damage factor is set to be 0.5, and the length of the flaw is $5.0H$. Figure 13.13 is for a load with a dimensionless frequency of 0.8 and the corresponding

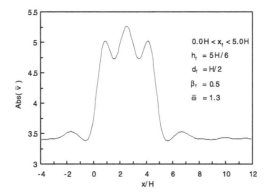

FIGURE 13.14
Scanning results of a sandwich plate with a rectangular flaw. The left boundary of the flaw is on the origin. The dimensionless frequency is 1.3. (From Liu, G. R. and Lam K. Y., *Composites Structure*, 34, 1996. With permission.)

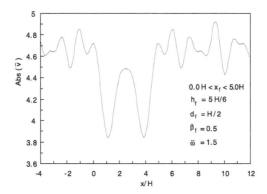

FIGURE 13.15
Scanning results of a sandwich plate with a rectangular flaw. The left boundary of the flaw is on the origin. The dimensionless frequency is 1.5. (From Liu, G. R. and Lam K. Y., *Composites Structure*, 34, 1996. With permission.)

shear wavelength in the face material is about $7.85H$. Figure 13.14 is for a load with a dimensionless frequency of 1.3 and the corresponding shear wavelength in the face material is about $4.83H$. Figure 13.15 is for a load with a dimensionless frequency of 1.5 and the corresponding shear wavelength in the face material is about $4.19H$. In comparing these figures, it is found that images of the flaw can be observed from the figures, and that the dimensionless frequency of 1.3 is the best to determine the location of the flaw. Hence, a clearer image of the flaw can be observed by choosing properly a frequency for the harmonic load. Figure 13.16 shows scanning results for a sandwich plate with a flaw of $3.0H$ in length. The dimensionless frequency of the load is 1.3. A clear image of the flaw can be observed from the figure. Hence, both the left and right boundaries of the flaw can be approximately determined.

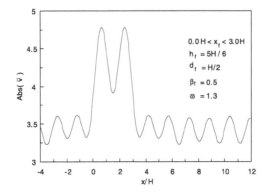

FIGURE 13.16
Scanning results of a sandwich plate with a rectangular flaw. The left boundary of the flaw is on the origin. The dimensionless frequency is $3.0H$. (From Liu, G. R. and Lam K. Y., *Composites Structure*, 34, 1996. With permission.)

13.10 Remarks

In this chapter, the SEM has been used to investigate the scattering of waves generated by a line harmonic load applied to an anisotropic laminate containing a flaw. A very simple rectangular flaw has been considered. From the computed results, it is possible to detect the presence and determine the position of the rectangular flaw, and to estimate its size by examining the absolute value of the displacement responses. SEM is a powerful tool for the investigation. The quantitative characterization of nonrectangular flaws is much more difficult and requires more detailed investigations and the use of inverse procedures.

We have also derived the SEM formulations for SH waves in orthotropic plates and used them to investigate the scattering of waves generated by a time-harmonic shear load applied to a sandwich plate containing a flaw. Using a method for the characterization of the flaws in sandwich plates, we can obtain the length of the flaw from the scanning results of responses of the plate, while a time-harmonic load scans over the flaw. We can qualitatively determine the damage factor of the flaw from the scattered wave field by the flaw. This technique is expected to be of practical use in the characterization of flaws in sandwich plates. More detailed and incisive studies on characterization of flaws, including determining the damage factor in sandwich plates, is given by Wang et al. (1996).

14

Bending Waves in Anisotropic Laminated Plates

14.1 Introduction

The previous chapters presented strip element methods for analyzing waves scattered by cracks and flaws in composite laminates. The formulation was in the framework of 2D or 3D solid mechanics. The strip element method can also be used to solve many other problems of wave motion and structural mechanics. This chapter uses the strip element method to analyze bending deformation and harmonic and transient bending waves in composite laminated plates.

In the analysis of laminated plates subjected to dynamic loading, the classical laminated plate theory (CLPT) is widely used because of its simplicity. More complex plate theories have also been used, such as the first order shear deformation theory (FSDT) (Whitney, 1969, 1970; Reissner, 1972), the third-order laminated theory (Reddy, 1984b), and many others (see, e.g., Reddy, 1984a, 1984b, 1985). The exact solution for the dynamic response of composite laminated plates is very difficult to obtain and is available only for plate theories applied to simply supported rectangular plates (Khdeir and Reddy, 1989). Therefore, approximate methods have been proposed for dynamic analysis of laminated plates. Most of the work available in the published literature is for plates with simple boundary conditions. Lu (1996) employed the Rayleigh-Ritz method and the method of superposition of normal modes to calculate the dynamic response of laminated angle-ply plates with clamped boundary conditions subjected to explosive loading. When the problem involves complex geometries and boundary conditions, one must resort to numerical methods.

Many numerical methods have been suggested to analyze the dynamic response of plates, in which the finite element method (FEM) has become a universally applicable technique for boundary and initial value problems. Reismann et al. (1968) and Lee et al. (1969) analyzed a simply supported rectangular isotropic plate subjected to a suddenly applied uniformly distributed load over a square area on the plate. Rock and Hinton (1974) presented a transient finite element analysis of thick and thin isotropic plates. Akay (1980) determined the large deflection transient response of isotropic plates using a mixed finite element. Reddy (1983) presented a finite element result for the transient analysis of layered composite plates based on the FSDT. Mallikarjuna and Kant (1988) presented an isoparametric finite element formulation based on a high-order displacement model for dynamic analysis of multilayer symmetric composite plates.

The formulation in this chapter is based on classic laminated plate theory or CLPT (Reddy 1997). We skip the details about the plate theory, as the concepts and equations related to CLPT are very clearly presented in Reddy (1997). Waves in infinite laminated plates and finite laminated plates with defined boundary conditions are considered. Note that topics for wave motion in finite laminated plates are very similar to the topics of vibration. Therefore, methods discussed in this chapter are applicable for vibration analysis of composite laminated plates.

The work introduced here was originally proposed by Wang and co-workers (1997, 1998, 2000a, 2000b, 2001). We first derive a set of ordinary differential equations for general composite laminated plates of rectangular shape by dividing the plates into strip elements and using the boundary conditions on two opposite edges of the laminated plate. These equations can be solved analytically to obtain the general solution for the equation system. The general solution contains constants that can be determined using boundary conditions on the other two edges of the laminated plate. Numerical examples are presented in application of the SEM formulation developed for analyzing bending deformation and bending waves in composite laminated plates in both frequency and time domains.

14.2 Governing Equation

Consider a rectangular laminated plate of overall thickness H made of K layers fiber-reinforced composites, as shown in Figure 14.1. The plate lies in the x-y plane and is bounded in the region $-a/2 \leq x \leq a/2$ and $0 \leq y \leq b$. The reference plane, $z = 0$, is chosen at the middle plane of the plate. This chapter considers only symmetrically stacked composite laminated plates. In this case, the geometrical middle plane is the neutral plane of bending where the normal stresses are zero. This also implies that there is no coupling of bending deflection and in-plane deformation. The fiber direction of the kth layer is given by the angle α_k as shown in Figure 14.2.

Using the classic laminated plate theory of bending (see Reddy 1997), the lateral mid-surface deflection $w(x, y)$ of the plate subjected to a distributed

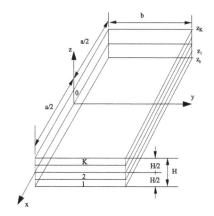

FIGURE 14.1
Composite Laminated plate and the coordinate system.

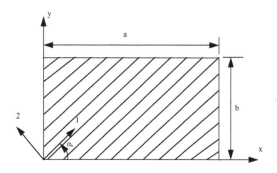

FIGURE 14.2
The kth layer with fiber orientation of α_k.

vertical force of $q(x,y)$ is governed by the following partial differential equation:

$$W = D_{11}\frac{\partial^4 w}{\partial x^4} + 4D_{16}\frac{\partial^4 w}{\partial x^3 \partial y} + 2(D_{12} + 2D_{66})\frac{\partial^4 w}{\partial x^2 \partial y^2} + 4D_{26}\frac{\partial^4 w}{\partial x \partial y^3}$$

$$+ D_{22}\frac{\partial^4 w}{\partial y^4} + I_0\frac{\partial^2 w}{\partial t^2} - I_2\frac{\partial^2}{\partial t^2}\left(\frac{\partial^2 w}{\partial x^2} + \frac{\partial^2 w}{\partial y^2}\right) - q(x,y,t) = 0 \qquad (14.1)$$

where D_{11}, D_{12}, D_{16}, D_{22}, D_{26}, and D_{66} are the coefficients of bending stiffness, which can be obtained using engineering constants of materials such as the modulus of elasticity parallel to the fiber orientation denoted by E_1, the modulus of elasticity perpendicular to the fiber orientation denoted by E_2, in plane shear modulus G_{12}, and the Poisson's ratio v_{12}. Equation (14.1) is established based on classic laminated plate theory of bending using the well-known

Kirchhoff assumption. Details lead to Eq. (14.1) and the formulae for the bending stiffness coefficients can be found in Reddy (1997). As for the moment terms, when all the layers have the same density, the mass moments of the inertia I_0 and rotatory inertia I_2 can be given as

$$I_0 = \int_{-H/2}^{H/2} \rho \, dz = \rho H, \quad I_2 = \int_{-H/2}^{H/2} \rho z^2 \, dz = \frac{1}{12} \rho H^3 \qquad (14.2)$$

where ρ is the mass density of the material. The following principle of minimum potential energy is used to establish the system equation for the strip element method:

$$\iint_\Omega W \delta w \, dx \, dy + \oint_\Gamma \left(\frac{\partial M_{ns}}{\partial s} + Q_n - \bar{q} \right) \delta w \, ds - \oint_\Gamma (M_n - \overline{M}_n) \frac{\partial \delta w}{\partial n} ds = 0 \qquad (14.3)$$

where M and Q denote moment and shear force, and the bar over them stands for prescribed value on the boundaries of the problem domain. Equation (14.3) provides a weak form of system equation for plates.

14.3 Strip Element Equation

We first consider a laminated plate of infinite length, shown in Figure 14.3. The plate occupies the region of $-\infty \le x \le \infty$ and $0 \le y \le b$, while the problem domain is bounded by $-a/2 \le x \le a/2$, $0 \le y \le b$, and $-H/2 \le z \le H/2$. The boundaries are denoted by S_1, S_2, S_3, and S_4. The infinite plate is divided in the y direction into N strip elements. The displacement field in an element is assumed in the form of

$$w(x, y, t) = \mathbf{N}(y) \mathbf{V}^e(x, t) \qquad (14.4)$$

where $\mathbf{N}(y)$ is the matrix of shape functions given by

$$\mathbf{N}(y) = [n_1(y) \quad n_2(y) \quad n_3(y) \quad n_4(y) \quad n_5(y) \quad n_6(y)] \qquad (14.5)$$

in which the elements in matrix $\mathbf{N}(y)$ are obtained using Hermite interpolation

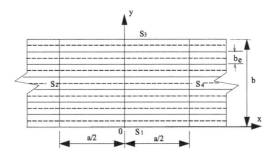

FIGURE 14.3
Infinite plate divided into strip elements in the y direction. The problem domain is bounded by S_1, S_2, S_3, and S_4.

functions. They are listed below.

$$n_1(y) = 1 - 23\frac{y^2}{b_e^2} + 66\frac{y^3}{b_e^3} - 68\frac{y^4}{b_e^4} + 24\frac{y^5}{b_e^5}$$

$$n_2(y) = y\left(1 - 6\frac{y}{b_e} + 13\frac{y^2}{b_e^2} - 12\frac{y^3}{b_e^3} + 4\frac{y^4}{b_e^4}\right)$$

$$n_3(y) = 16\frac{y^2}{b_e^2} - 32\frac{y^3}{b_e^3} + 16\frac{y^4}{b_e^4}$$

$$\hspace{8cm}(14.6)$$

$$n_4(y) = y\left(-8\frac{y}{b_e} + 32\frac{y^2}{b_e^2} - 40\frac{y^3}{b_e^3} + 16\frac{y^4}{b_e^4}\right)$$

$$n_5(y) = 7\frac{y^2}{b_e^2} - 34\frac{y^3}{b_e^3} + 52\frac{y^4}{b_e^4} - 24\frac{y^5}{b_{eb}^5}$$

$$n_6(y) = y\left(-\frac{y}{b_e} + 5\frac{y^2}{b_e^2} - 8\frac{y^3}{b_e^3} + 4\frac{y^4}{b_e^4}\right)$$

where b_e is the width of the strip element.

In Eq. (14.4), $\mathbf{V}^e(x, t)$ is the *generalized* displacement vector given by

$$\mathbf{V}^e(x, t) = \begin{cases} v_1 = w_1(x,t) \\ v_2 = \theta_1(x,t) \\ v_3 = w_2(x,t) \\ v_4 = \theta_2(x,t) \\ v_5 = w_3(x,t) \\ v_6 = \theta_3(x,t) \end{cases} \left.\begin{array}{l} \\ \end{array}\right\} \text{lower nodal line} \atop \left.\begin{array}{l} \\ \end{array}\right\} \text{middle nodal line} \atop \left.\begin{array}{l} \\ \end{array}\right\} \text{upper nodal line} \hspace{1cm}(14.7)$$

in which w_i ($i = 1, 2, 3$) are the lateral deflections on the nodal lines, and θ_i ($i = 1, 2, 3$) are the rotation angles on the nodal lines given by

$$\theta_i = \left.\frac{\partial w}{\partial y}\right|_{y=y_i} \tag{14.8}$$

Generalized is used for the fact that the displacement vector includes not only translational displacements but also angular displacements (rotations).

Applying the principle of minimum potential energy to a strip element in an infinite plate, Eq. (14.3) can be written as

$$
\int_0^{b_e} W \delta w \, dy - \left(\frac{\partial M_{yx}}{\partial x} + Q_y\right)\delta w \bigg|_{y=0} + \left(\frac{\partial M_{yx}}{\partial x} + Q_y\right)\delta w \bigg|_{y=b_e} + M_y \frac{\partial \delta w}{\partial y}\bigg|_{y=0}
$$
$$
- M_y \frac{\partial \delta w}{\partial y}\bigg|_{y=b_e} = -\bar{q}\,\delta w|_{y=0} + \overline{M}_y \frac{\partial \delta w}{\partial y}\bigg|_{y=0} + \bar{q}\,\delta w|_{y=b_e} - \overline{M}_y \frac{\partial \delta w}{\partial y}\bigg|_{y=b_e} \tag{14.9}
$$

or in the matrix form of

$$\int_0^{b_e} W \delta w \, dy + \mathbf{R}_y \delta \mathbf{V}^e(x,t) = \mathbf{T}^e \delta \mathbf{V}^e(x,t) \tag{14.10}$$

where $\mathbf{R}_y$ is the *generalized* force vector acting on the boundary lines of the strip element defined by

$$\mathbf{R}_y = \left[-\left(\frac{\partial M_{yx}}{\partial x} + Q_y\right)\bigg|_{y=0} \quad M_y|_{y=0} \quad 0 \quad 0 \quad \left(\frac{\partial M_{yx}}{\partial x} + Q_y\right)\bigg|_{y=b_e} \quad -M_y|_{y=b_e}\right] \tag{14.11}$$

Here again *generalized* is used for the fact that the force vector includes not only shear forces but also moments. In Eq. (14.10), $\mathbf{T}^e$ is the generalized external force vector acting on the boundary lines of the strip element with the form of

$$\mathbf{T}^e = \left[-\bar{q}|_{y=0} \quad \overline{M}_y|_{y=0} \quad 0 \quad 0 \quad \bar{q}|_{y=b_e} \quad -\overline{M}_y|_{y=b_e}\right] \tag{14.12}$$

The first term on the left side of Eq. (14.10) leads to

$$\int_0^{b_e} W\delta w\, dy$$

$$= \int_0^{b_e} \left(D_{11}\frac{\partial^4 w}{\partial x^4} + 4D_{16}\frac{\partial^4 w}{\partial x^3 \partial y} + 2(D_{12}+2D_{66})\frac{\partial^4 w}{\partial x^2 \partial y^2} + 4D_{26}\frac{\partial^4 w}{\partial x \partial y^3} \right.$$

$$\left. + D_{22}\frac{\partial^4 w}{\partial y^4} + I_0\frac{\partial^2 w}{\partial t^2} - I_2\frac{\partial^2}{\partial t^2}\left(\frac{\partial^2 w}{\partial x^2} + \frac{\partial^2 w}{\partial y^2}\right) - q \right)\delta w\, dy$$

$$= \int_0^{b_e} \left(D_{11}\mathbf{N}\frac{\partial^4 \mathbf{V}^e}{\partial x^4} + 4D_{16}\frac{d\mathbf{N}}{dy}\frac{\partial^3 \mathbf{V}^e}{\partial x^3} + 2(D_{12}+2D_{66})\frac{d^2\mathbf{N}}{dy^2}\frac{\partial^2 \mathbf{V}^e}{\partial x^2} + 4D_{26}\frac{d^3\mathbf{N}}{dy^3}\frac{\partial \mathbf{V}^e}{\partial x} \right.$$

$$\left. + D_{22}\frac{d^4\mathbf{N}}{dy^4}\mathbf{V}^e + I_0\mathbf{N}\frac{\partial^2 \mathbf{V}^e}{\partial t^2} - I_2\left(\mathbf{N}\frac{\partial^4 \mathbf{V}^e}{\partial x^2 \partial t^2} + \frac{d^2\mathbf{N}}{dy^2}\frac{\partial^2 \mathbf{V}^e}{\partial t^2}\right) - q \right)\delta w\, dy$$

$$= \int_0^{b_e} \left(D_{11}\sum_{j=1}^6 \left(n_j\frac{\partial^4 w_j}{\partial x^4}\right) + 4D_{16}\sum_{j=1}^6 \left(\frac{dn_j}{dy}\frac{\partial^3 w_j}{\partial x^3}\right) + 2(D_{12}+2D_{66})\sum_{j=1}^6 \left(\frac{d^2 n_j}{dy^2}\frac{\partial^2 w_j}{\partial x^2}\right) \right.$$

$$+ 4D_{26}\sum_{j=1}^6 \left(\frac{d^3 n_j}{dy^3}\frac{\partial w_j}{\partial x}\right) + D_{22}\sum_{j=1}^6 \frac{d^4 n_j}{dy^4}w_j + I_0\sum_{j=1}^6 \left(n_j\frac{\partial^2 w_j}{\partial t^2}\right)$$

$$\left. - I_2\sum_{j=1}^6 \left(n_j\frac{\partial^4 w_j}{\partial x^2 \partial t^2} + \frac{\partial^2 n_j}{\partial y^2}\frac{\partial^2 w_j}{\partial t^2}\right) - q \right)\sum_{i=1}^6 n_i\delta w_i\, dy$$

$$= \sum_{i=1}^6 \left\{ \sum_{j=1}^6 \left(A_{1ij}^e\frac{\partial^4 w_j}{\partial x^4} + A_{2ij}^e\frac{\partial^3 w_j}{\partial x^3} + A_{3ij}^e\frac{\partial^2 w_j}{\partial x^2} + A_{4ij}^e\frac{\partial w_j}{\partial x} + A_{5ij}^e w_j + A_{6ij}^e\frac{\partial^2 w_j}{\partial t^2} \right) \right.$$

$$\left. - A_{7ij}^e\frac{\partial^2 w_j}{\partial t^2} - A_{8ij}^e\frac{\partial^2}{\partial t^2}\left(\frac{\partial^2 w_j}{\partial x^2}\right) - p_i^e \right\}\delta w_i \tag{14.13}$$

where

$$A_{1ij}^e = D_{11}\int_0^{b_e} n_i n_j\, dy$$

$$A_{2ij}^e = 4D_{16}\int_0^{b_e} n_i\frac{dn_j}{dy}\, dy$$

$$A_{3ij}^e = 2(D_{12}+2D_{66})\int_0^{b_e} n_i\frac{d^2 n_j}{dy^2}\, dy$$

$$A_{4ij}^e = 4D_{26}\int_0^{b_e} n_i\frac{d^3 n_j}{dy^3}\, dy$$

$$A^e_{5ij} = D_{22} \int_0^{b_e} n_i \frac{d^4 n_j}{dy^4} dy$$

$$A^e_{6ij} = I_0 \int_0^{b_e} n_i n_j \, dy$$

$$A^e_{7ij} = I_2 \int_0^{b_e} n_i \frac{d^2 n_j}{dy^2} dy$$

$$A^e_{8ij} = I_2 \int_0^{b_e} n_i n_j \, dy$$

$$P^e_i = \int_0^{b_e} q n_i \, dy$$

with $i, j = 1, 2,...,6$. Carrying out the above integrals, we found in the following matrices.

$$\mathbf{A}^e_1 = D_{11} \begin{bmatrix} \dfrac{523b_e}{3465} & \dfrac{19b_e^2}{2310} & \dfrac{4b_e}{63} & \dfrac{-8b_e^2}{693} & \dfrac{131b_e}{6930} & \dfrac{-29b_e^2}{13860} \\[2mm] & \dfrac{2b_e^3}{3465} & \dfrac{2b_e^2}{315} & \dfrac{-b_e^3}{1155} & \dfrac{29b_e^2}{13860} & \dfrac{-b_e^3}{4620} \\[2mm] & & \dfrac{128b_e}{315} & 0 & \dfrac{4b_e}{63} & \dfrac{-2b_e^2}{315} \\[2mm] & & & \dfrac{32b_e^3}{3465} & \dfrac{8b_e^2}{693} & \dfrac{-b_e^3}{1155} \\[2mm] & \text{sym.} & & & \dfrac{523b_e}{3465} & \dfrac{-19b_e^2}{2310} \\[2mm] & & & & & \dfrac{2b_e^3}{3465} \end{bmatrix} \tag{14.14}$$

$$\mathbf{A}^e_2 = 4D_{16} \begin{bmatrix} \dfrac{1}{2} & \dfrac{23b_e}{630} & \dfrac{8}{21} & \dfrac{-32b_e}{315} & \dfrac{5}{42} & \dfrac{-b_e}{90} \\[2mm] & 0 & \dfrac{8b_e}{315} & \dfrac{-2b_e^2}{315} & \dfrac{b_e}{90} & \dfrac{-b_e^2}{1260} \\[2mm] & & 0 & \dfrac{64b_e}{315} & \dfrac{8}{21} & \dfrac{-8b_e}{315} \\[2mm] & \text{asy} & & 0 & \dfrac{32b_e}{315} & \dfrac{-2b_e^2}{315} \\[2mm] & & & & \dfrac{1}{2} & \dfrac{23b_e}{630} \\[2mm] & & & & & 0 \end{bmatrix} \tag{14.15}$$

$$\mathbf{A}_3^e = 2(D_{12}+2D_{66})\begin{bmatrix} \dfrac{-278}{105b_e} & \dfrac{-223}{210} & \dfrac{256}{105b_e} & \dfrac{-8}{21} & \dfrac{22}{105b_e} & \dfrac{1}{70} \\[2mm] \dfrac{-13}{210} & \dfrac{-2b_e}{45} & \dfrac{8}{105} & \dfrac{4b_e}{315} & \dfrac{-1}{70} & \dfrac{b_e}{126} \\[2mm] \dfrac{256}{105b_e} & \dfrac{8}{105} & \dfrac{-512}{105b_e} & 0 & \dfrac{256}{105b_e} & \dfrac{-8}{105} \\[2mm] \dfrac{-8}{21} & \dfrac{4b_e}{315} & 0 & \dfrac{-128b_e}{315} & \dfrac{8}{21} & \dfrac{4b_e}{315} \\[2mm] \dfrac{22}{105b_e} & \dfrac{-1}{70} & \dfrac{256}{105b_e} & \dfrac{8}{21} & \dfrac{-278}{105b_e} & \dfrac{223}{210} \\[2mm] \dfrac{1}{70} & \dfrac{b_e}{126} & \dfrac{-8}{105} & \dfrac{4b_e}{315} & \dfrac{13}{210} & \dfrac{-2b_e}{45} \end{bmatrix}$$

(14.16)

$$\mathbf{A}_4^e = 4D_{26}\begin{bmatrix} \dfrac{46}{b_e^2} & \dfrac{341}{35b_e} & \dfrac{-192}{7b_e^2} & \dfrac{688}{35b_e} & \dfrac{-130}{7b_e^2} & \dfrac{101}{35b_e} \\[2mm] \dfrac{79}{35b_e} & \dfrac{1}{2} & \dfrac{-48}{35b_e} & \dfrac{32}{35} & \dfrac{-31}{35b_e} & \dfrac{11}{70} \\[2mm] \dfrac{-32}{7b_e^2} & \dfrac{48}{35b_e} & 0 & \dfrac{-256}{35b_e} & \dfrac{32}{7b_e^2} & \dfrac{48}{35b_e} \\[2mm] \dfrac{-128}{35b_e} & \dfrac{-32}{35} & \dfrac{256}{35b_e} & 0 & \dfrac{-128}{35b_e} & \dfrac{32}{35} \\[2mm] \dfrac{130}{7b_e^2} & \dfrac{101}{35b_e} & \dfrac{192}{7b_e^2} & \dfrac{688}{35b_e} & \dfrac{-46}{b_e^2} & \dfrac{341}{35b_e} \\[2mm] \dfrac{-31}{35b_e} & \dfrac{-11}{70} & \dfrac{-48}{35b_e} & \dfrac{32}{35} & \dfrac{79}{35b_e} & \dfrac{-1}{2} \end{bmatrix}$$

(14.17)

$$\mathbf{A}_5^e = D_{22}\begin{bmatrix} \dfrac{-8768}{35b_e^3} & \dfrac{-1592}{35b_e^2} & \dfrac{448}{5b_e^3} & \dfrac{-960}{7b_e^2} & \dfrac{5632}{35b_e^3} & \dfrac{-808}{35b_e^2} \\[2mm] \dfrac{-472}{35b_e^2} & \dfrac{-88}{35b_e} & \dfrac{32}{5b_e^2} & \dfrac{-48}{7b_e} & \dfrac{248}{35b_e^2} & \dfrac{-32}{35b_e} \\[2mm] \dfrac{-512}{5b_e^3} & \dfrac{-128}{5b_e^2} & \dfrac{1024}{5b_e^3} & 0 & \dfrac{-512}{5b_e^3} & \dfrac{128}{5b_e^2} \\[2mm] \dfrac{384}{7b_e^2} & \dfrac{64}{7b_e} & 0 & \dfrac{256}{7b_e} & \dfrac{-384}{7b_e^2} & \dfrac{64}{7b_e} \\[2mm] \dfrac{5632}{35b_e^3} & \dfrac{808}{35b_e^2} & \dfrac{448}{5b_e^3} & \dfrac{960}{7b_e^2} & \dfrac{-8768}{35b_e^3} & \dfrac{1592}{35b_e^2} \\[2mm] \dfrac{-248}{35b_e^2} & \dfrac{-32}{35b_e} & \dfrac{-32}{5b_e^2} & \dfrac{-48}{7b_e} & \dfrac{472}{35b_e^2} & \dfrac{-88}{35b_e} \end{bmatrix}$$

(14.18)

$$
\mathbf{A}_6^e = I_0
\begin{bmatrix}
\dfrac{523b_e}{3465} & \dfrac{19b_e^2}{2310} & \dfrac{4b_e}{63} & \dfrac{-8b_e^2}{693} & \dfrac{131b_e}{6930} & \dfrac{-29b_e^2}{13860} \\[2ex]
 & \dfrac{2b_e^3}{3465} & \dfrac{2b_e^2}{315} & \dfrac{-b_e^3}{1155} & \dfrac{29b_e^2}{13860} & \dfrac{-b_e^3}{4620} \\[2ex]
 & & \dfrac{128b_e}{315} & 0 & \dfrac{4b_e}{63} & \dfrac{-2b_e^2}{315} \\[2ex]
 & & & \dfrac{32b_e^3}{3465} & \dfrac{8b_e^2}{693} & \dfrac{-b_e^3}{1155} \\[2ex]
 & \text{sym.} & & & \dfrac{523b_e}{3465} & \dfrac{-19b_e^2}{2310} \\[2ex]
 & & & & & \dfrac{2b_e^3}{3465}
\end{bmatrix}
\tag{14.19}
$$

$$
\mathbf{A}_7^e = I_2
\begin{bmatrix}
\dfrac{-278}{105b_e} & \dfrac{-223}{210} & \dfrac{256}{105b_e} & \dfrac{-8}{21} & \dfrac{22}{105b_e} & \dfrac{1}{70} \\[2ex]
\dfrac{-13}{210} & \dfrac{-2b_e}{45} & \dfrac{8}{105} & \dfrac{4b_e}{315} & \dfrac{-1}{70} & \dfrac{b_e}{126} \\[2ex]
\dfrac{256}{105b_e} & \dfrac{8}{105} & \dfrac{-512}{105b_e} & 0 & \dfrac{256}{105b_e} & \dfrac{-8}{105} \\[2ex]
\dfrac{-8}{21} & \dfrac{4b_e}{315} & 0 & \dfrac{-128b_e}{315} & \dfrac{8}{21} & \dfrac{4b_e}{315} \\[2ex]
\dfrac{22}{105b_e} & \dfrac{-1}{70} & \dfrac{256}{105b_e} & \dfrac{8}{21} & \dfrac{-278}{105b_e} & \dfrac{223}{210} \\[2ex]
\dfrac{1}{70} & \dfrac{b_e}{126} & \dfrac{-8}{105} & \dfrac{4b_e}{315} & \dfrac{13}{210} & \dfrac{-2b_e}{45}
\end{bmatrix}
\tag{14.20}
$$

$$
\mathbf{A}_8^e = I_2
\begin{bmatrix}
\dfrac{523b_e}{3465} & \dfrac{19b_e^2}{2310} & \dfrac{4b_e}{63} & \dfrac{-8b_e^2}{693} & \dfrac{131b_e}{6930} & \dfrac{-29b_e^2}{13860} \\[2ex]
 & \dfrac{2b_e^3}{3465} & \dfrac{2b_e^2}{315} & \dfrac{-b_e^3}{1155} & \dfrac{29b_e^2}{13860} & \dfrac{-b_e^3}{4620} \\[2ex]
 & & \dfrac{128b_e}{315} & 0 & \dfrac{4b_e}{63} & \dfrac{-2b_e^2}{315} \\[2ex]
 & & & \dfrac{32b_e^3}{3465} & \dfrac{8b_e^2}{693} & \dfrac{-b_e^3}{1155} \\[2ex]
 & \text{sym.} & & & \dfrac{523b_e}{3465} & \dfrac{-19b_e^2}{2310} \\[2ex]
 & & & & & \dfrac{2b_e^3}{3465}
\end{bmatrix}
\tag{14.21}
$$

For a uniformly distributed load in the y-direction, the force vector $\mathbf{P}^e$ is given as

$$\mathbf{P}^e = p\left[\frac{7b_e}{30} \quad \frac{b_e^2}{60} \quad \frac{8b_e}{15} \quad 0 \quad \frac{7b_e}{30} \quad -\frac{b_e^2}{60}\right]^T \tag{14.22}$$

Using the relationship between the displacement and the bending moment (Reddy, 1997), i.e.,

$$M_{yx} = M_{xy} = -\left(D_{16}\frac{\partial^2 w}{\partial x^2} + D_{22}\frac{\partial^2 w}{\partial y^2} + 2D_{66}\frac{\partial^2 w}{\partial x \partial y}\right) \tag{14.23}$$

$$M_y = -\left(D_{12}\frac{\partial^2 w}{\partial x^2} + D_{22}\frac{\partial^2 w}{\partial y^2} + 2D_{26}\frac{\partial^2 w}{\partial x \partial y}\right) \tag{14.24}$$

we obtain

$$Q_y = \frac{\partial M_{xy}}{\partial x} + \frac{\partial M_y}{\partial y} + I_2\frac{\partial^3 w}{\partial t^2 \partial x}$$

$$= -D_{16}\frac{\partial^3 w}{\partial x^3} - (D_{12} + 2D_{66})\frac{\partial^3 w}{\partial x^2 \partial y} - 3D_{26}\frac{\partial^3 w}{\partial x \partial y^2} - D_{22}\frac{\partial^3 w}{\partial y^3} + I_2\frac{\partial^3 w}{\partial t^2 \partial y} \tag{14.25}$$

$$\frac{\partial M_{yx}}{\partial x} + Q_y = -2D_{16}\frac{\partial^3 w}{\partial x^3} - (D_{12} + 4D_{66})\frac{\partial^3 w}{\partial x^2 \partial y} - 4D_{26}\frac{\partial^3 w}{\partial x \partial y^2}$$

$$- D_{22}\frac{\partial^3 w}{\partial y^3} + I_2\frac{\partial^3 w}{\partial t^2 \partial y} \tag{14.26}$$

From the above equations, the second term on the left side of Eq. (14.10) can be obtained from the following equations:

$$-\left(\frac{\partial M_{yx}}{\partial x} + Q_y\right) = 2D_{16}\mathbf{N}\frac{\partial^3 \mathbf{V}^e}{\partial x^3} + (D_{12} + 4D_{66})\frac{d\mathbf{N}}{dy}\frac{\partial^2 \mathbf{V}^e}{\partial x^2} + 4D_{26}\frac{d^2 \mathbf{N}}{dy^2}\frac{\partial \mathbf{V}^e}{\partial x}$$

$$+ D_{22}\frac{d^3 \mathbf{N}}{dy^3}\mathbf{V}^e - I_2\frac{d\mathbf{N}}{dy}\frac{\partial^2 \mathbf{V}^e}{\partial t^2} \tag{14.27}$$

$$-\left(\frac{\partial M_{yx}}{\partial x}+Q_y\right)\Bigg|_{y=0}=2D_{16}\frac{\partial^3 w_1}{\partial x^3}+(D_{12}+4D_{66})\frac{\partial^2 w_2}{\partial x^2}$$

$$+4D_{26}\left[-\frac{46}{b_e^2}\quad-\frac{12}{b_e}\quad\frac{32}{b_e^2}\quad-\frac{16}{b_e}\quad\frac{14}{b_e^2}\quad-\frac{2}{b_e}\right]\frac{\partial \mathbf{V}^e}{\partial x}$$

$$+D_{22}\left[\frac{396}{b_e^3}\quad\frac{78}{b_e^2}\quad-\frac{192}{b_e^3}\quad\frac{192}{b_e^2}\quad-\frac{204}{b_e^3}\quad\frac{30}{b_e^2}\right]\mathbf{V}^e$$

$$+I_2\frac{\partial^2 w_2}{\partial t^2}\tag{14.28}$$

which corresponds to virtual displacement δw_1, and the term with respect to virtual displacement δw_5 is given as

$$-\left(\frac{\partial M_{yx}}{\partial x}+Q_y\right)\Bigg|_{y=b_e}=2D_{16}\frac{\partial^3 w_5}{\partial x^3}+(D_{12}+4D_{66})\frac{\partial^2 w_6}{\partial x^2}$$

$$+4D_{26}\left[\frac{14}{b_e^2}\quad\frac{2}{b_e}\quad\frac{32}{b_e^2}\quad\frac{16}{b_e}\quad-\frac{46}{b_e^2}\quad\frac{12}{b_e}\right]\frac{\partial \mathbf{V}^e}{\partial x}$$

$$+D_{22}\left[\frac{204}{b_e^3}\quad\frac{30}{b_e^2}\quad\frac{192}{b_e^3}\quad\frac{192}{b_e^2}\quad-\frac{396}{b_e^3}\quad\frac{78}{b_e^2}\right]^T\mathbf{V}^e$$

$$+I_2\frac{\partial^2 w_6}{\partial t^2}\tag{14.29}$$

The terms correspond to virtual displacements δw_2 and δw_6 and are given by

$$-M_y=D_{12}\mathbf{N}\frac{d^2\mathbf{V}^e}{dx^2}+D_{22}\frac{d^2\mathbf{N}}{dy^2}\mathbf{V}^e+2D_{26}\frac{d\mathbf{N}}{dy}\frac{d\mathbf{V}^e}{dx}\tag{14.30}$$

where

$$-M_y\big|_{y=0}=D_{12}\frac{\partial^2 w_1}{\partial x^2}+2D_{26}\frac{\partial w_2}{\partial x}$$

$$+D_{22}\left[-\frac{46}{b_e^2}\quad-\frac{12}{b_e}\quad\frac{32}{b_e^2}\quad-\frac{16}{b_e}\quad\frac{14}{b_e^2}\quad-\frac{2}{b_e}\right]\mathbf{V}^e\tag{14.31}$$

$$-M_y\big|_{y=b_e}=D_{12}\frac{\partial^2 w_5}{\partial x^2}+2D_{26}\frac{\partial w_6}{\partial x}$$

$$+D_{22}\left[\frac{14}{b_e^2}\quad\frac{2}{b_e}\quad\frac{32}{b_e^2}\quad\frac{16}{b_e}\quad-\frac{46}{b_e^2}\quad\frac{12}{b_e}\right]\mathbf{V}^e\tag{14.32}$$

Substituting Eqs. (14.13) and (14.27)–(14.32) into Eq. (14.10), we can obtain the system of approximate differential equation for the strip element

$$\mathbf{B}_1^e \frac{\partial^4 \mathbf{V}^e}{\partial x^4} + \mathbf{B}_2^e \frac{\partial^3 \mathbf{V}^e}{\partial x^3} + \mathbf{B}_3^e \frac{\partial^2 \mathbf{V}^e}{\partial x^2} + \mathbf{B}_4^e \frac{\partial \mathbf{V}^e}{\partial x} + \mathbf{B}_5^e \mathbf{V}^e$$

$$+ (\mathbf{B}_6^e - \mathbf{B}_7^e) \frac{\partial^2 \mathbf{V}^e}{\partial t^2} - \mathbf{B}_8^e \frac{\partial^2}{\partial t^2}\left(\frac{\partial^2 \mathbf{V}^e}{\partial x^2}\right) = \mathbf{P}^e + \mathbf{T}^e \tag{14.33}$$

where

$$\mathbf{B}_1^e = \mathbf{A}_1^e$$

$$\mathbf{B}_6^e = \mathbf{A}_6^e$$

$$\mathbf{B}_8^e = \mathbf{A}_8^e$$

$$B_{211}^e = A_{211}^e + 2D_{16}$$

$$B_{255}^e = A_{255}^e - 2D_{16}$$

$$B_{312}^e = A_{312}^e + (D_{12} + 4D_{66})$$

$$B_{321}^e = A_{321}^e - D_{12}$$

$$B_{356}^e = A_{356}^e - (D_{12} + 4D_{66})$$

$$B_{365}^e = A_{365}^e + D_{12}$$

$$\mathbf{B}_{41}^e = \mathbf{A}_{41}^e + 4D_{26}\left[\frac{-46}{b_e^2} \quad \frac{-12}{b_e} \quad \frac{32}{b_e^2} \quad \frac{-16}{b_e} \quad \frac{14}{b_e^2} \quad \frac{-2}{b_e}\right]$$

$$\mathbf{B}_{45}^e = \mathbf{A}_{45}^e - 4D_{26}\left[\frac{14}{b_e^2} \quad \frac{2}{b_e} \quad \frac{32}{b_e^2} \quad \frac{16}{b_e} \quad \frac{-46}{b_e^2} \quad \frac{12}{b_e}\right]$$

$$B_{422}^e = A_{422}^e - 2D_{26}$$

$$B_{466}^e = A_{466}^e + 2D_{26}$$

$$\mathbf{B}_{51}^e = \mathbf{A}_{51}^e + D_{22}\left[\frac{396}{b_e^3} \quad \frac{78}{b_e^2} \quad \frac{-192}{b_e^3} \quad \frac{192}{b_e^2} \quad \frac{-204}{b_e^3} \quad \frac{30}{b_e^2}\right]$$

$$\mathbf{B}_{52}^e = \mathbf{A}_{52}^e - D_{22}\left[\frac{-46}{b_e^2} \quad \frac{-12}{b_e} \quad \frac{32}{b_e^2} \quad \frac{-16}{b_e} \quad \frac{14}{b_e^2} \quad \frac{-2}{b_e}\right]$$

$$\mathbf{B}^e_{55} = \mathbf{A}^e_{55} - D_{22}\left[\frac{204}{b_e^3} \quad \frac{30}{b_e^2} \quad \frac{192}{b_e^3} \quad \frac{192}{b_e^2} \quad \frac{-396}{b_e^3} \quad \frac{78}{b_e^2}\right]$$

$$\mathbf{B}^e_{56} = \mathbf{A}^e_{56} + D_{22}\left[\frac{14}{b_e^2} \quad \frac{2}{b_e} \quad \frac{32}{b_e^2} \quad \frac{16}{b_e} \quad \frac{-46}{b_e^2} \quad \frac{12}{b_e}\right]$$

$$B^e_{712} = A^e_{712} + I_2$$

$$B^e_{756} = A^e_{756} - I_2$$

In summary, we have matrices $\mathbf{B}^e_i (i = 1 \sim 8)$ in the form of

$$\mathbf{B}^e_1 = D_{11}\begin{bmatrix} \dfrac{523b_e}{3465} & \dfrac{19b_e^2}{2310} & \dfrac{4b_e}{63} & \dfrac{-8b_e^2}{693} & \dfrac{131b_e}{6930} & \dfrac{-29b_e^2}{13860} \\[2mm] & \dfrac{2b_e^3}{3465} & \dfrac{2b_e^2}{315} & \dfrac{-b_e^3}{1155} & \dfrac{29b_e^2}{13860} & \dfrac{-b_e^3}{4620} \\[2mm] & & \dfrac{12b_e}{315} & 0 & \dfrac{4b_e}{63} & \dfrac{-2b_e^2}{315} \\[2mm] & & & \dfrac{32b_e^3}{3465} & \dfrac{8b_e^2}{693} & \dfrac{-b_e^3}{1155} \\[2mm] & \text{sym.} & & & \dfrac{523b_e}{3465} & \dfrac{-19b_e^2}{2310} \\[2mm] & & & & & \dfrac{2b_e^3}{3465} \end{bmatrix} \quad (14.34)$$

$$\mathbf{B}^e_2 = 4D_{16}\begin{bmatrix} 0 & \dfrac{23b_e}{630} & \dfrac{8}{12} & \dfrac{-32b_e}{315} & \dfrac{5}{42} & \dfrac{-b_e}{90} \\[2mm] & 0 & \dfrac{8b_e}{315} & \dfrac{-2b_e^2}{315} & \dfrac{b_e}{90} & \dfrac{-b_e^2}{1260} \\[2mm] & & 0 & \dfrac{64b_e}{315} & \dfrac{8}{21} & \dfrac{-8b_e}{315} \\[2mm] & \text{asym} & & 0 & \dfrac{32b_e}{315} & \dfrac{-2b_e^2}{315} \\[2mm] & & & & 0 & \dfrac{23b_e}{630} \\[2mm] & & & & & 0 \end{bmatrix} \quad (14.35)$$

$$\mathbf{B}_3^e = 2D_{12}\begin{bmatrix}
\dfrac{-278}{105b_e} & \dfrac{-118}{210} & \dfrac{256}{105b_e} & \dfrac{-8}{21} & \dfrac{22}{105b_e} & \dfrac{1}{70} \\[2mm]
 & \dfrac{-2b_e}{45} & \dfrac{8}{105} & \dfrac{4b_e}{315} & \dfrac{-1}{70} & \dfrac{b_e}{126} \\[2mm]
 & & \dfrac{-512}{105b_e} & 0 & \dfrac{256}{105b_e} & \dfrac{-8}{105} \\[2mm]
 & & & \dfrac{-128b_e}{315} & \dfrac{8}{12} & \dfrac{4b_e}{315} \\[2mm]
 & \text{sym.} & & & \dfrac{-278}{105b_e} & \dfrac{118}{210} \\[2mm]
 & & & & & \dfrac{-2b_e}{45}
\end{bmatrix}$$

$$+\,4D_{66}\begin{bmatrix}
\dfrac{-278}{105b_e} & \dfrac{-13}{210} & \dfrac{256}{105b_e} & \dfrac{-8}{21} & \dfrac{22}{105b_e} & \dfrac{1}{70} \\[2mm]
 & \dfrac{-2b_e}{45} & \dfrac{8}{105} & \dfrac{4b_e}{315} & \dfrac{-1}{70} & \dfrac{b_e}{126} \\[2mm]
 & & \dfrac{-512}{105b_e} & 0 & \dfrac{256}{105b_e} & \dfrac{-8}{105} \\[2mm]
 & & & \dfrac{-128b_e}{315} & \dfrac{8}{21} & \dfrac{4b_e}{315} \\[2mm]
 & \text{sym.} & & & \dfrac{-278}{105b_e} & \dfrac{13}{210} \\[2mm]
 & & & & & \dfrac{-2b_e}{45}
\end{bmatrix} \tag{14.36}$$

$$\mathbf{B}_4^e = 4D_{26}\begin{bmatrix}
0 & \dfrac{-79}{35b_e} & \dfrac{32}{7b_e^2} & \dfrac{128}{35b_e} & \dfrac{-32}{7b_e^2} & \dfrac{31}{35b_e} \\[2mm]
 & 0 & \dfrac{-48}{35b_e} & \dfrac{32}{35} & \dfrac{-31}{35b_e} & \dfrac{11}{70} \\[2mm]
 & & 0 & \dfrac{-256}{35b_e} & \dfrac{32}{7b_e^2} & \dfrac{48}{35b_e} \\[2mm]
 & & & 0 & \dfrac{-128}{35b_e} & \dfrac{32}{35} \\[2mm]
 & \text{sym.} & & & 0 & \dfrac{-79}{35b_e} \\[2mm]
 & & & & & 0
\end{bmatrix} \tag{14.37}$$

$$
\mathbf{B}_5^e = D_{22}
\begin{bmatrix}
\dfrac{5092}{35b_e^3} & \dfrac{1138}{35b_e^2} & \dfrac{-512}{5b_e^3} & \dfrac{384}{7b_e^2} & \dfrac{-1058}{35b_e^3} & \dfrac{242}{35b_e^2} \\[2ex]
 & \dfrac{332}{35b_e} & \dfrac{-128}{5b_e^2} & \dfrac{64}{7b_e} & \dfrac{242}{35b_e^2} & \dfrac{38}{35b_e} \\[2ex]
 & & \dfrac{1024}{5b_e^3} & 0 & \dfrac{-512}{5b_e^3} & \dfrac{128}{5b_e^2} \\[2ex]
 & & & \dfrac{256}{7b_e} & \dfrac{-384}{7b_e^2} & \dfrac{64}{7b_e} \\[2ex]
 & \text{sym.} & & & \dfrac{5092}{35b_e^3} & \dfrac{-1138}{35b_e^2} \\[2ex]
 & & & & & \dfrac{332}{35b_e}
\end{bmatrix}
\tag{14.38}
$$

$$
\mathbf{B}_6^e = I_0
\begin{bmatrix}
\dfrac{523b_e}{3465} & \dfrac{19b_e^2}{2310} & \dfrac{4b_e}{63} & \dfrac{-8b_e^2}{693} & \dfrac{131b_e}{6930} & \dfrac{-29b_e^2}{13860} \\[2ex]
 & \dfrac{2b_e^3}{3465} & \dfrac{2b_e^2}{315} & \dfrac{-b_e^3}{1155} & \dfrac{29b_e^2}{13860} & \dfrac{-b_e^3}{4620} \\[2ex]
 & & \dfrac{12b_e}{315} & 0 & \dfrac{4b_e}{63} & \dfrac{-2b_e^2}{315} \\[2ex]
 & & & \dfrac{32b_e^3}{3465} & \dfrac{8b_e^2}{693} & \dfrac{-b_e^3}{1155} \\[2ex]
 & \text{sym.} & & & \dfrac{523b_e}{3465} & \dfrac{-19b_e^2}{2310} \\[2ex]
 & & & & & \dfrac{2b_e^3}{3465}
\end{bmatrix}
\tag{14.39}
$$

$$
\mathbf{B}_7^e = I_2
\begin{bmatrix}
\dfrac{-278}{105b_e} & \dfrac{-13}{210} & \dfrac{256}{105b_e} & \dfrac{-8}{21} & \dfrac{22}{105b_e} & \dfrac{1}{70} \\[2ex]
 & \dfrac{-2b_e}{45} & \dfrac{-8}{105} & \dfrac{4b_e}{315} & \dfrac{-1}{70} & \dfrac{b_e}{126} \\[2ex]
 & & \dfrac{-512}{105b_e} & 0 & \dfrac{256}{105b_e} & \dfrac{-8}{105} \\[2ex]
 & & & \dfrac{-128b_e}{315} & \dfrac{8}{12} & \dfrac{4b_e}{315} \\[2ex]
 & \text{sym.} & & & \dfrac{-278}{105b_e} & \dfrac{13}{210} \\[2ex]
 & & & & & \dfrac{-2b_e}{45}
\end{bmatrix}
\tag{14.40}
$$

$$\mathbf{B}_8^e = I_2 \begin{bmatrix} \dfrac{523b_e}{3465} & \dfrac{19b_e^2}{2310} & \dfrac{4b_e}{63} & \dfrac{-8b_e^2}{693} & \dfrac{131b_e}{6930} & \dfrac{-29b_e^2}{13860} \\[2mm] & \dfrac{2b_e^3}{3465} & \dfrac{2b_e^2}{315} & \dfrac{-b_e^3}{1155} & \dfrac{29b_e^2}{13860} & \dfrac{-b_e^3}{4620} \\[2mm] & & \dfrac{128b_e}{315} & 0 & \dfrac{4b_e}{63} & \dfrac{-2b_e^2}{315} \\[2mm] & & & \dfrac{32b_e^3}{3465} & \dfrac{8b_e^2}{693} & \dfrac{-b_e^3}{1155} \\[2mm] & \text{sym.} & & & \dfrac{523b_e}{3465} & \dfrac{-19b_e^2}{2310} \\[2mm] & & & & & \dfrac{2b_e^3}{3465} \end{bmatrix} \tag{14.41}$$

14.4 Assembly of Element Equations

The ordinary differential equations obtained in the last section are for a typical strip element. The continuous conditions of the generalized displacements v (deflection and rotation) at the element interfaces are used to assemble all the matrices together to form the system equation for the entire plate. These conditions are (see Eq. (14.7))

$$v_5^i = v_1^{i+1}, \quad v_6^i = v_2^{i+1} \tag{14.42}$$

The generalized reaction forces from the two adjacent strip elements at the common nodal line should be in equilibrium:

$$T_5^i + T_1^{i+1} = 0, \quad T_6^i + T_2^{i+1} = 0 \tag{14.43}$$

The assembly of all strip elements is based on the above two sets of continuous conditions for all the element interfaces. Since there are two equations for a nodal line of two adjacent strip elements, there will be a total of $2(2N + 1)$ equations if the plate is divided into N elements (see Figure 14.3). Assembling all the strip elements of the domain, a system of approximate ordinary differential equations for the entire plate can be obtained as follows:

$$\mathbf{B}_1 \frac{\partial^4 \mathbf{V}}{\partial x^4} + \mathbf{B}_2 \frac{\partial^3 \mathbf{V}}{\partial x^3} + \mathbf{B}_3 \frac{\partial^2 \mathbf{V}}{\partial x^2} + \mathbf{B}_4 \frac{\partial \mathbf{V}}{\partial x} + \mathbf{B}_5 \mathbf{V}$$

$$+ (\mathbf{B}_6 - \mathbf{B}_7) \frac{\partial^2 \mathbf{V}}{\partial t^2} - \mathbf{B}_8 \frac{\partial^2}{\partial t^2}\left(\frac{\partial^2 \mathbf{V}}{\partial x^2}\right) = \mathbf{P} \tag{14.44}$$

The matrices $\mathbf{B}_i$ ($i = 1 \sim 8$), $\mathbf{P}$ can be obtained by assembling the corresponding matrices and vectors of all the elements.

If the plate is divided into N strip elements, the dimension of matrix $\mathbf{B}_i$ ($i = 1 \sim 8$) will be $M \times M$ ($M = 4N + 2$) and that of vector $\mathbf{P}$ is $M \times 1$. It should be mentioned that Eq. (14.44) is an approximated system equation of Eq. (14.1); it is one dimension-reduced partial differential equation that is with respect to only x and t.

In order to solve these assembled system equations, the boundary conditions on the boundary lines at $y = 0$ and $y = b$ on the nodal lines should be imposed.

14.5 Static Problems for Orthotropic Laminated Plates

We use a simple static problem to demonstrate the procedures of SEM in solving Eq. (14.44) and issues concerning with the imposition of the boundary conditions.

14.5.1 System Equations

For static problems of orthotropic plates, the bending-twisting coefficients D_{16} and D_{26} and the dynamic related terms are zero. Hence, the lateral mid-surface deflection w of the plate is governed by the following partial differential equation reduced from Eq. (14.1).

$$D_{11}\frac{\partial^4 w}{\partial x^4} + 2(D_{12} + 2D_{66})\frac{\partial^4 w}{\partial x^2 \partial y^2} + D_{22}\frac{\partial^4 w}{\partial y^4} - q = 0 \qquad (14.45)$$

and the corresponding strip element equation can be obtained simply by removing the dynamic related terms and the terms related to D_{16} and D_{26} in Eq. (14.44).

$$\mathbf{B}_1\frac{d^4\mathbf{V}}{dx^4} + \mathbf{B}_3\frac{d^2\mathbf{V}}{dx^2} + \mathbf{B}_5\mathbf{V} = \mathbf{P} \qquad (14.46)$$

where $\mathbf{V}$ is the displacement vector at the nodal lines of the whole domain, which is a function of x only in static analysis. The matrices $\mathbf{B}_i$ ($i = 1, 3, 5$), $\mathbf{P}$ and $\mathbf{V}$ remain the same as those in Eq. (14.44). Since Eq. (14.46) is a one-dimensional differential equation system with constant matrices, it can be solved analytically as follows.

14.5.2 Complementary Solution

Equation (14.46) represents a set of fourth order differential equations with constant coefficients. Its general solution consists of two parts: the complementary solutions and the particular solutions. The complementary solutions can be obtained by solving the homogeneous equation of Eq. (14.46). Assuming

$$\mathbf{V} = \mathbf{d}_0 \exp(ikx) \tag{14.47}$$

where $\mathbf{d}_0$ is the displacement amplitude vector and k is the wavenumber. Substituting Eq. (14.47) into Eq. (14.46), we obtain the following eigenvalue equation:

$$(k^4 \mathbf{B}_1 - k^2 \mathbf{B}_3 + \mathbf{B}_5)\mathbf{d}_0 = 0 \tag{14.48}$$

In order to change Eq. (14.48) to a standard form of eigenvalue equation, the Cholesky decomposition is first applied to matrix $\mathbf{B}_1$ (it is easy to see that $\mathbf{B}_1$ is positive definite):

$$\mathbf{B}_1 = \mathbf{S}^T \mathbf{S} \tag{14.49}$$

where $\mathbf{S}$ is an upper triangle matrix. Multiplying $\mathbf{S}^{-T}$ to Eq. (14.48), we obtain

$$(k^4 \mathbf{I} - k^2 \bar{\mathbf{B}}_3 + \bar{\mathbf{B}}_5)\bar{\mathbf{d}}_0 = 0 \tag{14.50}$$

where

$$\bar{\mathbf{d}}_0 = \mathbf{S}\mathbf{d}_0 \tag{14.51}$$

$$\bar{\mathbf{B}}_i = \mathbf{S}^{-T}\mathbf{B}_i\mathbf{S}^{-1} \tag{14.52}$$

We can now rewrite Eq. (14.50) in the following standard eigenvalue equation:

$$\begin{bmatrix} 0 & \mathbf{I} \\ -\bar{\mathbf{B}}_5 & \bar{\mathbf{B}}_3 \end{bmatrix} \begin{Bmatrix} \bar{\mathbf{d}}_0 \\ \bar{\mathbf{d}}_1 \end{Bmatrix} = k^2 \begin{bmatrix} \mathbf{I} & 0 \\ 0 & \mathbf{I} \end{bmatrix} \begin{Bmatrix} \bar{\mathbf{d}}_0 \\ \bar{\mathbf{d}}_1 \end{Bmatrix} \tag{14.53}$$

where

$$\bar{\mathbf{d}}_1 = k^2 \bar{\mathbf{d}}_0 \tag{14.54}$$

Solving Eq. (14.53), $4M$ ($M = 4N + 2$, N-element number) eigenvalues k_j ($j = 1, 2,...,4M$) and eigenvectors can be obtained. If the part of the jth eigenvector corresponding to $\bar{\mathbf{d}}_0$ is denoted by $\bar{\Phi}_j$

$$\Phi_j^T = \{\varphi_{1j} \quad \varphi_{2j} \quad \cdots \quad \varphi_{Mj}\} \tag{14.55}$$

Then the eigenvector corresponding to $\mathbf{d}_0$ will be

$$\Phi_j = \mathbf{S}^{-1}\bar{\Phi}_j \tag{14.56}$$

and the complementary solution can be written by superposition of these eigenvectors as

$$\mathbf{V}_c = \sum_{j=1}^{4M} C_j\Phi_j \exp(ik_jx) = \mathbf{G}(x)\mathbf{C} \tag{14.57}$$

where C_j are constants that can be determined using the boundary conditions on S_2 and S_4.

14.5.3   Particular Solution

To obtain the particular solution of Eq. (14.46), we introduce the Fourier transformation with respect to the coordinate x as follows:

$$\tilde{\mathbf{V}}_p(k) = \int_{-\infty}^{\infty} \mathbf{V}_p(x)e^{-ikx}\,dx \tag{14.58}$$

where the subscript p indicates the particular solution. Applying Fourier transform to Eq. (14.46) leads to the following equation in the transform-domain:

$$\tilde{\mathbf{P}} = [k^4\mathbf{B}_1 - k^2\mathbf{B}_3 + \mathbf{B}_5]\tilde{\mathbf{V}}_p \tag{14.59}$$

This equation can be rewritten in the following form:

$$\left\{\begin{matrix} \mathbf{0} \\ -\tilde{\mathbf{P}} \end{matrix}\right\} = \left[\begin{matrix} \mathbf{0} & \mathbf{I} \\ -\mathbf{B}_5 & \mathbf{B}_3 \end{matrix}\right] - k^2\left[\begin{matrix} \mathbf{I} & \mathbf{0} \\ \mathbf{0} & \mathbf{B}_1 \end{matrix}\right]\left\{\begin{matrix} \tilde{\mathbf{V}}_p \\ k^2\tilde{\mathbf{V}}_p \end{matrix}\right\} \tag{14.60}$$

Let

$$\mathbf{p}^T = \{0 - \tilde{\mathbf{P}}\}^T, \quad \mathbf{d}^T = \{\tilde{\mathbf{V}}_p^T \quad k^2 \tilde{\mathbf{V}}_p^T\}^T \tag{14.61}$$

$$\mathbf{A} = \begin{bmatrix} 0 & \mathbf{I} \\ -\mathbf{B}_5 & \mathbf{B}_3 \end{bmatrix}, \quad \mathbf{B} = \begin{bmatrix} \mathbf{I} & 0 \\ 0 & \mathbf{B}_1 \end{bmatrix} \tag{14.62}$$

Then Eq. (14.60) can be rewritten as

$$\mathbf{p} = [\mathbf{A} - k^2 \mathbf{B}]\mathbf{d} \tag{14.63}$$

The eigenvalues k_m, left eigenvectors ϕ_m^L, and right eigenvectors ϕ_m^R that can be obtained from the corresponding homogeneous equation of Eq. (14.63) satisfy the following eigenvalue equations:

$$\phi_m^L[\mathbf{A} - k_m^2 \mathbf{B}] = 0, \quad [\mathbf{A} - k_m^2 \mathbf{B}]\phi_m^R = 0 \tag{14.64}$$

These eigenvectors satisfy the following orthogonal equations:

$$\phi_n^L \mathbf{A} \phi_m^R = k_m^2 B_n \delta_{nm}, \quad \phi_n^L \mathbf{B} \phi_m^R = B_n \delta_{nm} \tag{14.65}$$

where $\delta_{nm} = 1$ for $n = m$, and $\delta_{nm} = 0$ for $n \neq m$, and eigenvectors ϕ_m^L and ϕ_m^R can be written as

$$\phi_m^R = \begin{Bmatrix} \phi_{m1}^R \\ \phi_{m2}^R \end{Bmatrix}, \quad \phi_m^L = \{\phi_{m1}^L \quad \phi_{m2}^L\} \tag{14.66}$$

in which ϕ_{m1}^R, ϕ_{m2}^R, ϕ_{m1}^L, and ϕ_{m2}^L have the same dimension. Assuming

$$\mathbf{d} = \sum_{m=1}^{4M} C_m \phi_m^R, \quad \mathbf{p} = \sum_{m=1}^{4M} D_m \mathbf{B} \phi_m^R \tag{14.67}$$

From Eqs. (14.63), (14.65), and (14.67), we find

$$\mathbf{d} = \sum_{m=1}^{4M} \frac{\phi_m^L \mathbf{p} \phi_m^R}{(k_m - k) B_m} \tag{14.68}$$

and from Eqs. (14.61), (14.66), and (14.68), we obtain

$$\tilde{\mathbf{V}}_P = -\sum_{m=1}^{4M} \frac{\phi_{m4}^L \tilde{\mathbf{P}} \phi_{m1}^R}{(k_m - k) B_m} \tag{14.69}$$

It should be noted that the left eigenvectors can be obtained from right eigenvectors. From Eq. (14.64) and the associated equations, we can get two sets of equations. The first set of equations is given as

$$\phi_m^L[\mathbf{A} - k_m^2 \mathbf{B}] = [\phi_{mi}^L \quad \phi_{m2}^L] \left(\begin{bmatrix} \mathbf{0} & \mathbf{I} \\ -\mathbf{B}_5 & \mathbf{B}_3 \end{bmatrix} - k_m^2 \begin{bmatrix} \mathbf{I} & \mathbf{0} \\ \mathbf{0} & \mathbf{B}_1 \end{bmatrix} \right) = 0 \qquad (14.70)$$

that is

$$\begin{cases} -\phi_{m2}^L \mathbf{B}_5 - k_m^2 \phi_{m1}^L = 0 \\ \phi_{m1}^L + \phi_{m2}^L \mathbf{B}_3 - k_m^2 \phi_{m2}^L \mathbf{B}_1 = 0 \end{cases} \qquad (14.71)$$

from which we obtain

$$(-\mathbf{B}_5^T + k_m^2 \mathbf{B}_3^T - k_m^4 \mathbf{B}_1^T) \phi_{m2}^{LT} = 0 \qquad (14.72)$$

The second set of equations is

$$[\mathbf{A} - k_m^2 \mathbf{B}] \phi_m^R = \left(\begin{bmatrix} \mathbf{0} & \mathbf{I} \\ -\mathbf{B}_5 & \mathbf{B}_3 \end{bmatrix} - k_m^2 \begin{bmatrix} \mathbf{I} & \mathbf{0} \\ \mathbf{0} & \mathbf{B}_1 \end{bmatrix} \right) \begin{Bmatrix} \phi_{m1}^R \\ \phi_{m2}^R \end{Bmatrix} = 0 \qquad (14.73)$$

$$\begin{cases} \phi_{m2}^R - k_m^2 \phi_{m1}^R = 0 \\ -\mathbf{B}_5 \phi_{m1}^R + \mathbf{B}_3 \phi_{m2}^R - k_m^2 \mathbf{B}_1 \phi_{m2}^R = 0 \end{cases} \qquad (14.74)$$

from which we obtain

$$(-\mathbf{B}_5 + k_m^2 \mathbf{B}_3 - k_m^4 \mathbf{B}_1) \phi_{m1}^R = 0 \qquad (14.75)$$

It is known from Eqs. (14.72) and (14.75) that the left eigenvectors ϕ_{m2}^L are equal to the transposed right eigenvectors ϕ_{m1}^R if the coefficients matrices $\mathbf{B}_i$ in Eq. (14.46) is symmetric, which is true in our case.

Applying the inverse Fourier transformation to Eq. (14.69), the particular solution can be expressed by

$$\mathbf{V}_p(x) = \frac{1}{2\pi} \int_{-\infty}^{\infty} \tilde{\mathbf{V}}_p(k) e^{ikx} \, dk = \frac{-1}{2\pi} \int_{-\infty}^{\infty} \sum_{m=1}^{4M} \frac{\phi_{m2}^L \tilde{\mathbf{P}} \phi_{m1}^R}{(k_m^2 - k^2) B_m} e^{ikx} \, dk \qquad (14.76)$$

Then the general solution of Eq. (14.46) is given as

$$\mathbf{V} = \mathbf{V}_c + \mathbf{V}_p = \mathbf{G}(x)\mathbf{C} + \mathbf{V}_p \qquad (14.77)$$

where

$$\mathbf{G}(x) = \begin{bmatrix} \phi_{11}\exp(ik_1 x) & \phi_{21}\exp(ik_2 x) & \cdots & \phi_{4M1}\exp(ik_{4M}x) \\ \phi_{12}\exp(ik_1 x) & \phi_{22}\exp(ik_2 x) & \cdots & \phi_{4M2}\exp(ik_{4M}x) \\ \vdots & \vdots & & \vdots \\ \phi_{1M}\exp(ik_1 x) & \phi_{2M}\exp(ik_2 x) & \cdots & \phi_{4MM}\exp(ik_{4M}x) \end{bmatrix} \quad (14.78)$$

Thus, the displacements field can be determined once the constants C_i are determined using the boundary conditions.

14.5.4 Imposition of Boundary Conditions

As we have shown, Eq. (14.77) gives the fundamental solution for deformation of the infinite plate. The constant vector $\mathbf{C}$ with $4M$ elements can be determined using the boundary conditions on S_2 and S_4 to obtain the solution for a problem with defined boundary conditions. The boundary conditions include infinite and finite boundary conditions. Here we focus on finite boundary conditions. From the classical plate theory, it is known that the boundary conditions on $x = \pm a/2$ can have the following forms.

On clamped edges, the boundary conditions are

$$w = 0, \quad \frac{\partial w}{\partial x} = 0 \quad \text{at } x = -\frac{a}{2} \quad \text{or} \quad x = \frac{a}{2} \quad (14.79)$$

on simply supported edges

$$w = 0, \quad \frac{\partial^2 w}{\partial x^2} = 0 \quad \text{at } x = -\frac{a}{2} \quad \text{or} \quad x = \frac{a}{2} \quad (14.80)$$

and on free edges,

$$M_x = 0, \quad \frac{\partial M_{xy}}{\partial y} + Q_x = 0 \quad \text{at } x = -\frac{a}{2} \quad \text{or} \quad x = \frac{a}{2} \quad (14.81)$$

The above boundary equations should be imposed at all the crossing points of the element nodal lines and the plate boundary lines at $y = 0$ and $y = b$. Therefore, any equation set in Eq. (14.79), (14.80), or (14.81) gives $M(M = 4N + 2)$ boundary conditions on each boundary. We have, hence, a total of $2M$ boundary conditions on the two boundaries. However, there are $4M$ constants to be determined in Eq. (14.77); we need another $2M$ boundary condition.

Note that the satisfaction of any of the above three equation sets only at the nodal lines is not sufficient. The whole edges of the boundary ($x = \pm a/2$) should be imposed with a set of these three equation sets. In fact, the deflection for an element has been written as

$$w = n_1(y)w_1 + n_2(y)\theta_1 + n_3(y)w_2 + n_4(y)\theta_2 + n_5(y)w_3 + n_6(y)\theta_3 \quad (14.82)$$

and hence,

$$\frac{dw}{dx} = n_1(y)\frac{dw_1}{dx} + n_2(y)\frac{d\theta_1}{dx} + n_3(y)\frac{dw_2}{dx} + n_4(y)\frac{d\theta_2}{dx} + n_5(y)\frac{dw_3}{dx} + n_6(y)\frac{d\theta_3}{dx}$$

$$(14.83)$$

$$\frac{d^2w}{d^2x} = n_1(y)\frac{d^2w_1}{d^2x} + n_2(y)\frac{d^2\theta_1}{d^2x} + n_3(y)\frac{d^2w_2}{d^2x} + n_4(y)\frac{d^2\theta_2}{d^2x} + n_5(y)\frac{d^2w_3}{d_2x} + n_6(y)\frac{d^2\theta_3}{d^2x}$$

$$(14.84)$$

Since the $n_i(y)$ ($i = 1 \sim 6$) are independent functions of y, both w_i and θ_i must be zero in order to satisfy $w = 0$ at arbitrary y on the boundary. Also, both $\frac{dw_i}{dx}$ and $\frac{d\theta_i}{dx}$ should be zero to satisfy $\frac{dw}{dx} = 0$ at arbitrary y, and, similarly, if $\frac{d^2w}{dx^2} = 0$ holds at arbitrary y, then $\frac{d^2w_i}{dx^2}$ and $\frac{d^2\theta_i}{dx^2}$ should be zero. Therefore, additional boundary conditions can be introduced as follows.

For clamped edges, the introduced additional boundary conditions are

$$\frac{\partial w}{\partial y} = 0, \quad \frac{\partial}{\partial x}\left(\frac{\partial w}{\partial y}\right) = 0 \quad (14.85)$$

Combining Eqs. (14.79) and (14.85), the complete set of boundary conditions for clamped edges can be written in the form of displacement vector as

$$\mathbf{V} = 0, \quad \frac{d\mathbf{V}}{dx} = 0 \quad (14.86)$$

The additional boundary conditions for simply supported edges are found to be

$$\frac{\partial w}{\partial y} = 0, \quad \frac{\partial^2}{\partial x^2}\left(\frac{\partial w}{\partial y}\right) = 0 \quad (14.87)$$

and the complete boundary conditions are obtained by combining Eqs. (14.80) and (14.87) and can be expressed as follows:

$$\mathbf{V} = 0, \quad \frac{d^2\mathbf{V}}{dx^2} = 0 \quad (14.88)$$

The treatment of force boundary conditions is more difficult than the displacement boundary conditions. Writing the bending moment and the shear force in term of shape functions and nodal displacements as

$$
\begin{aligned}
-M_x &= D_{11}\frac{\partial^2 w}{\partial x^2} + D_{12}\frac{\partial^2 w}{\partial y^2} \\
&= D_{11}\sum_{i=1}^{n} n_i\frac{d^2 v_i}{dx^2} + D_{12}\sum_{i=1}^{6}\frac{d^2 n_i}{dy^2}v_i \\
&= \sum_{j=0}^{5} y^j \sum_{i=1}^{6}\left[D_{11}n_{ij}\frac{d^2 v_i}{dx^2} + D_{12}(j+2)(j+1)n_{ij+2}v_i\right] \quad (14.89)
\end{aligned}
$$

$$
\begin{aligned}
-\left(\frac{\partial M_{xy}}{\partial y} + Q_x\right) &= D_{11}\frac{\partial^3 w}{\partial x^3} + (D_{12}+4D_{66})\frac{\partial^3 w}{\partial x \partial y^2} \\
&= D_{11}\sum_{i=1}^{6} n_i\frac{d^3 v_i}{dx^3} + (D_{12}+4D_{66})\sum_{i=1}^{6}\frac{d^2 n_i}{dy^2}\frac{dv_i}{dx} \\
&= \sum_{j=0}^{5} y^j \sum_{i=1}^{6}\left[D_{11}n_{ij}\frac{d^3 v_i}{dx^3} + (D_{12}+4D_{66})(j+2)(j+1)n_{ij+2}\frac{dv_i}{dx}\right]
\end{aligned}
$$

$$
(14.90)
$$

It should be noted that in the above derivation, the shape functions have been written as

$$
n_i = \sum_{j=0}^{5} n_{ij}y^j \quad (14.91)
$$

and the coefficient n_{i6} and n_{i7} are set to zero. In order to satisfy the free boundary conditions of Eq. (14.81) at arbitrary y, the following conditions must be satisfied:

$$
\sum_{i=1}^{6}\left[D_{11}n_{ij}\frac{d^2 v_i}{dx^2} + D_{12}(j+2)(j+1)n_{ij+2}v_i\right] = 0, \quad \text{for } j = 0,1,\dots,5 \quad (14.92)
$$

$$
\sum_{i=1}^{6}\left[D_{11}n_{ij}\frac{d^3 v_i}{dx^3} + (D_{12}+4D_{66})(j+2)(j+1)n_{ij+2}\frac{dv_i}{dx}\right] = 0, \quad \text{for } j = 0,1,\dots,5
$$

$$
(14.93)
$$

which are equivalent with the following standard form of boundary conditions:

$$
\mathbf{M}_x = 0, \quad \frac{\partial \mathbf{M}_{xy}}{\partial y} + \mathbf{Q}_x = 0, \quad \frac{\partial \mathbf{M}_x}{\partial y} = 0, \quad \frac{\partial}{\partial y}\left(\frac{\partial \mathbf{M}_{xy}}{\partial y} + \mathbf{Q}_x\right) = 0 \quad (14.94)
$$

TABLE 14.1

Material Constants for Glass/Epoxy Plate

	Young's Modules 10³ MPa			Poisson's Ratio	Thickness mm
Glass/epoxy	E_1	E_2	G_{12}	ν_{12}	0.5
	30	5	10	0.25	
Aluminum	$E = 70$			$\nu = 0.3$	3

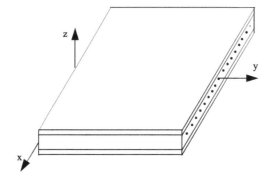

FIGURE 14.4
Sandwich plate constructed from a glass fiber-reinforced epoxy layer and two aluminum alloy face layers.

These equations give the exact number of boundary conditions needed to determine the constant vector C in Eq. (14.77).

14.5.5 Examples for Static Problems

Bending analysis for a square composite plate has been carried out using the SEM. The results are compared with the exact solution. The composite plate is constructed from a glass fiber-reinforced epoxy layer sandwiched between two aluminium alloy plates. The side length of the square plate is 150 mm. The fiber orientation of the unidirectional glass fiber-reinforced epoxy layer is along the y-direction (as shown in Figure 14.4). The material constants and the dimensions of the layers are given in Table 14.1.

The line load is first considered. If the line load is acting at $x = x_0$ along the y-axis, the Fourier transformation is given as

$$\tilde{P} = Pe^{-ikx_0} \tag{14.95}$$

$$
\begin{aligned}
V_p(x) &= \frac{-1}{2\pi}\int_{-\infty}^{\infty}\sum_{m=1}^{4M}\frac{\phi_{m2}^L P\phi_{m1}^R}{(k_m^2 - k^2)B_m}e^{ik(x-x_0)}\,dk \\
&= \sum_{m=1}^{4M}\frac{1}{2\pi}\frac{\phi_{m2}^L P\phi_{m1}^R}{B_m}\int_{-\infty}^{\infty}\frac{1}{k^2 - k_m^2}e^{ik(x-x_0)}\,dk
\end{aligned}
\tag{14.96}
$$

From Cauchy's theorem, the integration in Eq. (14.96) can be carried out as follows (see Chapters 1 and 12):

$$
\mathbf{V}_p =
\begin{cases}
\displaystyle\sum_{m=1}^{2M} i\,\frac{\phi_{m2}^{L+}\mathbf{P}\,\phi_{m1}^{R+}}{\mathbf{B}_m^+}\,\frac{e^{ik_m^+(x-x_0)}}{2k_m^+} & x \geq x_0 \\[20pt]
\displaystyle\sum_{m=1}^{2M} -i\,\frac{\phi_{m2}^{L-}\mathbf{P}\,\phi_{m1}^{R-}}{\mathbf{B}_m^-}\,\frac{e^{ik_m^-(x-x_0)}}{2k_m^-} & x < x_0
\end{cases}
\tag{14.97}
$$

where "+" denotes variables evaluated for the cases that the eigenvalues are positive real numbers or the images of the eigenvalues are positive, while "–" denotes the rest of the variables.

If the load is of uniform distribution along the x-axis, we can obtain the solution simply by integrating the solution of the line load:

$$
\mathbf{V}_p = \sum_{m=1}^{2M} \frac{\phi_{m2}^{L+}\mathbf{P}\,\phi_{m1}^{R+}}{\mathbf{B}_m^+}\,\frac{e^{ik_m^+(x+a/2)}-1}{2(k_m^+)^2} + \sum_{m=1}^{2M} \frac{\phi_{m2}^{L-}\mathbf{P}\,\phi_{m1}^{R-}}{\mathbf{B}_m^-}\,\frac{e^{ik_m^-(x-a/2)}-1}{2(k_m^-)^2}
\tag{14.98}
$$

The SEM program is used to demonstrate the efficiency and accuracy of the SEM for static problems. The plate is simply supported at two opposite sides, $y = 0$ and $y = b$. The Levy solutions are available for this problem. The calculation results of the deflections at the center of the plate have been given in Table 14.2 for the line load $p = 3$ N/mm acting at $y = b/2$ and the uniformly distributed load $q = 0.02$ N/mm^2. The distributions of deflections are shown in Figures 14.5 and 14.6, respectively.

From Table 14.2, it is found that maximum difference between the SEM solution and the Levy solution is about 0.7% for the line load and 0.38% for the uniform distributed load. The maximum difference occurs for the case of free boundary at $x = \pm a/2$, and minimum difference is found for the case of simply supported at four edges. This conclusion can also be found from Figures 14.5 and 14.6. In Figure 14.6, the curves of the SEM solution and Levy solution coincide, implying that the agreement between SEM and Levy solution for the case of uniformly distributed load is better than for the case of line load. This is because the Levy solution is represented in terms of a

TABLE 14.2

Deflections at the Center of a 150 × 150 Square Plate

Load	Line Load		Distributed Load	
Method	Levy	SEM	Levy	SEM
S-S-S-S	0.24174	0.24185	0.14632	0.14632
S-C-S-C	0.12650	0.12662	0.073215	0.073212
S-F-S-F	0.68010	0.68488	0.43296	0.43129

Unit: mm.

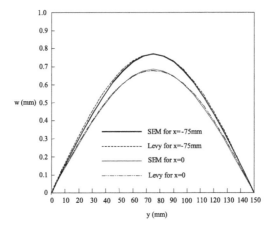

FIGURE 14.5
Deflection along y-axis given by the present SEM and Levy solutions. The plate is loaded by a uniform line load in the x direction acting at $y = b/2$. The plate is simply supported at $y = 0$ and $y = b$ and free at $x = \pm a/2$.

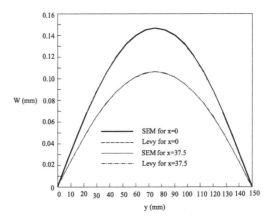

FIGURE 14.6
Deflection along y-axis given by the present SEM and Levy solutions. The plate is loaded by a uniform distributed load along x and y directions and simply supported at four edges.

single trigonometric series and the deformation of the plate subjected to uniformly distributed load is most close to a sine curve.

After obtaining the deflection, the stresses in the plate can be calculated. Since the stresses are not continuous through the thickness due to the layer-wise change in material property, the stresses should be calculated using the constitutive equation given by

$$
\begin{Bmatrix} \sigma_x \\ \sigma_y \\ \tau_{xy} \end{Bmatrix}^k = \begin{bmatrix} C_{11} & C_{12} & C_{16} \\ C_{12} & C_{22} & C_{26} \\ C_{26} & C_{26} & C_{66} \end{bmatrix}^k \begin{Bmatrix} \varepsilon_x \\ \varepsilon_y \\ \gamma_{xy} \end{Bmatrix}
\tag{14.99}
$$

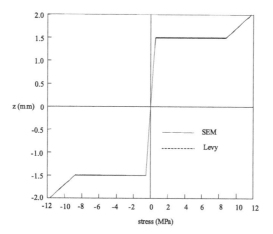

FIGURE 14.7
Distribution of normal stress σ_{xx} along z direction at center of plate. Comparison of SEM and Levy solutions. The plate is loaded by a uniform distributed load along x and y directions and simply supported at four edges.

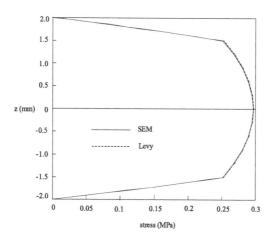

FIGURE 14.8
Distribution of shear stress τ_{xz} along z direction at center of plate. Comparison of SEM and Levy solutions. The plate is loaded by a uniform distributed load along x and y directions and simply supported at four edges.

The stress distribution along z-direction at the center of the simply supported plate have been shown in Figures 14.7 and 14.8, respectively, in which very good agreement for the stress distributions along the z direction is shown between SEM and Levy solutions.

14.6   Wave Motion in Anisotropic Laminated Plates

The SEM is now used to investigate the transient response of symmetric laminated plates. Transient responses of a rectangular symmetric laminated plate are studied for various loading and boundary conditions. The Fourier transform technique is then employed to obtain the time domain response and an exponential window method (see Chapter 4) is introduced to avoid singularities in the Fourier integration. Generally, the rotatory inertia is not considered in the CLPT that is applicable to thin plates only. However, the effect of the rotatory inertia cannot be ignored for thicker plates, especially for the transient analysis. One of the purposes of this chapter is to investigate the effect of rotatory inertia of laminated plates on the transient analysis. The effects of rotatory inertia and other material constants of laminated plates on the transient analysis are investigated.

14.6.1   System Equation in Frequency Domain

We start from Eq. (14.44). Note that the variables $\mathbf{V}$ and $\mathbf{P}$ in Eq. (14.44) are functions of coordinate x and time t. We introduce the Fourier transform from time t to frequency ω as

$$\tilde{\mathbf{V}}(x,\omega) = \int_{-\infty}^{\infty} \mathbf{V}(x,t)e^{-i\omega t}dt \tag{14.100}$$

$$\tilde{\mathbf{P}}(x,\omega) = \int_{-\infty}^{\infty} \mathbf{P}(x,t)e^{-i\omega t}dt \tag{14.101}$$

where "~" stands for a variable in the frequency domain. The application of the above Fourier transform to Eq. (14.44) leads to the following ordinary differential equation with respect to x:

$$\mathbf{B}_1\frac{d^4\tilde{\mathbf{V}}}{dx^4} + \mathbf{B}_2\frac{d^3\tilde{\mathbf{V}}}{dx^3} + (\mathbf{B}_3 + \omega^2\mathbf{B}_8)\frac{d^2\tilde{\mathbf{V}}}{dx^2} + \mathbf{B}_4\frac{d\tilde{\mathbf{V}}}{dx} + \mathbf{B}_5\tilde{\mathbf{V}} - \omega^2(\mathbf{B}_6 - \mathbf{B}_7)\tilde{\mathbf{V}} = \tilde{\mathbf{P}}$$

$$\tag{14.102}$$

The general solution of Eq. (14.102) consists of two parts: one is the complementary solution that satisfies the homogeneous equation corresponding to Eq. (14.102), and the other is the particular solution that satisfies Eq. (14.102).

14.6.2 Complementary Solution

The complementary solution can be obtained by solving the homogeneous equation corresponding to Eq. (14.102) ($\tilde{\mathbf{P}} = \mathbf{0}$). Assuming

$$\tilde{\mathbf{V}} = \mathbf{d}_0 \exp(ikx) \tag{14.103}$$

and substituting it into Eq. (14.102) with $\tilde{\mathbf{P}} = \mathbf{0}$, we obtain the following eigenvalue equation with respect to wavenumber k:

$$[k^4\mathbf{B}_1 - ik^3\mathbf{B}_2 - k^2(\mathbf{B}_3 + \omega^2\mathbf{B}_8) + ik\mathbf{B}_4 + \mathbf{B}_5 - \omega^2(\mathbf{B}_6 - \mathbf{B}_7)]\mathbf{d}_0 = 0 \tag{14.104}$$

For a given frequency, this equation can be changed to a standard eigenvalue equation as

$$\mathbf{A}\bar{\mathbf{d}} = k\mathbf{B}\bar{\mathbf{d}} \tag{14.105}$$

where

$$\mathbf{A} = \begin{bmatrix} \mathbf{0} & \mathbf{I} & \mathbf{0} & \mathbf{0} \\ \mathbf{0} & \mathbf{0} & \mathbf{I} & \mathbf{0} \\ \mathbf{0} & \mathbf{0} & \mathbf{0} & \mathbf{I} \\ -\mathbf{B}_5 + \omega^2(\mathbf{B}_6 - \mathbf{B}_7) & -i\mathbf{B}_4 & \mathbf{B}_3 + \omega^2\mathbf{B}_8 & i\mathbf{B}_2 \end{bmatrix} \tag{14.106}$$

$$\mathbf{B} = \begin{bmatrix} \mathbf{I} & 0 & 0 & 0 \\ 0 & \mathbf{I} & 0 & 0 \\ 0 & 0 & \mathbf{I} & 0 \\ 0 & 0 & 0 & \mathbf{B}_1 \end{bmatrix} \tag{14.107}$$

$$\bar{\mathbf{d}}^{\mathrm{T}} = \lfloor \mathbf{d}_0 \quad k\mathbf{d}_0 \quad k^2\mathbf{d}_0 \quad k^3\mathbf{d}_0 \rfloor \tag{14.108}$$

Note that if the Cholesky decomposition defined in Eq. (14.49) is performed to $\mathbf{B}_1$, matrix $\mathbf{B}$ in Eq. (14.107) will become an identity matrix.

Solving Eq. (14.105), $4M$ eigenvalues k_j ($j = 1, 2,...,4M$) and eigenvectors that are the functions of ω can be obtained, in which the first M elements in the eigenvectors are corresponding to $\mathbf{d}_0$. If the part of the jth eigenvector corresponding to $\mathbf{d}_0$ is denoted by $\Phi_j(\omega)$,

$$\Phi_j^T(\omega) = \{\phi_{j1} \quad \phi_{j2} \quad \cdots \quad \phi_{jM}\} \tag{14.109}$$

the complementary solution can then be written by superposition of these eigenvectors as

$$\tilde{\mathbf{V}}^c(x, \omega) = \sum_{j=1}^{4M} C_j(\omega)\Phi_j(\omega)\exp(ik_j x) = \mathbf{G}(x, \omega)\mathbf{C}(\omega) \qquad (14.110)$$

where the superscript c indicates the complementary solution and

$$\mathbf{G}(x, \omega) = \begin{bmatrix} \phi_{11}\exp(ik_1 x) & \phi_{21}\exp(ik_2 x) & \cdots & \phi_{L1}\exp(ik_L x) \\ \phi_{12}\exp(ik_1 x) & \phi_{22}\exp(ik_1 x) & \cdots & \phi_{L1}\exp(ik_L x) \\ \vdots & \vdots & \ddots & \vdots \\ \phi_{1M}\exp(ik_1 x) & \phi_{2M}\exp(ik_1 x) & \cdots & \phi_{LM}\exp(ik_L x) \end{bmatrix} \qquad (14.111)$$

where $L = 4M$. In Eq. (14.110), $\mathbf{C}$ is the constant vector that will be determined using the boundary conditions on S_2 and S_4 after the particular solution is obtained.

14.6.3   Particular Solution

The particular solution of Eq. (14.102) can be obtained by the seminumerical method (SNM), in which the Fourier transform spatial coordinate x to wavenumber k is introduced as the following:

$$\tilde{\tilde{\mathbf{V}}}^p(k, \omega) = \int_{-\infty}^{\infty} \tilde{\mathbf{V}}^p(x, \omega)e^{-ikx}\, dx \qquad (14.112)$$

$$\tilde{\tilde{\mathbf{P}}}^p(k, \omega) = \int_{-\infty}^{\infty} \tilde{\mathbf{P}}^p(x, \omega)e^{-ikx}\, dx \qquad (14.113)$$

where superscript p indicates the particular solution. The application of the Fourier transform to Eq. (14.102) leads to the following equation in the wavenumber domain:

$$\tilde{\tilde{\mathbf{P}}} = [k^4\mathbf{B}_1 - ik^3\mathbf{B}_2 - k^2(\mathbf{B}_3 + \omega^2\mathbf{B}_8) + ik\mathbf{B}_4 + \mathbf{B}_5 - \omega^2(\mathbf{B}_6 - \mathbf{B}_7)]\tilde{\tilde{\mathbf{V}}}^p \qquad (14.114)$$

This equation can be rewritten as

$$\mathbf{P} = [\mathbf{A} - k\mathbf{B}]\mathbf{d} \qquad (14.115)$$

where

$$\mathbf{p} = \{0 \ \ 0 \ \ 0 \ \ -\widetilde{\widetilde{\mathbf{P}}}^{T}\}^{T} \tag{14.116}$$

$$\mathbf{d} = \{\widetilde{\widetilde{\mathbf{V}}}^{pT} \ \ k\widetilde{\widetilde{\mathbf{V}}}^{pT} \ \ k^{2}\widetilde{\widetilde{\mathbf{V}}}^{pT} \ \ k^{3}\widetilde{\widetilde{\mathbf{V}}}^{pT}\}^{T} \tag{14.117}$$

Applying the modal analysis technique (see Chapter 8), we obtain

$$\mathbf{d} = \sum_{m=1}^{4M} \frac{\phi_m^L \mathbf{p} \phi_m^R}{(k_m - k) B_m} \tag{14.118}$$

where

$$B_m = \phi_m^L \mathbf{B} \phi_m^R \tag{14.119}$$

and k_m, ϕ_m^L, and ϕ_m^R are the mth eigenvalues and left and right eigenvectors which are obtained for a given frequency ω from

$$\phi_m^L \lfloor \mathbf{A} - k_m \mathbf{B} \rfloor = 0 \qquad \lfloor \mathbf{A} - k_m \mathbf{B} \rfloor \phi_m^R = 0 \tag{14.120}$$

It can be seen that k_m, ϕ_m^L, and ϕ_m^R are the same as that obtained in the previous subsection. The right eigenvectors ϕ_m^R and left eigenvectors ϕ_m^L can be written in the form of subvectors as

$$\phi_m^R = \begin{Bmatrix} \phi_{m1}^R \\ \phi_{m2}^R \\ \phi_{m3}^R \\ \phi_{m4}^R \end{Bmatrix}, \quad \phi_m^L = \{\phi_{m1}^L \ \ \phi_{m2}^L \ \ \phi_{m3}^L \ \ \phi_{m4}^L\} \tag{14.121}$$

where ϕ_{mi}^R ($i = 1, 2, 3, 4$) and ϕ_{mj}^R ($j = 1, 2, 3, 4$) have the same dimension. From Eqs. (14.117), (14.118), and (14.121), we obtain

$$\widetilde{\widetilde{\mathbf{V}}}^{p}(k, \omega) = -\sum_{m=1}^{4M} \frac{\phi_{m4}^L \widetilde{\widetilde{\mathbf{P}}} \phi_{m1}^R}{(k_m - k) B_m} \tag{14.122}$$

Once the external load is specified, the vector of the load Fourier transformation $\tilde{\mathbf{P}}$ can be obtained. Applying inverse Fourier transform to Eq. (14.122), the particular solution of Eq. (14.102) can be obtained.

$$\tilde{\mathbf{V}}^p(x,\omega) = \frac{1}{2\pi}\int_{-\infty}^{\infty} \tilde{\tilde{\mathbf{V}}}^p(k,\omega)e^{ikx}dk = \frac{1}{2\pi}\int_{-\infty}^{\infty}\sum_{m=1}^{4M} \frac{\phi_{m4}^L\tilde{\tilde{\mathbf{P}}}\phi_{m1}^R}{(k-k_m)B_m}e^{ikx}dk \quad (14.123)$$

For a line load in the x direction acting at $x = x_0$, the force vector in Eq. (14.101) can be written as

$$\mathbf{P} = \mathbf{P}_0(t)\delta(x - x_0) \quad (14.124)$$

and the Fourier transforms are as given below.

$$\tilde{\mathbf{P}} = \tilde{\mathbf{P}}_0(\omega)\delta(x - x_0) \quad (14.125)$$

$$\tilde{\tilde{\mathbf{P}}} = \tilde{\mathbf{P}}_0(\omega)e^{-ikx_0} \quad (14.126)$$

Hence, from Cauchy's theorem, the integral in Eq. (14.123) can be carried out to obtain

$$\tilde{\mathbf{V}}_p(x,\omega) = \begin{cases} \sum_{m=1}^{2M} i\frac{\phi_{m4}^{L+}\tilde{\mathbf{P}}_0\phi_{m1}^{R+}}{B_m^+}e^{ik_m^+(x-x_0)} & x \geq x_0 \\ \sum_{m=1}^{2M} -i\frac{\phi_{m4}^{L-}\tilde{\mathbf{P}}_0\phi_{m1}^{R-}}{B_m^-}e^{ik_m^-(x-x_0)} & (x < x_0) \end{cases} \quad (14.127)$$

where "+" denotes variables evaluated for the cases that the eigenvalues are positive real numbers or the imaginary parts of the eigenvalues are positive, which are corresponding to waves propagating in the positive x direction; and "−" denotes variables evaluated for the other eigenvalues that are corresponding to waves propagating in the negative x direction.

For a distributed load in the x direction, the solution can be obtained in the form of superposition integral over the solution of the line load. Such as, for the sinusoidally distributed load in the x direction,

$$q(x,y,t) = \sin\pi\left(\frac{1}{2} - \frac{x}{a}\right)p(p,t) \quad (14.128)$$

the particular solution can be obtained from the following integration:

$$\tilde{\mathbf{V}}^p = \int_{-\frac{a}{2}}^{x} \sin \pi \left(\frac{1}{2} - \frac{x}{a}\right) \sum_{m=1}^{2M} i \frac{\phi_{m4}^{L+} \tilde{\mathbf{P}}_0 \phi_{m1}^{R+}}{\mathbf{B}_m^{+}} e^{ik_m^{+}(x-x_0)} dx$$

$$+ \int_{x}^{\frac{a}{2}} \sin \pi \left(\frac{1}{2} - \frac{x}{a}\right) \sum_{m=1}^{2M} -i \frac{\phi_{m4}^{L-} \tilde{\mathbf{P}}_0 \phi_{m1}^{R-}}{\mathbf{B}_m^{-}} e^{ik_m^{-}(x-x_0)} dx \qquad (14.129)$$

That is evaluated to obtain

$$\tilde{\mathbf{V}}^p = \sum_{m=1}^{2M} i \frac{\phi_{m4}^{L-} \tilde{\mathbf{P}}_0 \phi_{m1}^{R-}}{\mathbf{B}_m^{-}} \frac{\frac{\pi}{a} \sin \frac{\pi x}{a} - ik_m^{-} \cos \frac{\pi x}{a} - \frac{\pi}{a} e^{ik_m^{-}(x-a/2)}}{-(k_m^{-})^2 + \left(\frac{\pi}{a}\right)^2}$$

$$+ \sum_{m=1}^{2} i \frac{\phi_{m4}^{L+} \tilde{\mathbf{P}}_0 \phi_{m1}^{R+}}{\mathbf{B}_m^{+}} \frac{\frac{\pi}{a} \sin \frac{\pi x}{a} - ik_m^{+} \cos \frac{\pi x}{a} + \frac{\pi}{a} e^{ik_m^{+}(x+a/2)}}{-(k_m^{+})^2 + \left(\frac{\pi}{a}\right)^2} \qquad (14.130)$$

14.6.4 The General Solution

The general solution of Eq. (14.102) can be given by adding the complementary solution given by Eq. (14.110) and the particular solution given by Eq. (14.123):

$$\tilde{\mathbf{V}}(x, \omega) = \tilde{\mathbf{V}}^c(x, \omega) + \tilde{\mathbf{V}}^p(x, \omega) = \mathbf{G}(x, \omega)\mathbf{C}(\omega) + \tilde{\mathbf{V}}^p(x, \omega) \quad (14.131)$$

Equation (14.131) gives the general solution with an unknown constant vector $\mathbf{C}$ of $4M$ elements. To obtain a special solution for the problem domain, the boundary conditions on S_2 and S_4 have to be used to determine the constant vector $\mathbf{C}$. Boundary conditions needed to determine $\mathbf{C}$ are discussed in detail in the previous section. For anisotropic laminated plates, Eqs. (14.89), (14.90), (14.92), and (14.93) need to be replaced, respectively, by the following four equations:

$$-M_x = D_{11}\frac{\partial^2 w}{\partial x^2} + D_{12}\frac{\partial^2 w}{\partial y^2} + 2D_{16}\frac{\partial^2 w}{\partial x \partial y}$$

$$= D_{11}\sum_{i=1}^{n} n_i \frac{d^2 v_i}{dx^2} + D_{12}\sum_{i=1}^{6} \frac{d^2 n_i}{dy^2} v_i + 2D_{16}\sum_{i=1}^{6} \frac{dn_i}{dy}\frac{dv_i}{dx}$$

$$= \sum_{j=0}^{5} y^j \sum_{i=1}^{6} \left[D_{11}n_{ij}\frac{d^2 v_i}{dx^2} + D_{12}(j+2)(j+1)n_{ij+2}v_i + 2D_{16}(j+1)n_{ij+1}\frac{dv_i}{dx} \right]$$

$$(14.132)$$

$$-\left(\frac{\partial M_{xy}}{\partial y} + Q_x\right) = D_{11}\frac{\partial^3 w}{\partial x^3} + 4D_{16}\frac{\partial^3 w}{\partial x^2 \partial y} + (D_{12} + 4D_{16})\frac{\partial^3 w}{\partial x \partial y^2} + 2D_{26}\frac{\partial^3 w}{\partial y^3}$$

$$= D_{11}\sum_{i=1}^{6} n_i \frac{d^3 v_i}{dx^3} + 4D_{16}\sum_{i=1}^{6} \frac{dn_i}{dy}\frac{d^2 v_i}{dx^2}$$

$$+ (D_{12} + 4D_{66})\sum_{i=1}^{6} \frac{d^2 n_i}{dy^2}\frac{dv_i}{dx} + 2D_{26}\sum_{i=1}^{6} \frac{d^3 n_i}{dy^3} v_i$$

$$= \sum_{j=0}^{5} y^j \sum_{i=1}^{6}\left[D_{11}n_{ij}\frac{d^3 v_i}{dx^3} + 4D_{16}(j+1)n_{ij+1}\frac{d^2 v_i}{dx^2}\right.$$

$$+ (D_{12} + D_{66})(j+2)(j+1)n_{ij+2}\frac{dv_i}{dx}$$

$$+ 2D_{26}(j+3)(j+2)(j+1)n_{ij+3}v_i]$$

$$\hspace{8cm}(14.133)$$

$$\sum_{i=1}^{6}\left[D_{11}n_{ij}\frac{d^2 v_i}{dx^2} + D_{12}(j+2)(j+1)n_{ij+2} + 2D_{16}(j+1)n_{ij+1}\frac{dv_i}{dx}\right] = 0 \quad (14.134)$$

$$\text{for } j = 0, 1, \ldots, 5$$

$$\sum_{i=1}^{6}\left[D_{11}n_{ij}\frac{d^3 v_i}{dx^3} + 4D_{16}(j+1)n_{ij+1}\frac{d^2 v_i}{dx^2} + (D_{12} + 4D_{66})(j+2)(j+1)n_{ij+2}\frac{dv_i}{dx}\right.$$

$$+ 2D_{26}(j+3)(j+2)(j+1)n_{ij+3}v_i] = 0 \quad \text{for } j = 0, 1, \ldots, 5 \hspace{2cm}(14.135)$$

Once the constant vector **C** is determined, the responses in the frequency domain can be obtained.

14.6.5   Solution in the Time Domain

The inverse Fourier transform to Eq. (14.131) gives the solution in the time domain.

$$\mathbf{V}(x,t) = \frac{1}{2\pi}\int_{-\infty}^{\infty} \tilde{\mathbf{V}}(x,\omega)e^{i\omega t}d\omega \hspace{3cm}(14.136)$$

The integral in the foregoing equation usually has to be carried out numerically. For the undamped plates, difficulties with the integration result from singularities of $\mathbf{V}(x,\omega)$ at $\omega = 0$ and at the cutoff frequencies ($k = 0$), as discussed in Chapters 1, 4, and 12. To overcome these difficulties, the exponential window method (EWM) needs to be used, which leads to

$$\tilde{\mathbf{V}}(x, \omega - i\eta) = \int_{-\infty}^{\infty} e^{-\eta t} \mathbf{V}(x, t) e^{-i\omega t} dt \tag{14.137}$$

$$\mathbf{V}(x, t) = \frac{e^{\eta t}}{2\pi} \int_{-\infty}^{\infty} \tilde{\mathbf{V}}(x, \omega - i\eta) e^{i\omega t} d\omega \tag{14.138}$$

The Fourier transform of external force in Eq. (10.4) should be

$$\tilde{\mathbf{P}}(x, \omega - i\eta) = \int_{-\infty}^{\infty} e^{-\eta t} \mathbf{P}(x, t) e^{-i\omega t} dt \tag{14.139}$$

In general, the external force is loaded on the plate only for a certain time duration in the transient analysis; therefore, the foregoing equation is often in the form of

$$\tilde{\mathbf{P}}(x, \omega - i\eta) = \int_{0}^{t_d} e^{-\eta t} \mathbf{P}(x, t) e^{-i\omega t} dt \tag{14.140}$$

where t_d is the duration of the external force.

14.6.6   Results for Anisotropic Laminated Plates

Example 1

Before we show the results for waves in anisotropic laminated plates, we show the results for a static example problem first, to verify the SEM formulation thoroughly. In this example, a five-layer laminated square plate of glass/epoxy with a stacking sequence of $[0°/45°/90°/45°/0°]$ is considered. The material properties are given in Table 14.1. The dimension of the plate is $a = b = 350$ mm, and the thickness of each layer is $h_1 = h_5 = 1.0$ mm, $h_2 = h_4 = 1.5$ mm, $h_3 = 2.0$ mm.

The convergence of the SEM is investigated first. Table 14.4 shows a comparison of SEM solutions with different strip elements numbers for a simply supported plate under uniformly distributed transverse load. It is clear from the results that the convergence of the SEM is very good. The bending analysis of the plate is carried out for a line load of $q = 7$ N/mm acting at $x = 0$. The deflection results at the center of the plate for different boundary conditions are presented in Table 14.5. When compared with Rayleigh-Ritz solutions, it is found that the maximum difference between SEM and Rayleigh-Ritz solutions is 0.62%. Next, the deflection results at the center of the plate subjected to a uniform lateral load $q = 0.02$ N/mm^2 are also given in Table 14.5. In this case, the maximum discrepancy is 0.66%. The displacement distributions

TABLE 14.3

Properties of Materials for Carbon/Epoxy Composites

Material	E_1(GPa)	E_2(GPa)	G_{12}(GPa)	ν_{12}	ρ(g/cm^2)
Carbon/epoxy-I	172.4	6.895	3.448	0.2500	1.603
Carbon/epoxy-II	131.7	8.550	6.670	0.3000	1.610

TABLE 14.4

Convergence of the SEM Solutions for a Simply Supported[0°/45°/90°/45°/0°] Plate of Glass/Epoxy Subjected to a Uniformly Distributed Transverse Load

Number of Element	2	4	6	8	10
w(mm)	2.2948	2.3033	2.3044	2.3048	2.3049
σ_x(MPa)	19.261	19.328	19.332	19.333	19.333
σ_y(MPa)	7.522	7.388	7.368	7.360	7.357
τ_{xy}(MPa)	4.866	4.913	4.942	4.952	4.957

Source: Wang, Y. Y., Lam, K. Y., and Liu, G. R., *Mechanics of Composite Materials and Structures*, 7, 225, 2000. With permission.

TABLE 14.5

Deflection at the Center of the [0°/45°/90°/45°/0°] Plate Made of Glass/Epoxy *

Method	Line Load (7 N/mm at x = 0)		Distributed Load (0.02 N/mm^2)	
	Rayleigh-Ritz	SEM**	Rayleigh-Ritz	SEM**
S-S-S-S	3.7549	3.7761	2.2905	2.3049
C-C-C-C	1.4511	1.4558	0.7181	0.7200
S-S-C-C	2.1494	2.1637	1.1960	1.2029
C-S-C-S	1.7864	1.7948	0.8949	0.8986

* Unit: mm.

** 10 elements are used.

Source: From Wang, Y. Y., Lam, K. Y., and Liu, G. R., *Mechanics of Composite Materials and Structures*, 7, 225, 2000. With permission.

along y-axis for different boundary conditions are shown simply supported at four edges in Figures 14.9 and 14.10.

The transient analysis of anisotropic composite laminated plates is carried out using a SEM program, and the results for two examples are presented below. We assume that the laminated plates are before the external excitation is applied.

Example 2

Consider a three-layer cross-ply [0°/90°/0°] square laminated plate made of carbon/epoxy-I composite, in which all layers are of the same thickness.

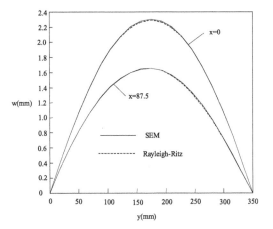

FIGURE 14.9
Deflection distributions along *y*-axis for the four edges simply supported plate subjected to a uniformly distributed loading. Comparison of SEM and Rayleigh-Ritz solutions. (From Wang, Y. Y., Lam, K. Y., and Liu, G. R., *Mechanics of Composite Materials and Structures*, 7, 225, 2000. With permission.)

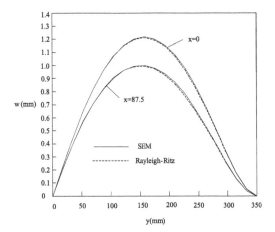

FIGURE 14.10
Deflection distributions along *y*-axis for the plate with two simply supported edges (at S1 and S2) and two clamped edges (at S3 and S4). The plate is subject to a uniformly distributed loading. Comparison of SEM and Rayleigh-Ritz solutions. (From Wang, Y. Y., Lam, K. Y., and Liu, G. R., *Mechanics of Composite Materials and Structures*, 7, 225, 2000. With permission.)

The material properties are given in Table 14.3. The total thickness of the plate is taken to be $H = 3.81$ cm, and the length of the plate is $a = b = 20H$.

The load is sinusoidally distributed on the whole surface of the plate:

$$q(x,y,t) = q_0 \sin \pi \left(\frac{1}{2} - \frac{x}{a} \right) \sin \frac{\pi y}{b} F(t) \tag{14.141}$$

Four time functions of the loads given in the following equation are used.

$$F(t) = \begin{cases} \begin{cases} \sin \pi t/t_1 & 0 \le t \le t_1 \\ 0 & t > t_1 \end{cases} & \text{sine loading} \\[2ex] \begin{cases} 1 & 0 \le t \le t_1 \\ 0 & t > t_1 \end{cases} & \text{step loading} \\[2ex] \begin{cases} 1 - t/t_1 & 0 \le t \le t_1 \\ 0 & t > t_1 \end{cases} & \text{triangular loading} \\[2ex] e^{-\gamma t} & \text{explosive blast loading} \end{cases} \qquad (14.142)$$

in which $t_1 = 0.006$ s and $\gamma = 3304$ s^{-1}. The intensity of the transverse load is taken to be $q_0 = 3.448$ Pa.

Figures 14.11–14.14 show the time history of the transverse deflection at the center of the plate for various loads. The comparisons of the SEM results with exact solutions (Khdeir and Reddy, 1989) are shown. From Figures 14.11–14.14, it is found that the SEM solutions agree very well with the exact solutions.

The stress responses in time domain can be obtained from the displacement distribution using a similar method as obtaining the deflection response. The normal stress response in time domain for sine loading and the comparison

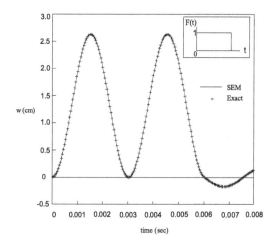

FIGURE 14.11
Time history of the center deflection for simple supported [0°/90°/0°] square laminated plate under step loading. Comparison of the SEM solution (solid line) and exact solution (cross marks). (From Wang, Y. Y., Lam, K. Y., and Liu, G. R., *Solids and Structures*, 38, 241, 2001. With permission.)

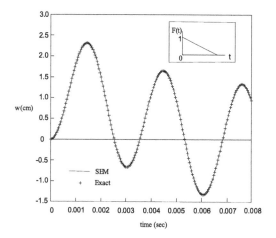

FIGURE 14.12
Same as Figure 10.1, but plate is subjected to triangular loading. (From Wang, Y. Y., Lam, K. Y., and Liu, G. R., *Solids and Structures*, 38, 241, 2001. With permission.)

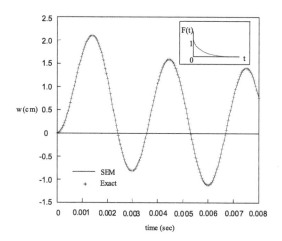

FIGURE 14.13
Same as Figure 10.1, but plate is subjected to explosive blast loading. (From Wang, Y. Y., Lam, K. Y., and Liu, G. R., *Solids and Structures*, 38, 241, 2001. With permission.)

of SEM solutions with exact solutions are given in Figure 14.15, in which the dimensionless normal stress are defined as

$$\bar{\sigma}_{xx} = \sigma_{xx}(a/2, b/2, H/2)/q_0 \qquad (14.143)$$

Also, very good agreements are observed. It can be seen that the stress response curve is very similar to the displacement response curve. This conclusion is valid also for other load cases.

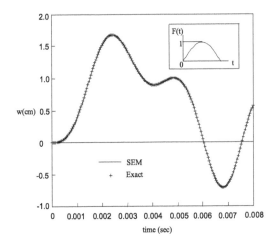

FIGURE 14.14
Same as Figure 10.1, but plate is subjected to sine loading. (From Wang, Y. Y., Lam, K. Y., and Liu, G. R., *Solids and Structures*, 38, 241, 2001. With permission.)

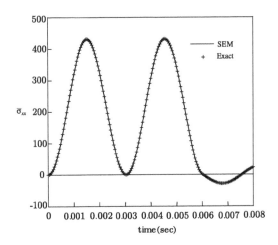

FIGURE 14.15
The time history of the dimensionless normal stress $\bar{\sigma}_{xx}$ versus time for simple supported $[0°/90°/0°]$ square laminated plate under step loading. Comparison of the SEM solution (solid line) and exact solution (cross). (From Wang, Y. Y., Lam, K. Y., and Liu, G. R., *Solids and Structures*, 38, 241, 2001. With permission.)

Example 3

A four-layer angle-ply square laminated plate made of carbon/epoxy-II with symmetrically stacking sequences $[30°/-30°/-30°/30°]$ is considered. All layers are assumed to have the same thickness and material properties are given in Table 14.3. The dimensions of the plate are

$$a_1 = 1.27 \text{ m} \qquad b_1 = 1.27 \text{ m} \qquad H = 0.0254 \text{ m}$$

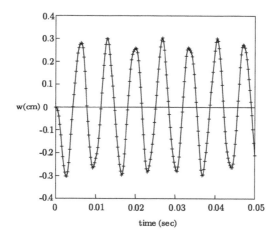

FIGURE 14.16
The time history of the deflection on the center of the [30°/−30°/−30°/30°] square laminated plate fully clamped at all edges. The plate is subjected to a conventional blast. Comparison of SEM solution (solid line) and Rayleigh-Ritz solution (cross). (From Wang, Y. Y., Lam, K. Y., and Liu, G. R., *Solids and Structures*, 38, 241, 2001. With permission.)

The plate is subjected to a conventional blast, while the pressure can be considered uniformly distributed over the plate and can be expressed as

$$q(x, y, t) = q_0\left(1 - \frac{t}{t_p}\right)e^{-\alpha t/t_p} \qquad (14.144)$$

where the parameters in Eq. (14.144) are taken as

$$\alpha = 1.98 \quad t_p = 4\,\text{ms} \quad q_0 = 68.95\,\text{kPa}$$

The data are taken from Lu (1996). For the uniformly distributed external load along the x-axis, the particular solution in Eq. (14.123) is given as

$$\tilde{\mathbf{V}}^p = \sum_{m=1}^{2M} \frac{\phi_{m4}^{L+}\tilde{\mathbf{P}}\phi_{m1}^{R+}}{\mathbf{B}_m^+}\frac{e^{ik_m^+(x+a/2)}-1}{k_m^+} + \sum_{m=1}^{2M} \frac{\phi_{m4}^{L-}\tilde{\mathbf{P}}\phi_{m1}^{R-}}{\mathbf{B}_m^-}\frac{e^{ik_m^-(x-a/2)}-1}{k_m^-} \qquad (14.145)$$

The results for the four edges clamped plate are shown in Figure 14.16. The comparison with Lu's results using Rayleigh-Ritz method shows a good agreement.

The effects of the boundary conditions of composite laminated plates have been shown in Figures 14.17 and 14.18. Figure 14.17 shows the transient response of the deflection of the square composite laminated plate for two types of boundary conditions: (1) S_1 and S_3 are clamped, and S_2 and S_4 are simply supported; (2) S_1 and S_3 are simply supported, and S_2 and S_4 are clamped. It is known that the deflection and frequency of a square isotropic plate are the

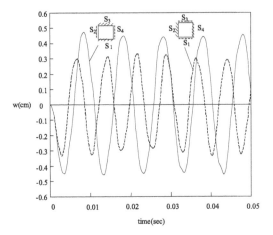

FIGURE 14.17
The time history of the center deflection of the [30°/−30°/−30°/30°] square laminated plate subjected to a conventional blast. The plate is clamped at two opposite edges and simply supported at the other two edges. (From Wang, Y. Y., Lam, K. Y., and Liu, G. R., *Solids and Structures*, 38, 241, 2001. With permission.)

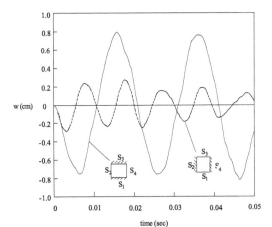

FIGURE 14.18
As in Figure 14.17, but the plate is clamped at two opposite edges and free at the other two edges. (From Wang, Y. Y., Lam, K. Y., and Liu, G. R., *Solids and Structures*, 38, 241, 2001. With permission.)

same for the above two types of boundary conditions. However, this is not true in the case of composite laminated plates, in which both the deflection and frequency are different for the two types of boundary conditions. This fact can also be found from Figure 14.18 in which results for two types of boundary conditions for the square composite laminated plate are presented: (1) clamped at S_1 and S_3, and free at S_2 and S_4; (2) free at S_1 and S_3, and clamped at S_2 and S_4.

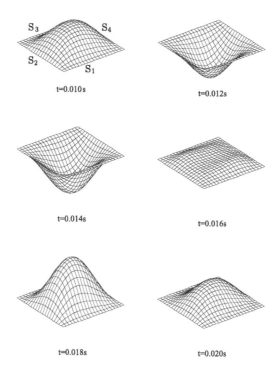

t=0.010s   t=0.012s

t=0.014s   t=0.016s

t=0.018s   t=0.020s

FIGURE 14.19
The deflection of the [30°/−30°/−30°/30°] square laminated plate subjected to a conventional blast. The plate is clamped on edges S_1 and S_3 and simply supported on edges S_2 and S_4. (From Wang, Y. Y., Lam, K. Y., and Liu, G. R., *Solids and Structures*, 38, 241, 2001. With permission.)

The deformation of the plate at different times is shown in Figures 14.19 and 14.20, in which the plate is clamped on two opposite edges and simply supported on the other two edges. In these figures, it is found that the deformation distribution is significantly different for the two different boundary conditions.

14.6.7 Effect of Rotatory Inertia

In the present study, zero initial conditions are again assumed. A three-layer cross-ply [0°/90°/0°] square laminated plate is considered. The plate is made of carbon/epoxy-I composite. The length of the plate is $a = b = 0.762$ m. All the layers are assumed to have the same thickness, and the material properties are shown in Table 14.3.

The loads acting on the plate are also given as in Eqs. (14.141) and (14.142), in which $t_1 = 0.006$ s and $\gamma = 330$ s^{-1}. The intensity of the transverse load is taken to be $q_0 = 68.9476$ MPa when $H = a/5$, and $q_0 = 689.476$ KPa when $H = a/10$ and $H = a/20$. These data are taken from Khdeir and Reddy (1989) for comparison purposes.

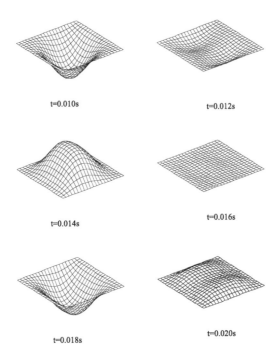

t=0.010s

t=0.012s

t=0.014s

t=0.016s

t=0.018s

t=0.020s

FIGURE 14.20
As in Figure 14.19, but the plate is simply supported on S_1 and S_3 and clamped on S_2 and S_4. (From Wang, Y. Y., Lam, K. Y., and Liu, G. R., *Solids and Structures*, 38, 241, 2001. With permission.)

Figures 14.21–14.24 show the time history of transverse deflection at the center of the plate for various loads. The SEM results are compared with exact solutions, which do not consider the rotatory inertia, and the exact solution given by Khdeir and Reddy (1989) who considered the rotatory inertia. From Figures 14.21–14.24, it is found that the SEM solutions have very good agreement with the exact solutions with and without the effect of the rotatory inertia. The stress responses in time domain can be obtained from the displacement distribution. The comparison of SEM solution with exact solution of normal stress response is given in Figure 14.25, in which the normal stress is normalized using Eq. (14.143). Also, very good agreements are observed.

The rotatory inertia has been found to have the effect of decreasing the frequency of the plate or increasing the wavelength slightly. It is also found that the rotatory inertia has less effect on the deflection amplitude when $t \leq t_1$, but has significant effect when $t > t_1$ (see, for example, Figures 14.21–14.25). This is because the plate will be in the state of free vibration after $t > t_1$ under the initial conditions (displacement and velocity) at $t = t_1$. A small change in natural frequency of the plate gives very different initial conditions at $t = t_1$. The different initial conditions result in a large difference in the amplitude of the free vibration after $t > t_1$.

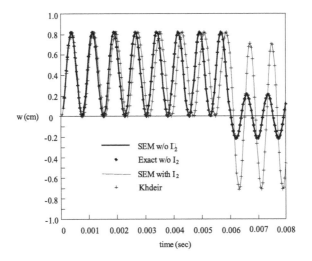

FIGURE 14.21
Center deflection versus time for simple supported symmetric cross-ply laminated plate subject to time-step loading. Comparison of the SEM solution and exact solution with and without rotatory inertia. (From Wang, Y. Y., Lam, K. Y., and Liu, G. R., *Composite Structures*, 48, 265–273, 2000. With permission.)

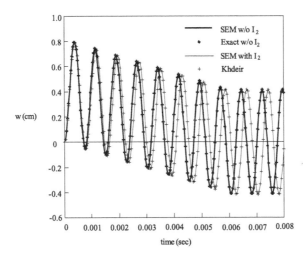

FIGURE 14.22
As in Figure 14.21, but the plate is subjected to a triangularly distributed loading. (From Wang, Y. Y., Lam, K. Y., and Liu, G. R., *Composite Structures*, 48, 265–273, 2000. With permission.)

Figures 14.23, 14.26, and 14.27 show, respectively, displacement response of plates with different thickness-length ratios of 1/5, 1/10, and 1/20. Comparison of these figures reveals that the smaller the thickness-length ratio, the smaller the effects of the rotatory inertia. When the thickness-length ratio is 1/20, as shown in Figure 14.27, there is nearly no observable difference

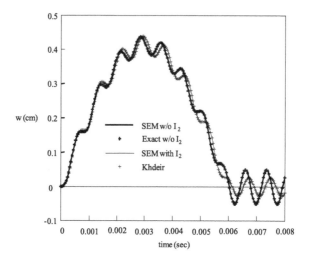

FIGURE 14.23
As in Figure 14.21, but the plate is subjected to a distributed loading of sine function. (From Wang, Y. Y., Lam, K. Y., and Liu, G. R., *Composite Structures*, 48, 265–273, 2000. With permission.)

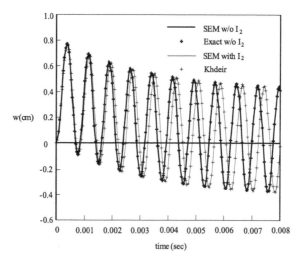

FIGURE 14.24
As in Figure 14.21, but the plate is subjected to an explosive blast loading. (From Wang, Y. Y., Lam, K. Y., and Liu, G. R., *Composite Structures*, 48, 265–273, 2000. With permission.)

between the results with and without the consideration of rotatory inertia. We can therefore conclude that the effect of rotatory inertia can be ignored only if the plate is very thin with a thickness-length ratio of less than 1/20. The rotatory inertia needs to be considered if the thickness-length ratio is between 1/5 and 1/20. For plates with thickness-length ratio of larger than 1/5, the CLPT is no longer valid, and a higher order plate theory should be used.

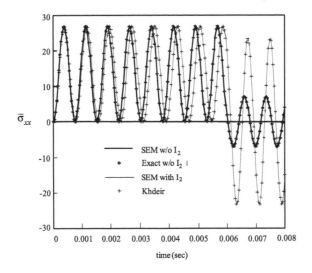

FIGURE 14.25
Transient response of the dimensionless normal stress $\bar{\sigma}_{xx}$ for simple supported symmetric cross-ply laminated plate under step loading. Comparison of the SEM and exact solution with and without rotatory inertia. (From Wang, Y. Y., Lam, K. Y., and Liu, G. R., *Composite Structures,* 48, 265–273, 2000. With permission.)

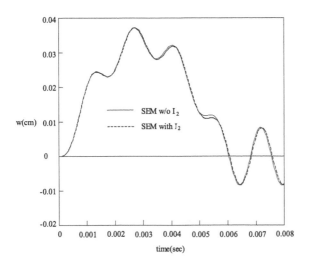

FIGURE 14.26
The effect of the thickness on the center of a square plate ($a = 0.762$ m, $H = a/10$) under sine loading. (From Wang, Y. Y., Lam, K. Y., and Liu, G. R., *Composite Structures,* 48, 265–273, 2000. With permission.)

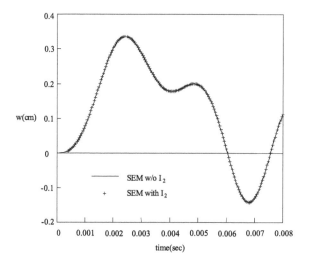

FIGURE 14.27
As in Figure 14.26, but $H = a/20$. (From Wang, Y. Y., Lam, K. Y., and Liu, G. R., *Composite Structures*, 48, 265–273, 2000. With permission.)

The effects of elastic constant E_1 and material density ρ have also been investigated and results are shown in Figures 14.28–14.31. In Figures 14.28 and 14.29, the thin line refers to $E_1 = 165.474$ GPa (decreased about 4%), the dash line refers to $\rho = 1550.16$ kg/m^3 (decreased about 3.3%), and all other parameters are unchanged. It is also found that the small change of the natural frequency results in a large difference in the amplitude of the free vibration after $t > t_1$. Comparing Figures 14.28 and 14.29, it is found that the difference of the amplitude after $t > t_1$ is not so significant when the rotatory inertia is considered, which shows that the rotatory inertia is very important in the analysis of thicker plate. Figures 14.30 and 14.31 show the response of the thin plate where $H/a = 1/20$. From Figures 14.28–14.30, it is found that the elastic constant affects not only the natural frequency but also the amplitude. The amplitudes increase as the elastic constant decreases when $t < t_1$, but it is not true $t > t_1$ in which the initial condition at $t = t_1$ will affect the response. The influence of the mass density of the material can be found from Figures 14.28, 14.29, and 14.31, which show that the mass density has less effect on the amplitude when $t < t_1$ but will affect the natural frequency and the amplitude when $t > t_1$.

The effect of the fiber orientation on the dynamic response of a symmetric laminated plate has been studied and the results are shown in Figures 14.32 and 14.33. It is found that the fiber orientation of the face layers affects significantly the response of the plate, and the fiber orientation of the inner layer has less effect on the dynamic response.

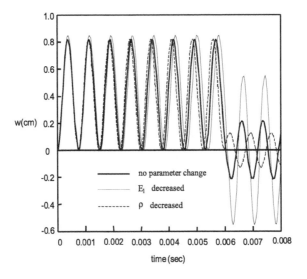

FIGURE 14.28
The effect of elastic constant E_1 and material density ρ on the center deflection of plate subject to time-step loading. The rotatory inertia effect is not considered. (From Wang, Y. Y., Lam, K. Y., and Liu, G. R., *Composite Structures*, 48, 265–273, 2000. With permission.)

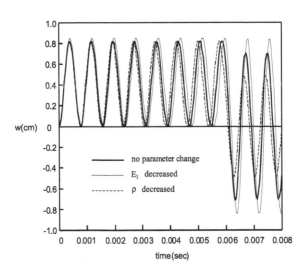

FIGURE 14.29
As in Figure 14.28, but the rotatory inertia effect is considered. (From Wang, Y. Y., Lam, K. Y., and Liu, G. R., *Composite Structures*, 48, 265–273, 2000. With permission.)

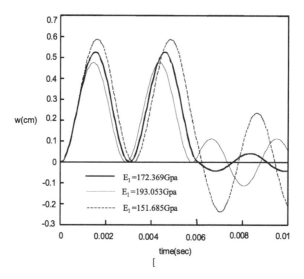

FIGURE 14.30
The effect of the elastic constant E_1 on the center deflection of a square plate subjected to time-step loading. (From Wang, Y. Y., Lam, K. Y., and Liu, G. R., *Composite Structures*, 48, 265–273, 2000. With permission.)

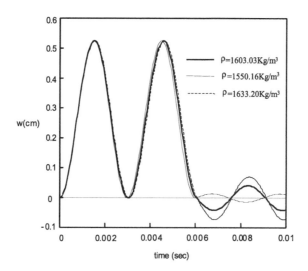

FIGURE 14.31
The effect of material density ρ on the center deflection of a square plate subjected to time-step loading. (From Wang, Y. Y., Lam, K. Y., and Liu, G. R., *Composite Structures*, 48, 265–273, 2000. With permission.)

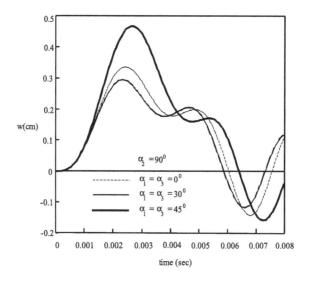

FIGURE 14.32
The effect of fiber orientation of the face layers on the centre deflection of a plate subjected to a loading of sine time function. (From Wang, Y. Y., Lam, K. Y., and Liu, G. R., *Composite Structures*, 48, 265–273, 2000. With permission.)

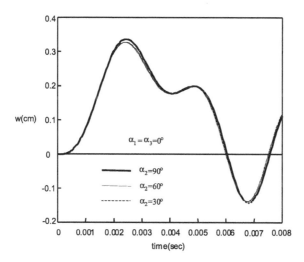

FIGURE 14.33
The effect of fiber orientation of the inner layers on the center deflection of a plate subjected to a loading of sine time function. (From Wang, Y. Y., Lam, K. Y., and Liu, G. R., *Composite Structures*, 48, 265–273, 2000. With permission.)

14.6 Concluding Remarks

This chapter introduced the strip element formulation for analyzing bending deformation and bending waves in composite laminated plates. The results obtained using SEM have been compared with many other variable solutions and good agreements are observed. The SEM method has the following advantages.

1. It requires only small memory in the computation.
2. It needs fewer elements to achieve the required accuracy. In above dynamic examples, the SEM results have good agreement with exact solutions for both thin plates ($H/a \leq 1/10$) and moderately thick plates ($H/a = 1/5$), using only four elements.
3. The SEM has high accuracy since both deflection and rotation angle are continuous on whole node lines.
4. Generally the integration of the particular solution given by Eq. (14.123) can be given analytically using superposition integral over the solution of the line load.
5. The boundary conditions in one direction are imposed only at the final stage of the solving procedure; it is, therefore, easy to get the responses for different boundary conditions in one running job. This will save much computing time when studies for many load cases are required.
6. The effects of rotatory inertia on the transient responses of the plate have been studied. It is found that the rotatory inertia has less effect on very thin plates with a thickness–length ratio of less than 1/20. The rotatory inertia effect significantly on the results for thicker plates whose thickness–length ratio is between 1/5 and 1/20. For plates with thickness–length ratio larger than 1/5, the CLPT is no longer valid, and a higher order plate theory should be used. SEM has not yet been developed based on those higher-order plate theories.

15

Characteristics of Waves in Composite Cylinders

15.1 Introduction

Anisotropic laminated cylinders or circular cylindrical shells are widely used in various engineering sectors. Because of the algebraic complexity, analytical solutions for waves in anisotropic cylinders are very difficult to obtain (Markus and Mead, 1995; Yuan and Hsieh, 1998). Numerical methods therefore need to be developed. Nelson et al. (1971) proposed a numerical-analytical method for analyzing waves in laminated orthotropic cylinders. In their analysis, the circumferential and axial displacements are represented by trigonometric functions, while the radial displacement is modeled by finite element approach. Huang and Dong (1988) later applied the technique to study the frequency dispersion of laminated composite cylinders. The method has proved expedient in the treatment of vibration and wave motion problems in anisotropic cylindrical shells (Rattanwangcharoen et al., 1994, 1997; Zhuang et al., 1997; Xi et al., 2000a; Han et al., 2001).

This chapter investigates such unique and important wave properties of anisotropic cylindrical shells as group velocity and characteristic wave surfaces. The techniques were originally presented by Xi et al. (2000a). The method used is formulated within the framework of the theory of 3D elasticity. The radial displacement of the shell is modeled by finite element techniques, while the axial and circumferential displacements are expanded as the complex exponentials. The associated discrete system equations are developed by applying Hamilton's principle. The body forces of the cylindrical shell are neglected, as we are interested in dealing with the characteristics of waves in cylinders. The dispersion relation of harmonic waves in cylinders is obtained by solving the eigenvalue equation derived from the system equations of free wave motion. The group velocity of the harmonic wave is obtained with the use of the Rayleigh quotient. Six representations of wave surfaces—the phase velocity, phase slowness, phase wave surfaces, group velocity, group slowness, and group wave surfaces—are formulated to visualize the effect of anisotropy on wave propagation. Parametric studies include the propagation direction and modes of waves, the ratio of the inner radius to the thickness of the shell, and the stacking sequence of the laminated shell.

15.2 Basic Equations

Suppose a helical harmonic wave propagates in a laminated composite circular cylindrical shell of inner radius R_i and outer radius R_o, where β is the angle of the propagation direction of the wave with respect to the z-axis (see Figure 15.1). Perfect bonding between plies is assumed. Deformations of the shell are assumed to be small to maintain a linear relationship between the strains and the displacements. Under these assumptions, the strain-displacement relations in the cylindrical coordinate system are given by

$$\boldsymbol{\varepsilon} = \mathbf{Lu} \tag{15.1}$$

where $\boldsymbol{\varepsilon} = [\varepsilon_z\ \varepsilon_\theta\ \varepsilon_r\ \gamma_{r\theta}\ \gamma_{rz}\ \gamma_{z\theta}]^T$ is the vector of strains and $\mathbf{u} = [u\ v\ w]^T$ is the vector of displacements. Here u, v, and w are the displacement components in the axial, circumferential, and radial directions, respectively. The differential operator matrix $\mathbf{L}$ is given by

$$\mathbf{L} = \begin{bmatrix} \dfrac{\partial}{\partial z} & 0 & 0 & 0 & \dfrac{\partial}{\partial r} & \dfrac{1}{r}\dfrac{\partial}{\partial \theta} \\[2ex] 0 & \dfrac{1}{r}\dfrac{\partial}{\partial \theta} & 0 & \dfrac{\partial}{\partial r} - \dfrac{1}{r} & 0 & \dfrac{\partial}{\partial z} \\[2ex] 0 & \dfrac{1}{r} & \dfrac{\partial}{\partial r} & \dfrac{1}{r}\dfrac{\partial}{\partial \theta} & \dfrac{\partial}{\partial z} & 0 \end{bmatrix}^T \tag{15.2}$$

or

$$\mathbf{L} = \mathbf{L}_1\frac{\partial}{\partial z} + \mathbf{L}_2\frac{1}{r}\frac{\partial}{\partial \theta} + \mathbf{L}_3\frac{\partial}{\partial r} + \mathbf{L}_4\frac{1}{r} \tag{15.3}$$

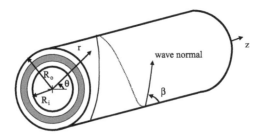

FIGURE 15.1
Configuration of a laminated composite circular cylindrical shell, where a helical harmonic wave is traveling. (From Xi et al., *J. Acoust. Soc. Am.*, 108, 2179, 2000a. With permission.)

where the matrices $\mathbf{L}_1$, $\mathbf{L}_2$, $\mathbf{L}_3$, and $\mathbf{L}_4$ are constant matrices, listed below.

$$\mathbf{L}_1 = \begin{bmatrix} 1 & 0 & 0 & 0 & 0 & 0 \\ 0 & 0 & 0 & 0 & 0 & 1 \\ 0 & 0 & 0 & 0 & 1 & 0 \end{bmatrix}^T \tag{15.4}$$

$$\mathbf{L}_2 = \begin{bmatrix} 0 & 0 & 0 & 0 & 0 & 1 \\ 0 & 1 & 0 & 0 & 0 & 0 \\ 0 & 0 & 0 & 1 & 0 & 0 \end{bmatrix}^T \tag{15.5}$$

$$\mathbf{L}_3 = \begin{bmatrix} 0 & 0 & 0 & 0 & 1 & 0 \\ 0 & 0 & 0 & 1 & 0 & 0 \\ 0 & 0 & 1 & 0 & 0 & 0 \end{bmatrix}^T \tag{15.6}$$

$$\mathbf{L}_4 = \begin{bmatrix} 0 & 0 & 0 & 0 & 0 & 0 \\ 0 & 0 & 0 & -1 & 0 & 0 \\ 0 & 1 & 0 & 0 & 0 & 0 \end{bmatrix}^T \tag{15.7}$$

A lamina under consideration is considered to be anisotropic, so the stresses are related to strains by

$$\boldsymbol{\sigma} = \mathbf{c}\boldsymbol{\varepsilon} \tag{15.8}$$

where

$$\boldsymbol{\sigma} = \begin{bmatrix} \sigma_z & \sigma_\theta & \sigma_r & \tau_{r\theta} & \tau_{rz} & \tau_{z\theta} \end{bmatrix}^T \tag{15.9}$$

is the vector of stresses and

$$\mathbf{c} = \begin{bmatrix} c_{11} & c_{12} & \cdots & c_{16} \\ & c_{22} & \cdots & c_{26} \\ & & \ddots & \vdots \\ \text{sym.} & & & c_{66} \end{bmatrix} \tag{15.10}$$

is the matrix of the off-principal-axis stiffness coefficients of the lamina whose expressions in terms of engineering constants are given in detail by Vinson and Sierakowski (1987).

15.3 Dispersion Relations

With the above basic equations, we apply Hamilton's principle to derive the system equation for the shell. We first use finite elements to model the radial displacement of the shell and then employs the complex exponentials to model the axial and circumferential displacements. In view of the heterogeneity of the laminated composite shell in the radial direction, an annular element shown in Figure 15.2 is used in the subdivision of the shell. The element has the inner, middle, and outer nodal surfaces denoted by i, m, o, and each nodal surface has three degrees of freedom, u, v, and w. Hence, the vector of the unknown displacement amplitudes of the element is expressed as

$$\mathbf{V}^e = [u_i \quad v_i \quad w_i \quad u_m \quad v_m \quad w_m \quad u_o \quad v_o \quad w_o]^T \tag{15.11}$$

The shell is now subdivided into N elements in the radial direction, and r_{j-1} and r_j represent, respectively, the inner and outer radii of element j. The displacements within an element are thus approximated as

$$\mathbf{u} = \mathbf{N}(r)\mathbf{V}^e \tag{15.12}$$

where

$$\mathbf{N}(r) = \lfloor (1 - 3\hat{r} + 2\hat{r}^2)\mathbf{I} \quad 4(\hat{r} - \hat{r}^2)\mathbf{I} \quad (-\hat{r} + 2\hat{r}^2)\mathbf{I} \rfloor \tag{15.13}$$

is the matrix of shape functions of the element. Here $\hat{r} = (r - r_{j-1})/(r_j - r_{j-1})$, $r_{j-1} \le r \le r_j$ and $\mathbf{I}$ is a 3×3 identity matrix.

The potential energy (or strain energy) of the element in the absence of body forces is given by

$$V = \frac{1}{2} \int_{-\infty}^{+\infty} \int_0^{2\pi} \int_{r_{j-1}}^{r_j} \boldsymbol{\varepsilon}^T \boldsymbol{\sigma} r \, dr \, d\theta \, dz \tag{15.14}$$

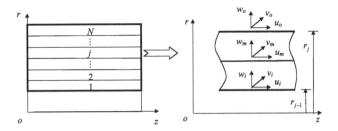

FIGURE 15.2
Annular element subdivision and the jth isolated annular element. (From Xi et al., *J. Acoust. Soc. Am.*, 108, 2179, 2000a. With permission.)

Invoking Eqs. (15.1) to (15.8), we obtain

$$
V = \frac{1}{2} \int_{-\infty}^{+\infty} \int_{0}^{2\pi} \int_{r_{j-1}}^{r_j} \left(\frac{\partial \mathbf{u}^T}{\partial z} \mathbf{D}_{11} \frac{\partial \mathbf{u}}{\partial z} + \frac{1}{r} \frac{\partial \mathbf{u}^T}{\partial z} \mathbf{D}_{12} \frac{\partial \mathbf{u}}{\partial \theta} + \frac{\partial \mathbf{u}^T}{\partial z} \mathbf{D}_{13} \frac{\partial \mathbf{u}}{\partial r} + \frac{1}{r} \frac{\partial \mathbf{u}^T}{\partial z} \mathbf{D}_{14} \mathbf{u} \right.
$$

$$
+ \frac{1}{r} \frac{\partial \mathbf{u}^T}{\partial \theta} \mathbf{D}_{12}^T \frac{\partial \mathbf{u}}{\partial z} + \frac{1}{r^2} \frac{\partial \mathbf{u}^T}{\partial \theta} \mathbf{D}_{22} \frac{\partial \mathbf{u}}{\partial \theta} + \frac{1}{r} \frac{\partial \mathbf{u}^T}{\partial \theta} \mathbf{D}_{23} \frac{\partial \mathbf{u}}{\partial r} + \frac{1}{r^2} \frac{\partial \mathbf{u}^T}{\partial \theta} \mathbf{D}_{24} \mathbf{u}
$$

$$
+ \frac{\partial \mathbf{u}^T}{\partial r} \mathbf{D}_{13}^T \frac{\partial \mathbf{u}}{\partial z} + \frac{1}{r} \frac{\partial \mathbf{u}^T}{\partial r} \mathbf{D}_{23}^T \frac{\partial \mathbf{u}}{\partial \theta} + \frac{\partial \mathbf{u}^T}{\partial r} \mathbf{D}_{33} \frac{\partial \mathbf{u}}{\partial r} + \frac{1}{r} \frac{\partial \mathbf{u}^T}{\partial r} \mathbf{D}_{34} \mathbf{u}
$$

$$
\left. + \frac{1}{r} \mathbf{u}^T \mathbf{D}_{14}^T \frac{\partial \mathbf{u}}{\partial z} + \frac{1}{r^2} \mathbf{u}^T \mathbf{D}_{24}^T \frac{\partial \mathbf{u}}{\partial \theta} + \frac{1}{r} \mathbf{u}^T \mathbf{D}_{34}^T \frac{\partial \mathbf{u}}{\partial r} + \frac{1}{r^2} \mathbf{u}^T \mathbf{D}_{44} \mathbf{u} \right) r \, dr \, d\theta \, dz \qquad (15.15)
$$

where $\mathbf{D}_{ij} = \mathbf{L}_i^T \overline{\mathbf{Q}} \mathbf{L}_j \ (i, j = 1, 2, 3, 4)$.

The kinetic energy of the element is expressed in terms of the displacement vector as

$$
T = \frac{1}{2} \int_{-\infty}^{+\infty} \int_{0}^{2\pi} \int_{r_{j-1}}^{r_j} \frac{\partial \mathbf{u}^T}{\partial t} \frac{\partial \mathbf{u}}{\partial t} \rho r \, dr \, d\theta \, dz \qquad (15.16)
$$

where ρ is the mass density of the material of the element.

We introduce now Hamilton's principle:

$$
\delta \int_{t_1}^{t_2} L \, dt = 0 \qquad (15.17)
$$

where L is the Lagrangian functional defined by the difference of kenetic and potential energies, i.e.,

$$
L = T - V \qquad (15.18)
$$

Hamilton's principle states

> Of all possible time histories of displacement states which satisfy (a) the compatibility equations, (b) the constraints or the kinematical boundary conditions and (c) the conditions at initial time (t_1) and final time (t_2), the history corresponding to the actual solution makes the Lagrangian functional a minimum.

Substituting Eqs. (15.15) and (15.16) into (15.18) and then (15.17), and performing a lengthy but simple algebraic manipulation, we obtain a set of partial differential equations of the element as follows:

$$
-\mathbf{A}_1^e \frac{\partial^2 \mathbf{V}^e}{\partial z^2} - \mathbf{A}_2^e \frac{\partial^2 \mathbf{V}^e}{\partial \theta \partial z} - \mathbf{A}_3^e \frac{\partial^2 \mathbf{V}^e}{\partial \theta^2} + \mathbf{A}_4^e \frac{\partial \mathbf{V}^e}{\partial z} + \mathbf{A}_5^e \frac{\partial \mathbf{V}^e}{\partial \theta} + \mathbf{A}_6^e \mathbf{V}^e + \mathbf{M}^e \frac{\partial^2 \mathbf{V}^e}{\partial t^2} = 0 \quad (15.19)
$$

where

$$\mathbf{A}_1^e = \int_{r_{j-1}}^{r_j} \mathbf{N}^T \mathbf{D}_{11} \mathbf{N} r \, dr \tag{15.20}$$

$$\mathbf{A}_2^e = \int_{r_{j-1}}^{r_j} \mathbf{N}^T (\mathbf{D}_{12} + \mathbf{D}_{12}^T) \mathbf{N} \, dr \tag{15.21}$$

$$\mathbf{A}_3^e = \int_{r_{j-1}}^{r_j} \frac{1}{r} \mathbf{N}^T \mathbf{D}_{22} \mathbf{N} \, dr \tag{15.22}$$

$$\mathbf{A}_4^e = \int_{r_{j-1}}^{r_j} \left[-\mathbf{N}^T \mathbf{D}_{13} \frac{d\mathbf{N}^T}{dr} + \frac{d\mathbf{N}^T}{dr} \mathbf{D}_{13}^T \mathbf{N} + \frac{1}{r} \mathbf{N}^T (\mathbf{D}_{14}^T - \mathbf{D}_{14}) \mathbf{N} \right] r \, dr \tag{15.23}$$

$$\mathbf{A}_5^e = \int_{r_{j-1}}^{r_j} \left[-\mathbf{N}^T \mathbf{D}_{23} \frac{d\mathbf{N}^T}{dr} + \frac{d\mathbf{N}^T}{dr} \mathbf{D}_{23}^T \mathbf{N} + \frac{1}{r} \mathbf{N}^T (\mathbf{D}_{24}^T - \mathbf{D}_{24}) \mathbf{N} \right] dr \tag{15.24}$$

$$\mathbf{A}_6^e = \int_{r_{j-1}}^{r_j} \left(\frac{d\mathbf{N}^T}{dr} \mathbf{D}_{33} \frac{d\mathbf{N}}{dr} + \frac{1}{r} \frac{d\mathbf{N}^T}{dr} \mathbf{D}_{34} \mathbf{N} + \frac{1}{r} \mathbf{N}^T \mathbf{D}_{34}^T \frac{d\mathbf{N}}{dr} + \frac{1}{r^2} \mathbf{N}^T \mathbf{D}_{44} \mathbf{N} \right) r \, dr \tag{15.25}$$

$$\mathbf{M}^e = \int_{r_{j-1}}^{r_j} \mathbf{N}^T \mathbf{N} \rho r \, dr \tag{15.26}$$

The axial and circumferential displacement vector $\mathbf{V}^e(\theta, z, t)$ can be expanded as the complex exponentials

$$\mathbf{V}^e(\theta, z, t) = \mathbf{U}^e \exp i(n\theta + k_z z - \omega t) \tag{15.27}$$

where ω is the angular frequency, n is the wavenumber in the circumferential direction, and $k_2 = k \cos \beta$ is the wavenumber in the z-axis direction. When a helical wave of wavenumber k propagates in the shell at an arbitrary β angle with respect to the z-axis, we have

$$n = R_o k \sin \beta, \quad k_z = k \cos \beta \tag{15.28}$$

Substituting Eq. (15.27) into Eq. (15.19), we find

$$\lfloor \mathbf{K}^e - \omega^2 \mathbf{M}^e \rfloor \mathbf{U}^e = 0 \tag{15.29}$$

where

$$\mathbf{K}^e = \mathbf{A}_1^e k_z^2 + \mathbf{A}_2^e n k_z + \mathbf{A}_3^e n^2 + i \mathbf{A}_4^e k_z + i \mathbf{A}_5^e n + \mathbf{A}_6^e \tag{15.30}$$

Assembling elements at the nodal surfaces in the same way for the laminated plates, we find the eigenvalue equation for the shell

$$\lfloor \mathbf{K} - \omega^2 \mathbf{M} \rfloor \mathbf{U} = 0 \tag{15.31}$$

During the process of assembling elements, we make use of the following interface and boundary conditions

$$\mathbf{p}_1^i = 0 \tag{15.32}$$

$$\mathbf{p}_j^o = \mathbf{p}_{j+1}^i, \quad \mathbf{U}_j^o = \mathbf{U}_{j+1}^i, \quad \text{for } 1 < j < N - 1 \tag{15.33}$$

$$\mathbf{p}_N^o = 0 \tag{15.34}$$

in which $\mathbf{p}$ is the vector of the surface tractions, the subscripts denote the element numbers, and the superscripts denote the inner and outer surfaces of the element.

When the wavenumber is specified, we can solve Eq. (15.31) for the circular frequency ω and, accordingly, obtain the relationship between the wavenumber and circular frequency, that is, the dispersion relationship for the shell. Subsequently, the group velocity and six characteristic wave surfaces can be calculated in a way similar to that in Chapter 5.

15.4 Examples

Now we present numerical examples for dispersion relations and characteristic surfaces of harmonic helical waves in a laminated composite cylindrical shell. In laminate codes used below, a lamina numbering increases from the inner to outer surface; letters C and G represent carbon/epoxy and glass/epoxy, respectively. The number following the letters indicates the azimuthal angle of the fiber orientation with respect to the z-axis; the subscript s denotes that the multilayered shell is symmetrically stacked with respect to the geometrical middle surface. The material properties of the shell are the same as before. The following dimensionless parameters are adopted:

$$\bar{k} = k(R_o - R_i), \quad \bar{\lambda} = \lambda/(R_o - R_i), \quad \bar{R} = R_i/(R_o - R_i), \quad \bar{\omega} = \omega(R_o - R_i)\sqrt{\rho/Q_{11}}$$

where the reference properties Q_{11} and ρ are the Young's modulus in the fiber direction and mass density of C0. Since the present method is based on the theory of 3D elasticity, it is applicable to both a circular cylindrical shell and a circular cylinder. The distinction between waves in a cylindrical shell and waves in a cylinder can be explored through studying the effect of the ratio of the inner radius to the thickness of the shell. To this end, two

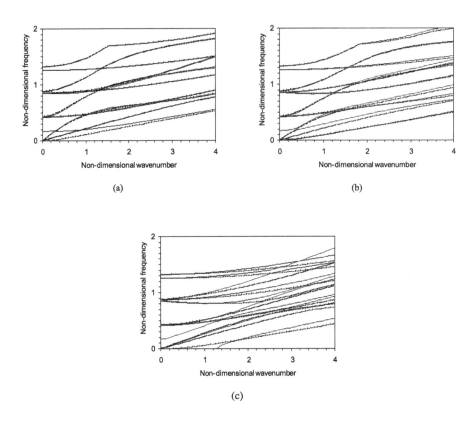

FIGURE 15.3
Dispersion curves for waves propagating in $(C0/G \pm 45)_s$ circular cylindrical shells. (a) $\beta = 0$; (b) $\beta = 30°$; (c) $\beta = 90°$. — $\bar{R} = 1$; — $\bar{R} = 100$. (From Xi et al., *J. Acoust. Soc. Am.*, 108, 2179, 2000a. With permission.)

ratios of radius to thickness, $\bar{R} = 1$ and 100, are used here. The former is for a thick cylinder, while the latter is for a cylindrical shell.

We first discuss the dispersion behavior of helical harmonic waves. In Figure 15.3(a)–(c), the dispersion curves for a $(C0/G \pm 45)_s$ shell are compared with those for a $(C0/G \pm 45)_s$ cylinder. The propagation directions of waves are chosen as $\beta = 0°$, $30°$, and $90°$, respectively. They are respectively referred to as the axial, helical, and circumferential waves. The distinction between the axial waves in the shell and cylinder can be found from Figure 15.3(a). The only curve for the first propagation mode in the case of the shell is straight lines passing through the origin. This reveals that the wave propagation for the first mode for the shell is nondispersive and all frequencies have the same phase velocity, whereas the wave propagation for higher order modes is dispersive. The wave propagation for all the modes in the case of the cylinder is dispersive. In addition, the ratio of radius to thickness affects only those curves for lower order modes and smaller wavenumbers. From Figure 15.3(b) it can be seen that for waves propagating

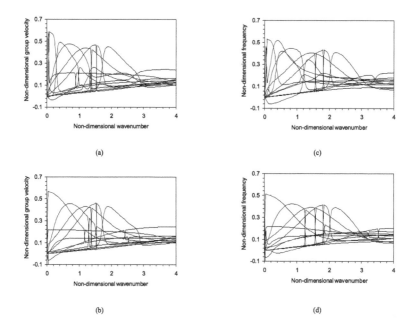

FIGURE 15.4

Group velocity spectra for waves propagating in $(C0/G \pm 45)_s$ circular cylindrical shells. (a) $\beta = 0$, $\bar{R} = 1$; (b) $\beta = 0$, $\bar{R} = 100$; (c) $\beta = 30°$, $\bar{R} = 1$; (d) $\beta = 30°$, $\bar{R} = 100$. (From Xi et al., *J. Acoust. Soc. Am.*, 108, 2179, 2000a. With permission.)

at an angle $\beta = 30°$ respective to the axial direction, the ratio of radius to thickness also affects only the curves for lower order modes, but this effect increases as the wavenumber increases. The results in Figure 15.3(c) indicate that the ratio of radius to thickness has considerable influence on the curves for all the modes, and that the larger the wavenumber, the stronger this effect.

We now turn to the computation of group velocity spectra. In Figure 15.4(a)–(d), the group velocity spectra for a $(C0/G \pm 45)_s$ cylindrical shell are compared with those for a $(C0/G \pm 45)_s$ cylinder. The propagation directions of waves are chosen as $\beta = 0°$ and $30°$, respectively. Similar to phenomena observed in Chapter 5 for laminated composite plates, negative group velocities occur at a range of smaller wavenumbers for both the shell and cylinder. The range of wavenumbers producing the negative group velocities for the wave with $\beta = 30°$ is larger than that for the axial wave. This indicates that the range varies with the wave normal. Another difference between the shell and cylinder waves lies in that the curves for the shell present more abrupt changes than those for the cylinder. It is thus evident that the ratio of radius to thickness can alter the pattern of the group velocity spectra. Besides, it is very clear that the velocities of the energy flux depend considerably on the wavenumber and propagation modes of waves.

Now, let us look at six characteristic wave surfaces of cylinders. Figure 15.5(a)–(d) illustrates six representation surfaces for the first two propagation modes

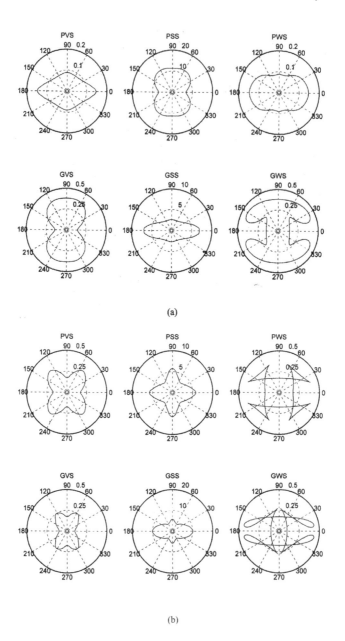

(a)

(b)

FIGURE 15.5
Characteristic wave surfaces for $(C0/G90/G0)_s$ circular cylindrical shells. (a) The first mode $\bar{R} = 1$; (a) The second mode $\bar{R} = 1$; (a) The first mode $\bar{R} = 100$; (d) The second mode $\bar{R} = 100$. (From Xi, Z. C., et al., *J. Acoust. Soc. Am.*, 108, 2179, 2000. With permission.)

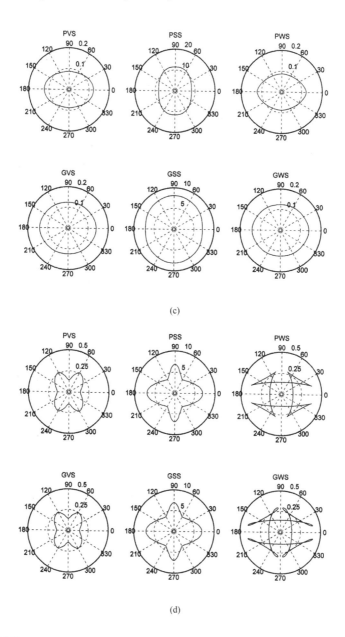

(c)

(d)

FIGURE 15.5
(Continued).

for a $(C0/G90/G0)_s$ cylindrical shell and a $(C0/G90/G0)_s$ cylinder. From these figures, we can see very clearly that the wave propagation in the shells depends strongly upon not only the wave propagation modes but also the anisotropy of composites and the ratio of radius to thickness. Energy flux deviation is observed for all the wave modes, as none of the slowness curves are circular.

15.5   Remarks

The frequency spectra, group velocity spectra, and characteristic surfaces of waves in a laminated composite circular cylindrical shell have been investigated. The method of approach is formulated within the framework of the theory of 3D elasticity. It is thus accurate in comparison with ones using various shell theories. The use of finite elements through the radial direction makes it easier to deal with composite circular cylindrical shells made of an arbitrary number of anisotropic layers, of arbitrary lay-ups, and of any type of materials. Furthermore, the method is capable of reducing the spatial dimensions of a problem by one and of omitting tedious preprocessors occupying a substantial part of finite element methods, and accordingly of reducing a great deal of computational labor. Based on the calculated results, the following conclusions can be drawn:

- The ratio of radius to thickness has stronger influence on the frequency spectra in the circumferential wave than on that in the axial wave.
- The ratio of radius to thickness can alter the pattern of the group velocity spectra.
- The negative group velocity appears at a range of smaller wavenumbers and the range varies as the wave normal.
- The characteristic wave surfaces vary with the propagation modes of waves, the ratio of radius to thickness, and the lay-ups of laminated composite cylindrical shells.

The present method has been extended to cases of laminated composite cylinders containing fluid or surrounded by fluid. These cases are frequently met in engineering applications. Numerical examples have shown that the wave behaviors of these wet cylinders are different from those of the corresponding dry cases. More details can be found in the reference (Xi et al., 2000a).

16

Wave Scattering by Cracks in Composite Cylinders

16.1 Introduction

The presence of cracks in any structure, including composite cylinders and cylindrical shells, is a big concern in operation. Crack detection in structures using elastic waves is, therefore, of great importance. Investigation of scattering problems associated with cylinders and circular cylindrical shells started with isotropic materials. Faran (1951) studied scattered sound of infinitely long elastic solid cylinders. Stanton (1988) treated scattering from cylinders of finite length using Faran's solution for the infinitely long elastic cylinder. Junger (1951, 1952) analyzed scattering from shells. These pioneering works were followed by a number of investigations, which are well reviewed by Gaunaurd (1989). Because advanced composite materials possess many properties superior to traditional metal materials, laminated composite circular cylinders and cylindrical shells have been widely used in lightweight structures. This situation has promoted the study of scattering problems related to layered cylinders and shells. Akay (1991) investigated scattering from a multiple concentric shell ensonified by a plane wave using a theory of resonance scattering. Nayfeh (1995) discussed scattering of horizontally polarized *SH* elastic waves from multilayered anisotropic cylinders embedded in isotropic solids. Because of the inherent complexities involved in the material itself, an analysis of wave propagation in layered composite cylinders needs to resort to numerical techniques. Dealing with propagating waves and edge vibration in anisotropic composite cylinders, Huang and Dong (1984) proposed an efficient numerical-analytical method (NAM) in which a composite cylinder was modeled by finite elements, triangular functions, and wave function expansions in the radial, circumferential, and axial directions, respectively. The salient features of the method are to be capable of reducing the spatial dimensions of a problem by one and to omit tedious preprocessors occupying a substantial part of a finite element method (FEM). Rattanwangcharoen et al. (1994) utilized the NAM to solve the reflection problem of waves at the free edge of a laminated circular cylinder. Rattanwangcharoen et al. (1997) combined the NAM and FEM to analyze scattering of axisymmetric guided waves by a weldment between two laminated cylinders. In their treatise, the NAM was employed to model the cylinders, and the FEM was used to model the weldment. This combinatory procedure was applied to axisymmetric guided wave scattering by cracks in welded steel pipes by Zhuang et al. (1997). The advantage of the combinatory procedure is to be able to treat complex local domains of a cylinder such as weldment, hole, and imperfection. The disadvantage is reduced efficiency of the NAM. Therefore, it is interesting to develop a NAM for analyzing waves in a composite cylinder containing a crack.

This chapter introduces a strip element method (SEM) for analyzing wave scattering by cracks in axisymmetric cross-ply laminated composite cylinders, subjected to a harmonic excitation of a line source along the circumferential direction. The techniques were originally presented by Xi et al. (2000c). The cylinder is first modeled using axisymmetric strip elements in the radial direction. Then the Hamilton variational principle is used to derive a system of governing ordinary differential equations of the system in frequency domain. A particular solution for the resulting equations is found using a modal analysis approach and techniques of Fourier transform. A general solution is obtained with axial stress boundary conditions. Last, numerical examples are presented for multilayered cylinders with outer surface-breaking and radial interior cracks.

16.2 Basic Equations

Consider an infinitely long cracked laminated composite cylinder of thickness H, inner radius R_i, and outer radius R_o as shown in Figure 16.1. The cylinder is made of an arbitrary number of linearly elastic cylinder-like laminae. The bonding between plies is perfect except in the region of the crack. Deformations of the cylinder are assumed small under a harmonic excitation. Let L represent the distance from the crack left tip to the axis r. The length and depth of the crack are designated by a and d, respectively. A radial line load of $q = q_0 \exp(-i\omega t)$ uniformly distributed along the circumferential direction is applied on the outer surface of the cylinder and at $z = 0$. Since the geometry of the cylinder and the load are independent of the circumferential direction, the problem is axisymmetric.

In the case of axisymmetry, the strain-displacement relations are given by

$$\boldsymbol{\varepsilon} = \mathbf{L}\mathbf{u} \tag{16.1}$$

where $\boldsymbol{\varepsilon} = [\varepsilon_z \ \varepsilon_\theta \ \varepsilon_r \ \gamma_{rz}]^T$ is the vector of strains, $\mathbf{u} = [u \ w]^T$ is the vector

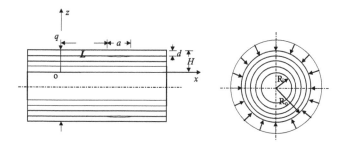

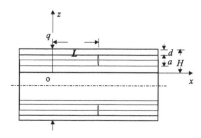

FIGURE 16.1
Geometry of a cracked laminate composite cylinder.

of displacements, and

$$
\mathbf{L} = \begin{bmatrix} \dfrac{\partial}{\partial z} & 0 & 0 & \dfrac{\partial}{\partial r} \\[2ex] 0 & \dfrac{1}{r} & \dfrac{\partial}{\partial r} & \dfrac{\partial}{\partial z} \end{bmatrix}^T \tag{16.2}
$$

is the operator matrix. The operator matrix can be rewritten in the form

$$
\mathbf{L} = \mathbf{L}_1 \frac{\partial}{\partial z} + \mathbf{L}_2 \frac{\partial}{\partial r} + \mathbf{L}_3 \frac{1}{r} \tag{16.3}
$$

where

$$
\mathbf{L}_1 = \begin{bmatrix} 1 & 0 & 0 & 0 \\ 0 & 0 & 0 & 1 \end{bmatrix}^T, \quad \mathbf{L}_2 = \begin{bmatrix} 0 & 0 & 0 & 1 \\ 0 & 0 & 1 & 0 \end{bmatrix}^T, \quad \mathbf{L}_3 = \begin{bmatrix} 0 & 0 & 0 & 0 \\ 0 & 1 & 0 & 0 \end{bmatrix}^T \tag{16.4}
$$

A lamina under consideration is transversely isotropic, and the stresses
are related to strains by

$$
\boldsymbol{\sigma} = \mathbf{Q}\boldsymbol{\varepsilon} \tag{16.5}
$$

where $\boldsymbol{\sigma} = [\sigma_z \ \sigma_\theta \ \sigma_r \ \tau_{rz}]^T$ is the vector of stresses and

$$
\mathbf{Q} = \begin{bmatrix} c_{11} & c_{12} & c_{13} & 0 \\ c_{12} & c_{22} & c_{23} & 0 \\ c_{13} & c_{23} & c_{33} & 0 \\ 0 & 0 & 0 & c_{55} \end{bmatrix}
\tag{16.6}
$$

is the matrix of the off-principal-axis stiffness coefficients of the lamina.

16.3   Axisymmetric Strip Element

A governing equation of the cylinder follows from the Hamilton variational principle that takes the form of

$$
\int_{t_0}^{t_1} \delta(T - V)\, dt = 0
\tag{16.7}
$$

where t_0 and t_1 are arbitrary time instants, V and T are the potential energy and kinetic energy of the cylinder, respectively. The potential energy of the cylinder in the absence of body forces is given by

$$
V = \int_{R_i}^{R_0} \boldsymbol{\varepsilon}^T \boldsymbol{\sigma} r\, dr - \pi R_o q \bar{w}
\tag{16.8}
$$

where $\bar{w}$ is the radial displacement at the loaded position which is expressible as $\bar{w} = \mathbf{n}\mathbf{u}$. Here $\mathbf{n} = [0 \ 1]$.

Substitution of Eqs. (16.1) to (16.5) into Eq. (16.8), in view of Eq. (16.3), gives

$$
V = \pi \int_{R_i}^{R_0} \left(\frac{\partial \mathbf{u}^T}{\partial z} \mathbf{D}_{11} \frac{\partial \mathbf{u}}{\partial z} + \frac{\partial \mathbf{u}^T}{\partial z} \mathbf{D}_{12} \frac{\partial \mathbf{u}}{\partial r} + \frac{1}{r}\frac{\partial \mathbf{u}^T}{\partial z} \mathbf{D}_{13} \mathbf{u} + \frac{\partial \mathbf{u}^T}{\partial r} \mathbf{D}_{12}^T \frac{\partial \mathbf{u}}{\partial z} + \frac{\partial \mathbf{u}^T}{\partial r} \mathbf{D}_{22} \frac{\partial \mathbf{u}}{\partial r} \right.
$$

$$
\left. + \frac{1}{r}\frac{\partial \mathbf{u}^T}{\partial r} \mathbf{D}_{23} \mathbf{u} + \frac{1}{r} \mathbf{u}^T \mathbf{D}_{13}^T \frac{\partial \mathbf{u}}{\partial z} + \frac{1}{r} \mathbf{u}^T \mathbf{D}_{23}^T \frac{\partial \mathbf{u}}{\partial r} + \frac{1}{r^2} \mathbf{u}^T \mathbf{D}_{33} \mathbf{u} \right) r\, dr - \pi \mathbf{u}^T \mathbf{n}^T R_o q
\tag{16.9}
$$

where

$$
\begin{array}{lll}
\mathbf{D}_{11} = \mathbf{L}_1^T \overline{\mathbf{Q}} \mathbf{L}_1 & \mathbf{D}_{12} = \mathbf{L}_1^T \overline{\mathbf{Q}} \mathbf{L}_2 & \mathbf{D}_{13} = \mathbf{L}_1^T \overline{\mathbf{Q}} \mathbf{L}_3 \\
\mathbf{D}_{22} = \mathbf{L}_2^T \overline{\mathbf{Q}} \mathbf{L}_2 & \mathbf{D}_{23} = \mathbf{L}_2^T \overline{\mathbf{Q}} \mathbf{L}_3 & \mathbf{D}_{33} = \mathbf{L}_3^T \overline{\mathbf{Q}} \mathbf{L}_3
\end{array}
\tag{16.10}
$$

The kinetic energy of the system is expressed in terms of the displacement vector as

$$T = \pi \int_{R_i}^{R_o} \frac{\partial \mathbf{u}^T}{\partial t} \frac{\partial \mathbf{u}}{\partial t} \rho r \, dr \tag{16.11}$$

where ρ is the mass density of the material.

Assume that the cylinder is divided into N axisymmetric strip elements in the radial direction. And let R_{m-1} and R_m be, respectively, the inner and outer radii of any element m. Then the displacements within an element are approximated as

$$\mathbf{u}(r, z, t) = \mathbf{N}(r)\mathbf{U}(z)\exp(-i\omega t) \tag{16.12}$$

where $\mathbf{U}(z)$ is the vector of unknown displacement amplitudes, and

$$\mathbf{N}(r) = \lfloor (1 - 3\hat{r} + 2\hat{r}^2)\mathbf{I} \quad 4(\hat{r} - \hat{r}^2)\mathbf{I} \quad (-\hat{r} + 2\hat{r}^2)\mathbf{I} \rfloor \tag{16.13}$$

is the matrix of shape functions of the element. Here $\hat{r} = (r - r_{j-1})/(r_j - r_{j-1})$, $r_{j-1} \le r \le r_j$, and $\mathbf{I}$ is a 2×2 identity matrix.

Substituting Eqs. (16.9) and (16.11) into Eq. (16.7), in view of Eq. (16.14), and then taking variation with respect to $\mathbf{U}$ leads to the following governing ordinary differential equations of the cylinder

$$-\mathbf{A}_2 \frac{d^2\mathbf{U}}{dz^2} + \mathbf{A}_1 \frac{d\mathbf{U}}{dz} + (\mathbf{A}_0 - \omega^2\mathbf{M})\mathbf{U} = \mathbf{q} \tag{16.14}$$

where

$$\mathbf{A}_0 = \sum_{m=1}^{N} \int_{R_{m-1}}^{R_m} \left(\frac{\partial \mathbf{N}^T}{\partial r} \mathbf{D}_{22} \frac{\partial \mathbf{N}}{\partial r} + \frac{1}{r} \frac{\partial \mathbf{N}^T}{\partial r} \mathbf{D}_{23}\mathbf{N} + \frac{1}{r} \mathbf{N}^T \mathbf{D}_{23}^T \frac{\partial \mathbf{N}}{\partial r} + \frac{1}{r^2} \mathbf{N}^T \mathbf{D}_{33}\mathbf{N} \right) r \, dr \tag{16.15}$$

$$\mathbf{A}_1 = \sum_{m=1}^{N} \int_{R_{m-1}}^{R_m} \left(-\mathbf{N}^T \mathbf{D}_{12} \frac{\partial \mathbf{N}^T}{\partial r} - \frac{1}{r} \mathbf{N}^T \mathbf{D}_{13}\mathbf{N} + \frac{\partial \mathbf{N}^T}{\partial r} \mathbf{D}_{12}^T \mathbf{N} + \frac{1}{r} \mathbf{N}^T \mathbf{D}_{13}^T \mathbf{N} \right) r \, dr \tag{16.16}$$

$$\mathbf{A}_2 = \sum_{m=1}^{N} \int_{R_{m-1}}^{R_m} \mathbf{N}^T \mathbf{D}_{11}\mathbf{N} r \, dr \quad \mathbf{M} = \sum_{m=1}^{N} \int_{R_{m-1}}^{R_m} \mathbf{N}^T \mathbf{N} \rho r \, dr \quad \mathbf{q} = \mathbf{N}^T \mathbf{n}^T R_o q_0 \tag{16.17}$$

As can be seen from Eq. (16.14), the original 2D problem has simplified to a 1D problem through the strip element formulation in the radial direction. Tedious preprocessors in FEMs are omitted here. Consequently, computational labor can be greatly reduced. In addition, the SEM requires a nodal line numbering along the r-axis only. This yields the minimum matrix bandwidth, and, therefore, the SEM requires much less computer memory and time compared with FEMs.

Equation (16.14) is a set of inhomogeneous ordinary differential equations in the frequency domain. A combination of modal analysis and inverse Fourier transform techniques gives its particular solution (see Chapter 12) of

$$
\mathbf{U}_p = \begin{cases} -i\sum_{m=1}^{M} \dfrac{\varphi_{m2}^{+L}\mathbf{q}_0 R_o \varphi_{m1}^{+R}}{B_m^+} e^{-ik_m^+ z}, & \text{for } z \geq 0 \\[4mm] i\sum_{m=1}^{M} \dfrac{\varphi_{m2}^{-L}\mathbf{q}_0 R_o \varphi_{m1}^{-R}}{B_m^-} e^{-ik_m^- z}, & \text{for } z < 0 \end{cases}
\tag{16.18}
$$

where $B_m = \varphi_m^L \mathbf{B} \varphi_m^R$, k_m, φ_{m1}^L, φ_{m2}^L φ_{m1}^R, and φ_{m2}^R are, respectively, the eigenvalues, the left and right eigenvectors obtained from the following characteristic equations:

$$
\left[\begin{bmatrix} 0 & I \\ \omega^2 \mathbf{M} - \mathbf{A}_0 & -i\mathbf{A}_1 \end{bmatrix} - k_m \begin{bmatrix} I & 0 \\ 0 & \mathbf{A}_2 \end{bmatrix}\right] \begin{Bmatrix} \varphi_{m1}^R \\ \varphi_{m2}^R \end{Bmatrix} = 0
\tag{16.19}
$$

$$
\begin{Bmatrix} \varphi_{m1}^L \\ \varphi_{m2}^L \end{Bmatrix}^T \left[\begin{bmatrix} 0 & I \\ \omega^2 \mathbf{M} - \mathbf{A}_0 & -i\mathbf{A}_1 \end{bmatrix} - k_m \begin{bmatrix} I & 0 \\ 0 & \mathbf{A}_2 \end{bmatrix}\right] = 0
\tag{16.20}
$$

The complementary solution of the associated homogeneous equation of Eq. (16.14) can be expressed by superposition of the right eigenvectors φ_m^R:

$$
\mathbf{U}_c = \sum_{m=1}^{2M} C_m \varphi_m^R \exp(ik_m z) = \mathbf{G}(z)\mathbf{C}
\tag{16.21}
$$

where the subscript c denotes the complementary solution and the coefficient vector $\mathbf{C}$ is to be determined. The addition of the particular and complementary solutions yields the general solution of Eq. (16.14) in the form

$$
\mathbf{U} = \mathbf{U}_c + \mathbf{U}_p = \mathbf{G}(z)\mathbf{C} + \mathbf{U}_p
\tag{16.22}
$$

Thus the coefficient vector $\mathbf{C}$ can be expressed in terms of particular and general solutions at radial boundaries:

$$
\mathbf{C} = \mathbf{G}_b^{-1}(\mathbf{U}_b - \mathbf{U}_{pb})
\tag{16.23}
$$

where the subscript b denotes boundaries. Substitution of Eq. (16.23) into Eq. (16.22) gives

$$\mathbf{U} = \mathbf{G}(z)\mathbf{G}_b^{-1}(\mathbf{U}_b - \mathbf{U}_{pb}) + \mathbf{U}_p \qquad (16.24)$$

The stress boundary conditions at the crack tips are given by

$$\mathbf{R}_b = \mathbf{KU}_b + \mathbf{S}_p \qquad (16.25)$$

where

$$\mathbf{R}_b = \left\{ \begin{matrix} \mathbf{R}_b^L \\ \mathbf{R}_b^R \end{matrix} \right\} \quad \mathbf{U}_b = \left\{ \begin{matrix} \mathbf{U}_b^L \\ \mathbf{U}_b^R \end{matrix} \right\} \qquad (16.26)$$

$$\mathbf{K} = \begin{bmatrix} \mathbf{R}_1 & 0 \\ 0 & \mathbf{R}_1 \end{bmatrix} + \begin{bmatrix} \mathbf{R}_2 \dfrac{\partial \mathbf{G}^L}{\partial z}\mathbf{G}_b^{-1} \\ \mathbf{R}_2 \dfrac{\partial \mathbf{G}^R}{\partial z}\mathbf{G}_b^{-1} \end{bmatrix} \qquad (16.27)$$

$$\mathbf{S}_p = \begin{bmatrix} \mathbf{R}_2 & 0 \\ 0 & \mathbf{R}_2 \end{bmatrix} \left\{ \begin{matrix} \dfrac{\partial \mathbf{U}_p^L}{\partial z} \\ \dfrac{\partial \mathbf{U}_p^R}{\partial z} \end{matrix} \right\} - \begin{bmatrix} \mathbf{R}_2 \dfrac{\partial \mathbf{G}^L}{\partial z}\mathbf{G}_b^{-1} \\ \mathbf{R}_2 \dfrac{\partial \mathbf{G}^R}{\partial z}\mathbf{G}_b^{-1} \end{bmatrix} \left\{ \begin{matrix} \mathbf{U}_p^L \\ \mathbf{U}_p^R \end{matrix} \right\} \qquad (16.28)$$

are the external traction and displacement vectors on the radial boundaries, the stiffness matrix, and the equivalent external force acting on the radial boundaries, respectively. The superscripts L and R represent the left and right sides of the crack, respectively.

16.4 Examples

In this section, numerical examples are given for $(C90/G0)_s$ and $(C0/G90)_s$ laminated composite cylinders. In the laminate codes, lamina numbering increases from the inner to outer surface. C and G represent carbon/epoxy and glass/epoxy, respectively. The number following the letters indicates the azimuthal angle of the fiber orientation with respect to the z-axis. The subscript s denotes that the multilayered cylinders are symmetrically stacked. The material properties of the cylinders are taken from Takahashi and Chou (1987). Reference properties Q_{44} and ρ are the material constant and mass density of C0.

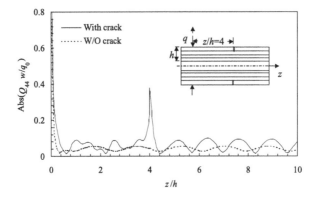

FIGURE 16.2

Distribution of the radial displacement on the outer surface of a $(C90/G0)_s$ cylinder with an outer surface-breaking crack ($\omega h \sqrt{\rho/Q_{44}} = 3.14, R_i/h = 1$). (From Xi et al., *ASME Journal of Applied Mechanics*, 67, 427, 2000c. With permission.)

Figure 16.2 shows the distributions of the radial displacement on the outer surface of a $(C90/G0)_s$ composite cylinder with an outer surface-breaking crack. For comparison, the results for the corresponding uncracked case are also plotted in the same figure with dotted lines. It can be seen from the figure that the presence of the crack causes a singularity in the displacement at the position of the crack. This phenomenon is of practical importance, from which the position of the crack can be clearly identified. It should be noted that the displacements at the left and right tips of the crack are discontinuous although this is not visible in the figure. Because of superposition of the incident and scattered wave fields, the absolute value of the displacement between the loaded point and the crack becomes irregular.

The present method may also be employed to detect a radial interior crack in a composite cylinder. Figure 16.3 shows the distributions of the radial displacement on the outer surface of a $(C0/G90)_s$ composite cylinder with a radial interior crack. It is apparent that when a wave strikes the radial interior crack in the cylinder, it generates scattering and causes an irregular oscillation of the absolute value of the displacement between the load point of application and crack. From the different patterns of wave fields at the left and right sides of the crack, the position of the interior crack can be easily identified although the crack is at a distance from the outer surface.

16.5   Remarks

This chapter introduced a SEM for analyzing wave scattering by cracks in laminated composite cylinders and circular cylindrical shells. This method can readily be used to deal with cylinders and shells made of an arbitrary number of anisotropic layers, of arbitrary lay-ups, and of any types of materials.

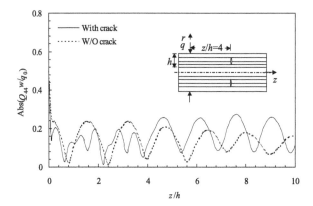

FIGURE 16.3

Distribution of the radial displacement on the outer surface of a $(C0/G90)_s$ cylinder with a radial interior crack ($\omega h \sqrt{\rho/Q_{44}} = 6.28$, $R_i/h = 1$). (From Xi et al., *ASME Journal of Applied Mechanics*, 67, 427, 2000c. With permission.)

Numerical results indicate that either a radial or axial crack in laminated composite cylinders and shells can be identified by using the present method.

The SEM reduces the spatial dimensions of the original problem by one. Therefore it does not need tedious preprocessors, which occupy a substantial part of the finite element method (FEM). As it adopts unidirectional nodal line numbering, it yields the minimum matrix bandwidth and requires much less computer memory and time. As strip elements are used, the number of the elements can be very small compared with those of the conventional FEM. Moreover, the strip elements can be used as infinite elements. The ease of treatment of infinite problems is an additional advantage of the SEM.

In engineering practice, there are a large number of composite laminated cylinders containing fluid or surrounded by fluid. Because of the fluid presence, simulation of wave scattering in cylinders becomes much more complicated. On the basis of the present method, Xi et al. (2000b) have further developed SEMs for analyzing fluid-loaded composite laminated cylinders containing cracks.

17

Inverse Identification of Impact Loads Using Elastic Waves

17.1 Introduction

A major concern in applying fiber-reinforced composite laminates in engineering applications is their relatively poor damage tolerance to localized surface impact loads. Even low-velocity impact can often easily cause delaminations. The delaminations could severely reduce the load-carrying capability and the structural reliability of the laminates. It can also lead to catastrophic failure of the entire structural system. Because of this, a better understanding of impact loading, including its spatial distribution and time history, in actual service situations is very important for designing composite laminates that are strong against impact loads. However, it is very difficult to obtain the loading profile in an actual impact situation. The traditional ways of direct sensing require direct access to the impact point; therefore, it is technically difficult and sometimes impossible to perform due to the complex nature of impact loading or the danger that arises in actual service situations.

Using elastic waves is a promising way to solve the problem, as one can measure the response at a distance from the impact point. Waves generated by the impact load will propagate to the point of observation. It is also routinely convenient for us to accurately measure structural response at a distant point away from the impact point. If one has a model of wave propagation that can provide the wave response at a point distant to the impact load, a properly formulated inverse procedure can reconstruct the impact loading using the wave response measured at the distant point.

Techniques have been proposed for reconstructing loadings of impact using responses of structures. Most of these considered, however, cases of either isotropic structures or single layer anisotropic structures. Doyle (1984a, 1984b, 1987) determined the impact force for beam and plate types of structures using the phase difference of the signals measured at two different sensors. Chang and Sun (1989) employed an experimentally generated Green's function along with signal deconvolution to recover the dynamic impact force. No governing equations of the structure were needed because

Green's function was obtained experimentally for actual structure under the actual boundary conditions. Michaels and Pao (1985, 1986) presented a general procedure of deconvolution for the case where the kernel of the convolution time integral is a linear combination of Green's function with unknown coefficients. Double iterative method of disconsolation was applied to solve an inverse source problem for an oblique impact force on an elastic plate. Chang and Sachse (1985) proposed a method to analyze both the forward and inverse problem of an extended, finite source of elastic wave in a thick plate. An iterative procedure of disconsolation was developed based on the signals detected at two receiving points to recover extended sources. Yen and Wu (1995) presented a method to identify both the impact location and the time history of the impact force from the strain responses at certain points on a rectangle isotropic plate. A relationship among an arbitrary pair of strain responses was used to find the impact location without knowing in advance the impact force history. The impact force history was then subsequently determined. An optimization method was used to search for the optimal impact location as well as the time history of the force that fits best to the measured data.

The investigation on inverse problems of force profile recovery for composite laminates with many anisotropic layers is also reported. An inverse procedure has been proposed to deal with force identification problems of composite laminates for 2D cases (Liu and Ma, 2001). Both the time history and the spatial distribution function of line loads applied on the surface of composite laminates are reconstructed from the dynamic displacement response at one receiving point. The procedure has been tested computationally and proven effective and reliable. In that procedure, the HNM method developed in Chapter 8 is introduced as a forward problem solver to analyze the wave response of composite laminates with many anisotropic layers subjected to a line load of excitation on the surface. The HNM combines the fast Fourier transform technique (FFT) and the concept of the finite element method (FEM), and it can efficiently compute the dynamic displacement response of composite laminates. The outstanding efficiency paves the way for the inverse analysis of loading applied on composite laminates.

In this chapter, an inverse procedure is introduced for 2D and 3D problems of impact loading identification for anisotropic laminates with any number of layers. The procedure was originally proposed by Liu and Ma in 2001. By employing the HNM method, the time history is reconstructed for line and point loads as well as distributed loads on the surface of composite laminates. A kernel signal function—the displacement response of composite laminates subjected to a line or point time-step load—is derived using the HNM. By applying the kernel function, the displacement response of composite laminates subjected to a line or point load with an arbitrary time function is expressed in a form of convolution. Deconvolution is then performed in time domain to recover the loading time history. Distribution loads with separable time and spatial dependencies are treated in a similar way as a line or point load. For 3D cases, we focus on reconstruction of time history of the force. Distributed forces can be treated as a point force, if the spatial distribution of the loads is very small compared with the distance from the load to the receiving point. The displacement response can be approximately modeled as a convolution integral of the loading time history with a kernel function subjected to a point step-impact load at the center of the locally distributed load. By performing the same method of deconvolution as the point load case, the time history can be reconstructed.

The effects of measurement noises on reconstructed or identified results of impact loading are also investigated. The measurement noises are simulated by adding Gaussian noises directly to the computer-generated displacement responses. Numerical verifications are performed for line, point, and distribution loads to verify the effectiveness and robustness of the proposed procedure to the added noise.

17.2 Two-Dimensional Line Load

As shown in Figure 17.1, a composite laminate is excited by a line load at $x = 0$. The load is applied in the thickness direction of the laminate and independent of y-axis. Since both the load and the displacement field are independent of y, the problem is 2D. In the x-z plane, the loading is simplified as a point load. The composite laminate is assumed to be stationary before the load is applied.

In the following, we use two methods to identify the loading time function of a line load using two kernel functions: Green's function and time-step response function.

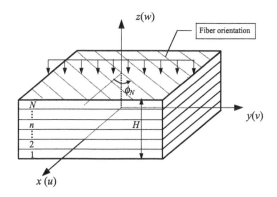

FIGURE 17.1
Composite laminate subjected to a line load on the surface.

17.2.1   Green's Function

The displacement response at any receiving point can be expressed as a convolution integral of the time function of the force and corresponding Green's function:

$$u(x, t) = \int_0^t G(x, t - \tau)s(\tau)\, d\tau \qquad (17.1)$$

where $u(x, t)$ is the displacement in x direction at point x, $s(t)$ is the time function of the line load, and $G(x, t)$ is the corresponding Green's function.

By discretizing this convolution integral into m equally spaced sample points in the time domain, Eq. (17.1) is transformed into a matrix form of

$$\begin{Bmatrix} u_1 \\ u_2 \\ \vdots \\ u_m \end{Bmatrix} = \begin{bmatrix} G_1 & 0 & \cdots & 0 \\ G_2 & G_1 & \cdots & 0 \\ \vdots & \vdots & \ddots & 0 \\ G_m & G_{m-1} & \cdots & G_1 \end{bmatrix} \begin{Bmatrix} s_0 \\ s_1 \\ \vdots \\ s_{m-1} \end{Bmatrix} \Delta t \qquad (17.2)$$

where Δt is the discrete time step, and u_j, G_j, and s_j are displacement, Green's function, and impact force at time $t = j\Delta t$, respectively. Note that since the plate is stationary before impact, u_0 and G_0 are equal to zero. Equation (17.2) can be rewritten in a concise matrix form of

$$\mathbf{U} = \mathbf{GS} \qquad (17.3)$$

The special triangular form of the Green's function matrix reflects the characteristic of the convolution integral.

Note that the Green's function $G(x, t)$ for composite laminates can be obtained using HNM described in Chapter 8, by computing the displacement at point x and time t for a point excitation of Dirac delta pulse at $x = 0$. If the displacement $u(x, t)$ can be obtained experimentally, the time history $s(t)$ can then be obtained by solving Eq. (17.3). However, because of the Green's function, matrix $\mathbf{G}$ in Eq. (17.2) is generally ill-conditioned; it cannot be solved by directly using conventional methods for solving algebraic equation systems. An optimization scheme must be used to minimize the object function that is an error function defined as

$$E = \|\mathbf{GS} - \mathbf{U}\|^2 \qquad (17.4)$$

where $\|\bullet\|$ denotes a square norm of the vector. The subject now turns to find a force function S, which minimizes E. Many least-square fit methods are available to this end. We use here the conjugate gradient method.

17.2.2 Time-Step Response Function

It has been found that the response to a point-pulse excitation is extremely oscillatory. Response to a Heaviside time-step excitation behaves, however, much more smoothly. It is therefore recommended to replace the Green's function in Eq. (17.3) with the time-step response function that can also be computed using the HNM. Formulation is briefed as follows. Using a Heaviside step function $H(t)$, an arbitrary force $F(t)$ can be approximately expressed as

$$F(t) = s_0 H(t) + (s_1 - s_0)H(t - t_1) + \cdots + (s_i - s_{i-1})H(t - t_i) \quad (t_i < t < t_{i+1})$$

(17.5)

where s_j is the impact force function value at time $t = j\Delta t$. Let $h(t)$ be the *time-step response function* that is defined as the displacement response of the composite laminate subjected to Heaviside time-step load at time t. According to superposition principle, the response of the laminate to the impact force $F(t)$ can be expressed as

$$u(t) = s_0 h(t) + (s_1 - s_0)h(t - t_1) + \cdots + (s_i - s_{i-1})h(t - t_i) \quad (t_i < t < t_{i+1}) \quad (17.6)$$

or

$$u(t) = s_0[h(t) - h(t - t_1)] + s_1[h(t - t_1) - h(t - t_2)]$$
$$+ \cdots + s_{i-1}[h(t - t_i) - h(t - t_i)] + s_i h(t - t_i) \quad (17.7)$$

where $h(t - t_j)$ is the time-step response function at time $t - t_j$ for a Heaviside time-step excitation at time $t = 0$.

Let u_j, h_j be the displacements at time $t = j\Delta t$ subjected to the impact force and the Heaviside time-step force applied at time $t = 0$, respectively. Equation (17.7) can be rewritten as

$$u_{m+1} = \sum_{i=0}^{m} s_i[h_{m-i+1} - h_{m-i}] \quad (17.8)$$

in which h_0 is set to zero since the plate is assumed stationary before excitation. We note

$$g_i = h_i - h_{i-1} \quad (17.9)$$

which is the difference in the time-step response of the laminate observed at $t = i\Delta t$ and $t = (i - 1)\Delta t$. Substituting Eq. (17.9) into Eq. (17.8) gives

$$u_{m+1} = \sum_{i=0}^{m} s_i g_{m-i+1} \quad (17.10)$$

Equation (17.10) is a standard convolution using a time-step response function. It can be rewritten in a matrix form similar to Eq. (17.2) as

$$
\begin{Bmatrix} u_1 \\ u_2 \\ \vdots \\ u_m \end{Bmatrix}
=
\begin{bmatrix}
g_1 & 0 & \cdots & 0 \\
g_2 & g_1 & \cdots & 0 \\
\vdots & \vdots & \ddots & 0 \\
g_m & g_{m-1} & \cdots & g_1
\end{bmatrix}
\begin{Bmatrix} s_0 \\ s_1 \\ \vdots \\ s_{m-1} \end{Bmatrix}
\tag{17.11}
$$

Since Eqs. (17.11) and (17.2) have similar convolution form, we can follow a procedure similar to that in the last section to recover **S**, i.e., searching for **S** that minimize the error function defined by Eq. (17.4). The minimization is performed using the conjugate gradient method.

The same form in Eqs. (17.11) and (17.2) gives us the convenience to code an inverse algorithm using both Green's function method and the time-step response function. Numerical computations have proven that both methods always lead to similar good results, but the time-step response function allows the use of a longer time step and hence requires less computational effort, as discussed later in this chapter. In the following, $g(x, t)$, refers to either Green's function or the time-step response function defined by Eq. (17.9).

17.3 Two-Dimensional Extended Line Load

Assume that the line load extends along the x-axis from ζ_1 to ζ_2, as shown in Figure 17.2. Since the extended line load is independent of y, it is a 2D problem. We denote the loading function as $F(x, t)$ and assume that the time and spatial dependencies of $F(x, t)$ are separable; i.e., we can have $F(x, t) =$

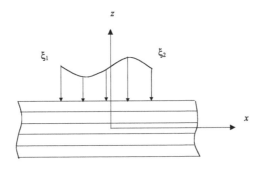

FIGURE 17.2
An extended line load in the x-z plane.

$s(t)p(x)$, where $s(t)$ is loading time function and $p(x)$ is the loading distribution function. The displacement at receiving point x is given by

$$u(x, t) = \int_0^t s(t')\, dt' \int_{\zeta_1}^{\zeta_2} p(x')g(x|x', t - t')\, dx' \qquad (17.12)$$

We model the extended loads as a collection of point loads covering the loading region. The displacement response at a receiving point is a superposition of contributions from all these point loads. We discretize loading region (ζ_1, ζ_2) into N discrete point loads separated equally with an interval of $\Delta x'$ such that $X^n = n\Delta x'$ while the time domain is also discretized evenly. Equation (17.12) can then be rewritten as

$$u_m(x) = \sum_{k=1}^m s(k\Delta t)\Delta t \sum_{n=1}^N p(X^n)\Delta x' g[x|X^n, (m-k)\Delta t] \qquad (17.13)$$

where $u_m(x)$ is the displacement response at receiver point x at time $T^m = m\Delta t$.

The inverse problem for extended source is to recover the time function $s(t)$ as well as the distribution function $p(x)$. In the following section, we consider three different cases of loading reconstruction.

17.3.1 Identification of Loading Time Function

We assume distribution function $p(x)$ is known and our task is to find time function $s(t)$. Considering the right side of Eq. (17.13), we introduce

$$G_m(x) = \sum_{n=1}^N p(X^n)\Delta x' g[x|X^n, (m-k)\Delta t] \qquad (17.14)$$

which can be rewritten in a matrix form of

$$[G] = \begin{Bmatrix} G_1 \\ G_2 \\ \vdots \\ G_M \end{Bmatrix} = \begin{Bmatrix} \begin{bmatrix} g_1 \\ g_2 \\ \vdots \\ g_M \end{bmatrix}_{\Delta x'} \begin{bmatrix} g_1 \\ g_2 \\ \vdots \\ g_M \end{bmatrix}_{2\Delta x'} \cdots \begin{bmatrix} g_1 \\ g_2 \\ \vdots \\ g_M \end{bmatrix}_{N\Delta x'} \end{Bmatrix} \begin{bmatrix} p(\Delta x') \\ p(2\Delta x') \\ \vdots \\ P(N\Delta x') \end{bmatrix} \Delta x' \qquad (17.15)$$

Then substituting Eq. (17.15) into Eq. (17.13), we arrive at the following formula for time function reconstruction of impact load with a given distribution function:

$$
[U] = \left\{ \begin{matrix} u_1 \\ u_2 \\ \vdots \\ u_m \end{matrix} \right\} = \left\{ \begin{bmatrix} G_1 & & & \\ G_2 & G_1 & & \\ \vdots & \vdots & \ddots & \\ G_m & G_{m-1} & \cdots & G_1 \end{bmatrix} \left\{ \begin{matrix} s_0 \\ s_1 \\ \vdots \\ s_{m-1} \end{matrix} \right\} \right\} \Delta t \tag{17.16}
$$

or in concise form of

$$
\mathbf{U} = \mathbf{GS} \tag{17.17}
$$

The objective function can also be defined as $E = \|\mathbf{GS} - \mathbf{U}\|^2$. The conjugate gradient method can be used to find $\mathbf{S}$, which minimize E.

17.3.2   Identification of Loading Distribution Function

If time function $s(t)$ is known, the distribution function $p(x)$ can be recovered through the following procedure. We rewrite Eq. (17.13) as

$$
u_m(x) = \sum_{n=1}^{N} p(X^n) \Delta x' \sum_{k=1}^{m} s(k\Delta t) g[x|X^n, (m-k)\Delta t] \Delta t \tag{17.18}
$$

Let

$$
R^n = \sum_{k=1}^{m} s(k\Delta t) g[x|X^n, (m-k)\Delta t] \Delta t \tag{17.19}
$$

Then Eq. (17.18) has matrix form of

$$
\mathbf{R}^n = \left\{ \begin{matrix} r_1 \\ r_2 \\ \vdots \\ r_m \end{matrix} \right\}_{x'=n\Delta x} = \left[\begin{matrix} g_1 & & & \\ g_2 & g_1 & & \\ \vdots & \vdots & \ddots & \\ g_m & g_{m-1} & \cdots & g_1 \end{matrix} \right]_{x'=n\Delta x} \left\{ \begin{matrix} s_0 \\ s_1 \\ \vdots \\ s_{m-1} \end{matrix} \right\} \Delta t \tag{17.20}
$$

Substituting Eq. (17.20) into Eq. (17.18), we obtain

$$
\left[\begin{matrix} u_1 \\ u_2 \\ \vdots \\ u_m \end{matrix} \right] = \left\{ \left\{ \begin{matrix} r_1 \\ r_2 \\ \vdots \\ r_m \end{matrix} \right\}_{x'=\Delta x} \left\{ \begin{matrix} r_1 \\ r_2 \\ \vdots \\ r_m \end{matrix} \right\}_{x'=2\Delta x} \cdots \left\{ \begin{matrix} r_1 \\ r_2 \\ \vdots \\ r_m \end{matrix} \right\}_{x'=N\Delta x} \right\} \left\{ \begin{matrix} p_1 \\ p_2 \\ \vdots \\ p_N \end{matrix} \right\} \Delta x' \tag{17.21}
$$

or in concise matrix form of

$$\mathbf{U} = \mathbf{RP} \tag{17.22}$$

We introduce an objective function in Eq. (17.4) and use an optimization method to solve for $\mathbf{P}$, which minimize E.

17.3.3   Identification of Both Time History and Distribution Functions

If both the time function and the distribution function are unknown and need to be recovered from the displacement measured at one receiving point, Eq. (17.13) can be used in a double loop iterative procedure. First, an initial guess of the distribution function $p(x)$ is made to initiate the iterative process at outer loop. Based on the procedure described in section 17.3.1, a time function $s(t)$ can be obtained in the inner loop. Then the $s(t)$ is used as input to the outer loop to evaluate the new $p(x)$. These iterations are repeated until a desired accuracy is attained.

17.4   Three-Dimensional Concentrated Load

Consider a composite laminate subjected to a transient point load on its upper surface, as shown in Figure 8.1. By introducing a Heaviside step function $H(t)$, an arbitrary force $F(t)$ can be expressed as

$$F(t) = f_0 H(t) + (f_1 - f_0)H(t - t_1) + \cdots + (f_i - f_{i-1})H(t - t_i) \quad (t_i < t < t_{i+1}) \tag{17.23}$$

where f_j is the load value at time $t_j = j\Delta t$. Let $h(t)$ be the displacement response of the laminate subjected to a step-impact load on the surface, which can be obtained using the HNM for 3D transient wave field. According to superposition principle, the response of the impact force $F(t)$ can be expressed as

$$F(t) = f_0 h(t) + (f_1 - f_0)h(t - t_1) + \cdots + (f_i - f_{i-1})h(t - t_i) \quad (t_i < t < t_{i+1}) \tag{17.24}$$

$$u(t) = f_0[h(t) - h(t - t_1)] + f_1[h(t - t_1) - h(t - t_2)]$$
$$+ \cdots + f_{i-1}[h(t - t_i) - h(t - t_i)] + f_i h(t - t_i) \tag{17.25}$$

where $h(t - t_j)$ is the displacement at time t subject to a Heaviside step load applied at time t_j. It is easy to notice that its value is equal to the displacement response at time $t - t_j$ subjected to a step force applied at time $t = 0$.

In discrete time domain, we use u_j, h_j to denote the displacements at time $t_j = j\Delta t$ subjected to the impact force and the Heaviside step force applied at time $t = 0$, respectively. Equation (17.25) can be transformed to

$$u_{m+1} = \sum_{i=0}^{m} f_i[h_{m-i+1} - h_{m-i}] \tag{17.26}$$

in which h_0 is set to zero since the laminate is assumed stationary before excitation.

We introduce

$$g_i = h_i - h_{i-1} \tag{17.27}$$

Substituting Eq. (17.27) into Eq. (17.26) leads to

$$u_{m+1} = \sum_{i=0}^{m} f_i g_{m-i+1} \tag{17.28}$$

Equation (17.28) is a convolution, which can be written in the following familiar form of

$$\begin{Bmatrix} u_1 \\ u_2 \\ \vdots \\ u_m \end{Bmatrix} = \begin{bmatrix} g_1 & 0 & \cdots & 0 \\ g_2 & g_1 & \cdots & 0 \\ \vdots & \vdots & \ddots & 0 \\ g_m & g_{m-1} & \cdots & g_1 \end{bmatrix} \begin{Bmatrix} f_0 \\ f_1 \\ \vdots \\ f_{m-1} \end{Bmatrix} \tag{17.29}$$

The matrix in Eq. (17.29) is generally ill-conditioned. An optimization scheme should be used with an objective function (error function) defined as

$$E = \sum_{j=1}^{m} \left(u_{j+1} - \sum_{i=0}^{j} f_i g_{j-i+1} \right)^2 \tag{17.30}$$

Our task in this case is to find a time function f, which minimizes the error function E. Standard least-square fit methods can be employed to achieve the goal.

17.5 Examples

Having established the formulation for different cases of inverse problems of impact load reconstruction for composite laminates, numerical investigation has been conducted to verify the effectiveness of the procedure. We considered a composite laminate made of two layers of carbon/epoxy and four layers of glass/epoxy composites. The stacking sequence of the laminate is denoted by [C90/G+45/G–45]$_s$ that is symmetrically stacked.

In the numerical computation, the following dimensionless parameters are used.

$$\bar{x} = x/H, \bar{t} = tc_s/H, \quad c_s = (c(4,4)/\rho)^{1/2}, \quad \bar{u} = u/u_0, \quad u_0 = q_0/c(4,4)$$

where u is the displacement in the x direction, and ρ, $c(4,4)$ and c_s are the density, the elastic constant, and the velocity of shear wave (in the direction of fiber orientation) in the carbon/epoxy composite. q_0 is a constant related to the force vector. H is the overall thickness of the laminate.

17.5.1 Identification of Time-Function for Concentrated Line Load

Assume a line load with time function

$$f(\bar{t}) = \sin(\pi\bar{t}), \quad 0 < \bar{t} < 2 \tag{17.31}$$

which is one cycle of sine function. Green's function method is used to identify the time function of the load. The experimental measurements are simulated using the HNM with the actual input of time function for the impact load. The conjugate gradient method is adopted to perform optimization. The result of reconstructed force function are shown in Figure 17.3,

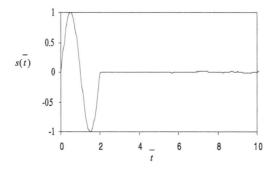

FIGURE 17.3
Identified time function of the line load applied on the surface of the composite laminate.

from which it is seen that the reconstructed time history function agrees very well with the true input function except some small ripples in the long time period.

The Green's function oscillates very frequently because of the nature of Dirac delta pulse excitation. Therefore in time discretization, the sampling rate should be high enough to avoid the aliasing phenomenon, meaning that if the time interval Δt is too large, the high frequency of the Green's function will appear as lower frequency. $\Delta t = 0.04 \sim 0.05$ was used in constructing results given in Figure 17.3.

Theoretically, one can choose sampling points of response in any time period on the response curve for reconstruction. However, our experience indicates that the time period chosen should be within the main response period to achieve better accuracy.

It has also been tested for differents points of observation of wave response, such as at $\bar{x} = 0.6$ and $\bar{x} = 2$. Identified results of reconstructed force were found, and both agree well with the true functions. Reconstruction has also been performed using the time-step response function instead of Green's function, and the result obtained is almost the same as that we get using the Green's function method. Since the time-step response curve varies much more smoothly than the Green's function, it is less sensitive to the density of the time sample points in convolution integral. Larger time step of $\Delta t = 0.1$ is therefore used, and the result is as good as those using Green's function with $\Delta t = 0.04 \sim 0.05$. Obviously the computational efficiency is higher using time-step response function compared to the use of Green's function.

When the loading time function is of long duration, the time-step response function is even more efficient to use, and much less time sample points are needed compared to using Green's function.

17.5.2   Identification for Extended Line Loads

In the following numerical computation, we choose

$$p(\bar{x}) = 2.5\sin((10\bar{x} + 1)\pi/2 + 1), \quad (-0.2 < \bar{x} < 0.2) \qquad (17.32)$$

as the true spatial distribution function, and

$$f(\bar{t}) = \sin(\pi\bar{t}), \quad 0 < \bar{t} < 4 \qquad (17.33)$$

as the true time functions of loading.

If distribution function $p(x)$ is known, the identified time function is recovered and shown in Figure 17.4, together with the true function. The recovered and true functions agree very well, and they are almost identical in the diagram.

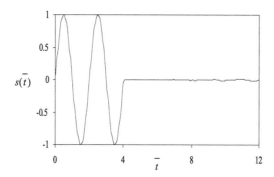

FIGURE 17.4
Comparison of the identified and the true time functions for an extended line load.

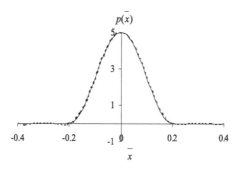

FIGURE 17.5
Comparison of the identified and the true distribution functions for an extended line load. Dashed line: identified function. Solid line: true function.

If $s(t)$ is known and $p(x)$ is to be recovered, we usually need to choose a larger area where the force may be distributed. In our numerical computation, we have chosen $(-0.4 < \bar{x} < 0.4)$ to recover the distributed load that doubles its actual range of $(-0.2 < \bar{x} < 0.2)$. The distribution function $p(x)$ is recovered and the comparison between the identified and the true function is shown in Figure 17.5. The accuracy of the recovered distribution function is excellent.

For recovering both time and distribution functions, a two-layer iterative procedure has been performed. The comparison of identified and the true distribution functions are shown in Figure 17.6. The identified time function has the same result as shown in Figure 17.3. It is found that the reconstructed distribution function is less accurate than the time function reconstructed.

The conjugate gradient method is used in the double loop iterative process. The numerical testing shows that the algorithm has fast convergence when recovering the loading time function. It gives satisfactory and stable results of time function after several double layer iterations. Comparatively, the

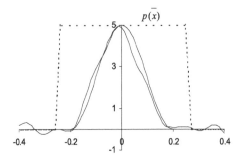

FIGURE 17.6
Comparison of the identified and the true distribution functions for an extended line load: double iteration case. Dotted line: the initial guess for distribution function. Thick line: the identified function after 30 double iterations. Thin line: the true function.

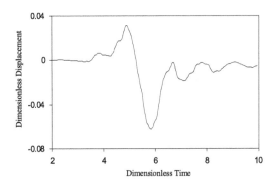

FIGURE 17.7
The displacement response at $(\sqrt{8}H, \sqrt{8}H)$ on the upper surface of composite laminate [C90/G+45/G–45]$_s$ subjected to a point load at (0, 0) with time history of one cycle of sine function.

distribution function identification is more sensitive to the choice of parameter used in the optimization program.

17.5.3   Concentrated Three-Dimensional Loads

The composite laminate is excited at (0, 0) by a point load with time function of $f(t) = \sin(\pi t), 0 < t < 2$ that is one cycle of sine function. A receiving point is chosen at $(\sqrt{8}H, \sqrt{8}H)$ on the upper surface of laminate. The dynamic displacement response at the receiving point is computed using the HNM method and shown in Figure 17.7. This computer-generated displacement response is employed as noise-free input data instead of the experimentally recorded data for the inverse identification. The displacement response of composite laminate at the same receiving point, but subjected to a point time-step impact load, is also computed by HNM and given in Figure 17.8. With these two sets of data obtained, the objective function is established.

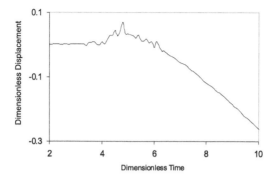

FIGURE 17.8

Time history of the vertical displacement at $(\sqrt{8H}, \sqrt{8H})$ on the upper surface of composite laminate $[C90/G+45/G-45]_s$ subjected to a time-step point load acting on the upper surface at $(0, 0)$.

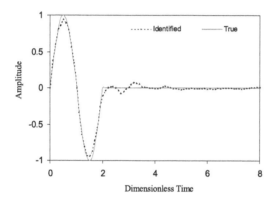

FIGURE 17.9

Comparison of identified time history to the true time history of the point load.

The reconstructed result of time function of load is shown in Figure 17.9. The identified time history function agrees well with the true force function except for some small ripples.

Noise effects are investigated by adding Gaussian noise directly to the computer-generated displacement readings. A Gauss random number generator is used to generate a series of random numbers with a standard deviation of

$$\sigma = b \times \left(1/N \sum_{j=1}^{N} (u_j^r)^2\right)^{1/2} \tag{17.34}$$

where u_j^r is the computer-generated displacement reading at the jth time sample point, N is the total number of time sample points, and b is a scale coefficient. By varying the coefficient b, one can generate Gauss noises of

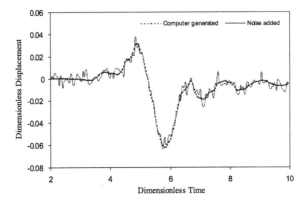

FIGURE 17.10

Comparison of noise-free and noise-contaminated displacement response at $(\sqrt{8H}, \sqrt{8H})$ on the upper surface of composite laminate $[C90/G+45/G-45]_s$ subjected to a point load at $(0, 0)$ with time history of one cycle of sine function.

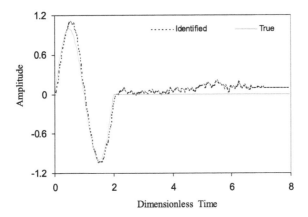

FIGURE 17.11

Comparison of the true force history and identified result based on noise-contaminated displacement.

different levels. In this numerical example, we choose b as 0.2 to generate a set of noise, which appears to be a quite large measurement error. The noise-contaminated displacement is generated and shown in Figure 17.10. This noise-added displacement is then used as input data for the inverse operation and the force time history is reconstructed. The results are shown in Figure 17.11. Compared with the identification result in Figure 17.9, which is based on the noise-free displacement, the result in Figure 17.11 exhibits many more ripples. However, both the shape and the magnitude of the identified curve agree satisfactorily well with the true time function of the load. The satisfactory identification result demonstrates the robustness of the present procedure to the given type of noise.

Numerical verification is also performed for a point Gaussian-pulse of excitation. The time history of a Gaussian pulse can be expressed as

$$f(t) = \frac{1}{\sqrt{2\pi}\sigma}\exp(-(t-t_0)^2/2\sigma^2), \quad (0 < t < \infty) \tag{17.35}$$

where t is time, σ is a parameter that controls the duration of the pulse, and t_0 determines the time delay of the pulse. In this numerical computation, t_0 and σ is taken as 1.5 and 0.4, respectively. The identification result based on the noise-free displacement response is shown in Figure 17.12, from which we observe a satisfactory agreement between the identified result and the true time function of the load. The noise effect is also considered by adding a Gauss noise with b equal to 0.1; the identification result is shown in Figure 17.13. Similar to the previous numerical example, the identified result exhibits more ripples, but overall the result is satisfactory from an engineering viewpoint.

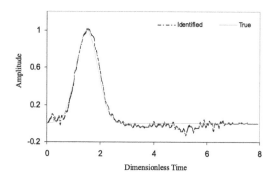

FIGURE 17.12
Comparison of the identified result to the true time history of a Gaussian pulse.

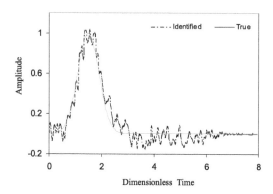

FIGURE 17.13
Comparison of identified result based on noise-contaminated displacement to the true time history of Gaussian pulse.

It is clearly shown that the inverse procedure is stable, meaning that it does not increase the error in the original input data.

The inverse identifications of loads with small distribution on the surface of the composite laminate are also investigated. We denote the loading function as $F(x, y, t)$ and assume that the time and spatial dependencies of force function are separable:

$$F(x, y, t) = s(t)p(x, y) \tag{17.36}$$

where $s(t)$ is loading time function, and $p(x, y)$ is the loading distribution function. We model the loading as a distribution of point loads covering over the loading region, and then the displacement response of the distribution load can be evaluated by superimposing the contributions from all the point loads. If the receiving point is located far enough from the loading compared to the dimension of the loading distribution range, the effect of the spatial distribution function to the displacement response becomes insignificant. The displacement can be roughly modeled as a convolution integral of force history with the displacement function of a point time-step load at the center of loading range. By applying deconvolution in the same way as we dealt with the point load, the loading history of the distribution force can be approximately reconstructed.

Numerical calculations are performed for loadings with different spatial distributions. Considering a dynamic load distributing in a rectangular region on the surface of the composite laminate, as shown in Figure 17.14. The time history of the loading is one cycle of sine function. The spatial distribution function can be expressed as

$$p(x, y) = \begin{cases} 6.25, & -0.2H < x < 0.2H \quad \text{and} \quad -0.2H < y < 0.2H \\ 0, & \text{elsewhere} \end{cases} \tag{17.37}$$

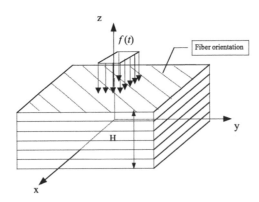

FIGURE 17.14
A composite laminate subjected to a load distributed in a rectangular region.

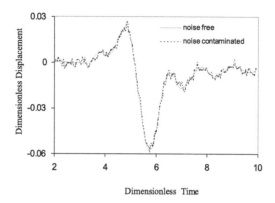

FIGURE 17.15
Displacement response at ($\sqrt{8}H$, $\sqrt{8}H$) on the surface of composite laminate [C90/G45/G–45]$_s$ subjected to a load of rectangular distribution and time history of one cycle sine function .

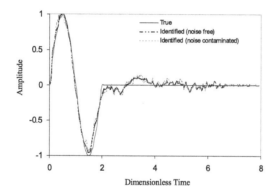

FIGURE 17.16
Identification result based on the displacement response subjected to a load of rectangle distribution.

Eight point sources that uniformly distribute in the loading region are used to simulate this distribution load. The dynamic displacement response of a composite laminate subjected to the distribution force, as shown in Figure 17.15, is a superposition of all the responses excited by the eight point sources. By adding 10% Gaussian noise to this computer-generated displacement, we obtain a simulated noise-contaminated displacement response, which is also given in Figure 17.15. The displacement response is approximately modeled as a convolution integral of the force time function with the displacement function subjected to a point time-step load at the center of loading range. By performing the same method of deconvolution as we did for the point loading cases, the loading history is reconstructed. The identified results based on the noise-free as well as the noise-contaminated displacements are given in Figure 17.16. Similar to the point load cases, both the identification results agree well with the true function, while the identification result based on the noise-contaminated displacement exhibits more ripples. The inverse procedure again demonstrates its stability to error contamination.

Loads with other spatial distributions, such as Gaussian distribution, are also investigated. We state, without showing further data, that the results for all cases are satisfactory and stable to error contamination.

We therefore conclude that inverse procedures discussed in this chapter are found to be computationally effective and robust for 2D and 3D loading reconstruction. This inverse procedure is expected to be applied in practice.

18

Inverse Determination of Material Constants of Composite Laminates

18.1 Introduction

Analysis and design of composite laminates depend directly upon accurate knowledge of the properties of composite materials. It is obviously valuable to develop a reliable and convenient method for nondestructively measuring properties of composite laminates. Inverse procedures are therefore being developed. We have seen from the previous chapter that the impact load identification can be performed very efficiently using elastic waves. The identification problem has been nicely formulated using Green's function, which fortunately can be obtained efficiently using the HNM (Chapter 8).

For parameter identification problems, sophisticated procedures are required because of the complex nature of the problem. Considerable work has been done in the inverse identification of the properties of composite materials. Chu and Rokhlin (1994) studied the elastic constants determination from ultrasonic bulk wave velocity data taken in a plane of symmetry. The problem is well posed with a unique and stable solution. Chu, Degtyar, and Rokhlin (1994) analyzed the reconstruction of elastic constants from ultrasonic velocity measurements in nonsymmetry planes of orthotropic composite materials. Balasubramaniam and Rao (1998) investigated the reconstruction of material stiffness properties of unidirectional fiber-reinforced composites from obliquely incident ultrasonic bulk wave data. Genetic algorithms (GAs) were used as the inversion technique and detail discussion in advantage as well as disadvantage of GAs for the identification problem over conventional methods were presented.

In the above references, elastic constants of composite materials were identified based on the ultrasonic bulk wave velocity data. The Christoffel equation was adopted to establish the relationship between material properties and bulk wave velocity. Complicated techniques were needed to measure the phase velocity of ultrasonic bulk waves in anisotropic materials, and only single-ply materials were considered in their works. Mota Soares, et al. (1993) presented a technique to predict the mechanical properties of composite plate specimens using measured eigenfrequencies. A finite element model based on the Mindlin plate theory was used for the laminate analysis. Later, Rikards et al. (1998, 1999) used the same laminate analysis model as Mota Soares et al. (1993), but, instead of using direct minimization of identification function, the method of experiment design was introduced. In Rikards et al. (1999), physical experiments were performed on the sample plates to measure the eigenfrequencies by a real-time television holography. The finite element method was used to obtain the numerical data in the reference points, and response surfaces were then determined. Based on the response surfaces and experimental data of eigenfrequency, the identification of material properties was performed. Frederiksen (1997) identified the elastic constants of thick orthotropic plates, focusing on the experiment's technique and the application of the method in actual tests. A mathematical model based on the higher-order shear deformation theory and higher mode frequencies was used to obtain reliable estimates of the two transverse shear moduli.

Genetic algorithms (GAs) have been applied in wide varieties of optimization problems in scientific research as well as engineering fields for the last decade and achieved great successes. In engineering structural optimization problems, GAs are being widely used as optimization techniques for dealing with complex, multimodal space optimization problems. GAs are based on the principles of genetics and natural selection to constitute search and optimization procedures. GA searches are stochastic in nature and, therefore, are capable of obtaining the global optimum without using gradient information, and no initial guesses are required. Furthermore, GAs work on a population of points in the search space rather than on a single point, which enables it to avoid being trapped at local optima points. In contrast, the gradient-based techniques have the obvious drawback of depending heavily on the initial guesses and often being trapped at a local optimum. This advantage of GAs over the conventional optimization technique makes GAs much more promising in the complicated, nonlinear, discrete, and multimodal optimization and search problems, such as the problem under consideration in this chapter. More details are available in Goldberg (1989).

Despite the above advantages, genetic algorithms usually require a large number of analyses and have high computational cost. For the inverse study using GAs, it is usually required to call the subprogram of forward computation for each chromosome. Therefore, the total number of calling the forward subprogram can be very large; usually several thousands or more depending on the complexity of the problem. Therefore, a fast forward solver is a prerequisite. An uncertainty using GAs for inverse problems is that its satisfactory performances from an optimization perspective do not obviously guarantee its success for multiparameter identification problems. For both optimal design and inversion problems of composite laminates, there are usually many near-optimal points or even sometimes many global optimum points in the whole search space. For optimization problems such as optimal design, generally every near-optimal point can lead to a satisfactory result, but, for identification problems, what we expect only the global optimum, which is obviously more difficult to be certain than the optimum obtained at a stage, is the result we are looking for.

In this chapter, we propose a procedure for inverse determination of the material constants of a composite laminate from the wave response of displacement recorded at only one receiving point on the surface of the laminate. The present inverse procedure identifies the material constants of composite laminates made of any anisotropic layers with different fiber orientations. The data needed for inverse identification are the time history of displacement observed at only one receiving point on the surface of the laminate excited by a transient line load, which can be easily obtained by a simple experiment. The study of an inverse problem heavily depends on the development of a forward problem solver. In general, the inverse problem study can be explored only after we have developed enough proper and efficient methods for forward computation. The HNM developed in Chapter 8 is used here as a forward approach to establish the relationship between the displacement response and the material constants. A GA search is employed for the present optimization process, and it is proven to be effective for this identification problem. Noise effects are also investigated by adding Gaussian noise to the time history of computer-generated displacement responses.

18.2   Inverse Operation

It is aimed to inversely determine the material constants of a composite laminate with many anisotropic layers, as shown in Figure 17.1. The laminate is made of one type anisotropic material and composed of many layers with different fiber orientations. The laminate is loaded by a dynamic line load at $x = 0$. The line load is in the thickness direction of the laminate and independent of the y-axis. Since the load and the displacement field are independent of y, it is a two-dimensional problem. The receiving point is arbitrarily chosen on the surface of laminated plate, and the displacement response subjected to the line load excitation can be obtained using the HNM method for the given load.

The displacement readings at the receiving point are used as input to inversely determine the material constants of the laminate. We discretize the whole time domain of data recording into N time sample points, the displacement readings at the jth time sample point is denoted as u_j^r, where the superscript means that the data is actually recorded.

In a GA run, each individual chromosome represents a candidate combination of material constants, denoted by $\mathbf{C}$. For each candidate combination, forward calculation has to be performed using the HNM to obtain u^c that is the displacement response subjected to the given dynamic loading. These calculated displacements are used to obtain the fitness value of the candidate combination. The fitness value will determine the probability of the candidate's being chosen as a future parent.

The object function for the inverse problem can be established as follows:

$$\text{ERR}(\mathbf{C}, \mathbf{C}^r) = \sum_{j=1}^{N} [u_j^r(\mathbf{C}^r) - u_j^c(\mathbf{C})]^2 \tag{18.1}$$

where $\mathbf{C}^r$ denotes the real set of material constants, and $\mathbf{C}$ denotes a candidate set of material constants. Displacement u_j^r is the recorded response at the jth time point on the actual composite laminate with material constant of $\mathbf{C}^r$, and u_j^c is the calculated displacement response at jth time point at the same receiving point using the candidate set material constant $\mathbf{C}$. Obviously the true set of material constants leads to a zero error. Therefore the object function needs to be minimized. A GA is employed to seek for a better candidate set of $\mathbf{C}$ that minimizes the objection function defined by Eq. (18.1). The search domain for $\mathbf{C}$ is predefined based on the existing knowledge and experience on the composite laminate of interest. Any combination of the material constants within the searching domain provides a set of $\mathbf{C}$. One should always provide the smallest possible searching domain, so as to reduce the possible combinations and hence the number of sets.

Equation (18.1) is the actual sum of squared errors between the actual measured data and the numerically calculated data on displacement using HNM. Instead of displacement, other valuables such as velocity or acceleration or strain can also be used to construct the error function. This sum is also the fitness value of the candidate.

18.3   Uniform-Micro Genetic Algorithms

A uniform-micro GA (or μGA) is used in this study. As per Krishnakumar's study (1989), a micro-GA can avoid premature convergence and demonstrated faster convergence to the near optimal region than does a simple GA for the multimodal problems. Carroll (1996) found that uniform-μGA is more robust in handling an order-3 deceptive function than traditional GA methods, and pointed out that the robustness of uniform-μGA lies in the constant infusion of new genetic information as the micro-population restart as well as the uniform crossover operator's characteristic of being no position biased.

Similar as a simple GA, a μGA cannot search in an infinite search space. The search domain defined gives the bounds on the parameters. Discretized points in the domain form the searching spaces. The density of the points depends on the accuracy requirement of the parameters to be identified.

A μGA starts with a very small population (usually 5 or 6 sets) randomly sampled in the searching space. The population evolves in normal GA fashion such as performing binary coding, selection, and crossover. The elitism operator is adopted to replicate the best individual of current generation into the next generation. The small population evolves and will soon converge after a few evolutions (converge criteria need to be preset). At this point, a new population is randomly generated while keeping the best individual from the previously converged generation, and the evolution process then restarts. This process repeats until the preset stopping criterion is met.

There is no mutation operation for the population evaluation in μGA. Since new chromosomes with the new genetic information keep flowing when the micro population is reborn, the μGA shows better robustness.

18.4 Examples

18.4.1 Inverse Procedure

Numerical investigations are conducted to verify the effectiveness of the method. Three composite laminated plates that are made of layers of glass/epoxy and carbon/epoxy materials, respectively, are studied. The ply fiber orientations of laminated plates are arbitrarily chosen.

First, we consider a composite laminate consisting of twelve glass/epoxy layers. The stacking sequence of the laminated layer is denoted by [G0/+45/ −45/90/−45/+45]$_s$, where G stands for glass/epoxy composite, and the digital numbers stand for the angles of fiber-orientation of each ply to the x-axis. The subscript s means that the plate is symmetrically stacked. Although we use a symmetrical laminate for the numerical verification, the method is applicable to laminates with arbitrary layer stacking sequence.

The laminate is loaded by a dynamic line load at $x = 0$. The line load is in z-axis direction and the time history is one cycle of sine function. The receiving point is set at $x = 1.5H$, where H denotes the thickness of the laminate. The displacement response in the z direction at the receiving point is computed using HNM and is shown in Figure 18.1. The letters P and S mark the estimated arrival times of the dilatational and shear waves. From the figure we can see the longitudinal wave has much smaller magnitude and is more vulnerable noise, while the shear wave has a much larger magnitude and dominates the main response of the plate. Figure 18.1 also designates the time range selected for the inverse operation in this study, which locates in the shear wave dominated area. The measurement response data is simulated by HNM using the actual material property. This computer-simulated displacement response is considered noise-free data and used to identify the material constants.

A uniform micro-GA (μGA) with binary parameter coding, tournament selection, uniform crossover, and elitism is adopted. The population size of each generation is set to be 6 and the probability of uniform crossover is set to be 0.5. The population convergence criterion is 5%, which means when less than 5% of the total bits of other individuals in a generation are different

TABLE 18.1

GA Search Space for Glass/Epoxy Laminates

Parameter	True Data	Search Range	Possibilities	Binary Digits
E_1 (GPa)	38.48	9.00–58.00	1024	10
E_2 (GPa)	9.38	4.00–14.00	512	9
G_{12} (GPa)	3.41	1.00–6.00	512	9
ν_{12}	0.292	0.100–0.481	128	7
ν_{23}	0.507	0.250–0.760	128	7

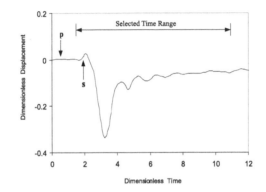

FIGURE 18.1

Displacement response at $x = 1.5H$ on the upper surface of a laminated plate [G0/−45/+45/90/−45/45]$_s$ excited by a line load with time history of one cycle sine function.

than the best individual, the convergence occurs. New random population, in which the best individual of the last generation is replicated, will flow in and the evolution process restarts. Knuth's subtractive method is used to generate random numbers, which makes the result of each GA run repeatable. With a different negative number initialization, the Knuth's algorithm generates a different series of a random number. The stopping criterion is imposed to limit each GA run to a maximum of 600 generations.

The GA search space defined for this numerical example is listed in Table 18.1. The bounds on the five parameters, two Young's modulus, one shear modulus, and two Poisson's ratios, are set to approximate ±50% of the true values; then the five parameters are discretized and translated into a chromosome of length 42. In the whole search space, there are approximately 2^{42} (or 4.4 trillion) possible combinations of the five parameters. If we would have to exhaust all the possible combinations, it would take about 140,000 years to find the optimum, even if one forward run takes only 1 second. Obviously, without an efficient searching scheme, it is not possible to solve an inverse problem of this kind of complexity.

Since the elitism technique is adopted, the best individual is automatically copied into the next generation, so in each generation (except the initial

TABLE 18.2

Identification Results for Glass/Epoxy Laminate [G0/+45/−45/90/−45/+45]$_s$

Material		Noise free		Gauss Noise	
Constant	True Data	Mean	% Error	Mean	%Error
E_1 (GPa)	38.48	38.34	0.36	37.75	1.89
E_2 (GPa)	9.38	9.24	1.49	9.41	0.32
G_{12} (GPa)	3.41	3.48	2.05	3.31	2.93
v_{12}	0.292	0.293	0.34	0.301	3.08
v_{23}	0.507	0.497	1.97	0.479	5.52

generation in which all six individuals have to be evaluated), only five individuals need to call the forward subprogram (HNM) for evaluating their fitness. The total number of calling forward calculation subprograms is 3001 in a GA run with maximum 600 generations.

The total computation time of the GA run directly depends on the single run time of the forward calculation subprogram since any time increase for forward calculation will be amplified by 3001 times. Only a high computation efficient forward solver can lead to a tolerable level of the computer time for each GA run. The HNM method has outstanding computation efficiency for the wave analysis of composite laminate; a subprogram developed using the HNM method only needs about 40 seconds for each forward calculation on an SGI Origin 2000 computer. A GA run still needs 40s × 3001 ≅ 33 hours of computer time, which show the GA's characteristic of high demand on computational power. Note that the total population is about 4.4 trillion, and 3001 forward computations are only 1.5 billionth of the total population. The power of GA is indeed astonishing.

Since genetic algorithms use random numbers, the result of a single GA run can be a matter of chance. Furthermore, GA's performances near global optima are relatively slow and imprecise compared with conventional gradient-based technique. The solution of each GA run approaches the global optima.

Eight GA runs with different random number series generated by Knuth's algorithm are performed, and the mean values of identification results are shown in Table 18.2. The variation of error value against the number of generation for a GA run is plotted in Figure 18.2.

18.4.2 Effects of Noise

Noise effects are also investigated by adding Gaussian noise directly to the computer-generated displacement readings. A Gauss random number generator is used to generate a series of random numbers with the standard deviation of

$$\sigma = 0.01 \times \left(1/N \sum_{j=1}^{N} (u_j^r)^2 \right)^{1/2} \tag{18.2}$$

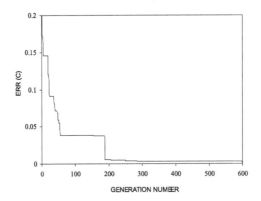

FIGURE 18.2
Variation of error against generation number in a GA run.

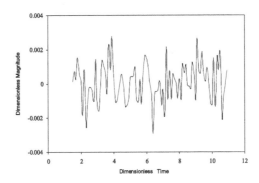

FIGURE 18.3
Gauss noise added to the displacement response of a laminated plate $[G0/+45/-45/90/-45/+45]_s$.

where N is the total number of time sample points. A typical Gauss noise generated is shown in Figure 18.3. The noise-added displacement readings are then used as input data for the inverse identification operation. The mean values of eight runs are listed also in Table 18.2.

A laminate made of the same material but with different layer orientations is also considered. The stacking sequence is arbitrarily chosen as $[0/-30/60/45/-70/90]_s$. A same search space defined in Table 18.1 is used for GA computation, and identified results, based on both noise-free response and Gauss noise contaminated response, are obtained and shown in Table 18.3. The identified results are as satisfactory as those obtained for $[G0/+45/-45/90/-45/45]_s$ laminate. This fact is very important because it shows that the procedure results are independent of the stacking sequence of the laminates.

Numerical verification is also performed on a laminate made of carbon/epoxy layers, whose anisotropy is much stronger than the glass/epoxy layers. The stacking sequence of the carbon/epoxy laminate is $[0/45/-45/90/-45/45]_s$. Table 18.4 lists the identified results based on both a noise-free

TABLE 18.3

Identification Results for Glass/Epoxy Laminate [G0/+30/−60/+45/−70/90]$_s$

Material		Noise Free		Gauss Noise	
Constant	True Data	Mean	% Error	Mean	% Error
E_1 (GPa)	38.48	38.81	0.86	38.80	0.83
E_2 (GPa)	9.38	9.28	1.07	9.13	2.66
G_{12} (GPa)	3.41	3.45	1.17	3.51	2.93
v_{12}	0.292	0.289	1.03	0.275	5.82
v_{23}	0.507	0.493	2.76	0.485	4.34

TABLE 18.4

Identification Results for Carbon/Epoxy Laminate [C0/+45/−45/90/−45/+45]$_s$

Material		Noise Free		Gauss Noise	
Constant	True Data	Mean	% Error	Mean	% Error
E_1 (GPa)	142.2	142.7	0.35	140.2	1.41
E_2 (GPa)	9.26	9.24	0.22	9.44	1.94
G_{12} (GPa)	4.80	4.85	1.04	4.83	0.63
v_{12}	0.33	0.333	0.91	0.350	6.06
v_{23}	0.49	0.491	0.20	0.463	5.51

response and a Gauss noise-contaminated response. The good results indicate that the present procedure also works for materials with very strong anisotropy.

From Tables 18.2 to 18.4, it is clear that the procedure is stable to the contamination of the noise. We also tested for a larger level of noise, and there are times when the inverse procedure fails. The robustness of the inverse procedure can be improved by having the problem being heavily over-determined, meaning increasing the sampling points N. The minimum sampling points is the number of parameters to be identified. In our inverse analysis, the parameter number is 5, but we use approximately 70 sampling points in the time history curve of response as inputs. Error functions using such over-determined data gives very little chance for the inverse problem's being ill-posed and unstable to noise.

It is found from the tables that the identified results of Young's modulus as well as shear modules are accurate and more stable to the contamination of noise, while the results of Poisson's ratio have relatively larger variation when the noised is added. In general, the results obtained are satisfactory for engineering applications.

In practice, the material properties for composite laminates are measured by testing a unidirectional composite specimen made in the same manufacturing process as the composite laminates. Using the present procedure, the material properties can be directly measured on the actual composite laminates with any stacking sequence, as long as the stacking sequence is known. The procedure of measurement is very simple and requires only standard testing equipment. All one needs is a small impact ball and a transient waveform recorder, plus, of course, a PC with a HNM code. There is no need to prepare a separate unidirectional specimen for material properties measurement.

18.5   Remarks

A procedure to inversely determine the material constants of composite laminates using elastic waves has been discussed. The HNM method is used as a forward solver for computing the wave response. The displacement response at one receiving point on the surface of the laminate is used as input for inverse identification. The GA technique is adopted in the inverse operation and is proven to be effective. Good identification results demonstrated the effectiveness of the procedure and its robustness to noise contamination, which is added to the computer-generated displacement response to simulate actual measurements.

The predicated results are independent of the stacking sequence of laminates and, therefore, can be applied directly to the actual laminates. The present method works also for composites of weak anisotropy, such as glass/epoxy, and of strong anisotropy, such as carbon/epoxy.

Note that since HNM is used, the laminate should not be bounded. Therefore, in practical application, the excitation and the observation points should be far away from the boundary of the laminate to avoid the contamination from the reflection of waves from the boundaries.

Other examples using elastic waves and neural networks can be found in papers by Xu et al. (2001) and Liu, Han, and Lau (2001b).

References

Aalami, B., Waves in prismatic guides of arbitrary cross section, *J. Appl. Mech.*, 40, 1067, 1973.

Abramowitz, M. and Stegun, I. A., *Handbook of Mathematical Functions with Formulas, Graphs, and Mathematical Tables,* U. S. Govt. Print, Washington, D.C., 1964.

Achenbach, J. D., *Wave Propagation in Elastic Solids,* North-Holland, Amsterdam, 1973.

Achenbach, J. D. and Rajapakse, Y., *Solid Mechanics Research for Quantitative Nondestructive Evaluation,* Martinus Nijhoff Publishers, 1987.

Ahn, V. S., Achenbach, J. D., and Kim, J. O., Numerical modeling of the V(z) curve for a thin-layer/substrate configuration, *Res. Nondestr. Eval.*, 3, 183–200, 1991.

Akay, A., Scattering of sound from concentric cylindrical shells, *J. Acoust. Soc. Am.*, 89, 1572, 1991.

Akay, H. U., Dynamic Large Deflection Analysis of Plates Using Mixed Finite Elements. *Computers and Structures*, 11, 1–11, 1980.

Al-Nassar, Y. N., Datta, S. K., and Shah, A. H., Scattering of Lamb waves by a normal rectangular strip weldment, *Ultrasonics*, 29, 125, 1991.

Ash, E. A. and Paige, E. G. S., Eds., *Rayleigh Wave Theory and Application,* Springer, Berlin, 1985.

Balasubramaniam, K. and Rao, N. S., Inversion of composite material elastic constants from ultrasonic bulk wave phase velocity data using genetic algorithms. *Composites,* 29B, 171, 1998.

Bamberger, A., Chalinder, B., Joly, P., Roberts, J. E., and Teron, J. L., Absorbing boundary conditions for Rayleigh waves, *SIAM J. Sci. Statistical Computing,* 9(6), 1016–1049, 1988.

Bathe, K. J. and Wilson, E. L., *Numerical Methods in Finite Element Analysis,* Prentice-Hall, Englewood Cliffs, NJ, 1976.

Beskos, D. E., Ed., *Boundary Element Method in Mechanics,* North-Holland, 1987.

Bleustein, J. F., A new surface wave in piezoelectric materials, *Appl. Phys. Lett.*, 13–12, 412, 1968.

Booth, D. C. and Crampin, S., The anisotropic reflectivity technique: theory, *Geophys. J. R. Astr. Soc.*, 72, 755, 1983.

Booth, D. C. and Crampin, S., The anisotropic reflectivity technique: anomalous reflected arrival from an anisotropic upper mantle, *Geophys. J. R. Astr. Soc.*, 72, 767, 1983.

Brekhovskikh, L. M., *Waves in Layered Media,* Academic Press, 79, 1960.

Carroll, D. L., Chemical laser modeling with genetic algorithms, *AIAA J.*, 34, 338, 1996.

Carroll, D. L., Genetic algorithms and optimizing chemical oxygen-iodine lasers, *Dev. Theoretical Appl. Mech.*, XVIII, 411, 1996.

Ceranoglu, A. N. and Pao, Y. H., Propagation of elastic pulses and acoustic emission in a plate, *ASME J. Appl. Mech.*, 48, 125, 1981.

Chang, C. and Sachse, W., Analysis of elastic wave signals from an extended source in a plate, *J. Acoust. Soc. Am.*, 77, 1335, 1985.

Chang, C. and Sun, C. T., Determining transverse impact force on a composite laminate by signal deconvolution, *Experimental Mech.*, 29, 414, 1989.

Cheung, Y. K., *Finite Strip Method in Structural Analysis*, Pergamon Press, 1976.

Chu, Y. C. and Rokhlin, S. I., Stability of determination of composite moduli from velocity data in planes of symmetry for weak and strong anisotropies, *J. Acoust. Soc. Am.*, 95, 213, 1994.

Chu, Y. C., Degtyar, A. D., and Rokhlin, S. I., On determination of orthoropic material moduli from ultrasonic velocity data in nonsymmetry planes, *J. Acoust. Soc. Am.*, 95, 3191, 1994.

Clayton, R. and Engquist, B., Absorbing boundary conditions for acoustic and elastic wave, *Bull. Seismol. Soc., Am.*, 67(6), 1529–1540, 1977.

Datta, S. K., Ju, T. H., and Shah, A. H., Scattering of an impact wave by a crack in a composite plate, *ASME J. Appl. Mech.*, 59, 596, 1992.

Dong, S. B. and Nelson, R. B., On natural vibrations and waves in laminated orthotropic plates, *ASME J. Appl. Mech.*, 39, 739, 1972.

Dong, S. B., A Block-Stodola eigensolution technique for large algebraic systems with non-symmetrical matrices, *Int. J. Numer. Methods Eng.*, 11, 247, 1977.

Doyle, J. F., An experimental method for determining the dynamic contact law, *Experimental Mech.*, 24, 10, 1984.

Doyle, J. F., Determining the contact force during the transverse impact of plates, *Experimental Mech.*, 27, 68, 1987.

Doyle, J. F., Further developments in determining the dynamic contact law, *Experimental Mech.*, 24, 265, 1984.

Engquist, B. and Majda, A., Radiation boundary conditions for acoustic and elastic wave calculations, *Commun. Pure Appl. Amth.*, 32, 313–357, 1979.

Erdelyi, A. et al., *Higher Transcendental Functions*, McGraw-Hill, New York, 1953.

Faran, J. J., Sound scattering by solid cylinders and spheres, *J. Acoust. Soc. Am.*, 23, 405, 1951.

Felippa, C. A., Solution of linear equations with skyline-stored symmetric matrix., *Computers and Structures*, 5, 13–30, 1975.

Frederiksen, P. S., Application of an improved model for the identification of material parameters, *Mech. Composite Mater. Structures*, 4, 297, 1997.

Fryer, G. J. and Frazer, L. N., Seismic waves in stratified anisotropic media, *Geophys. J. R. Astr. Soc.*, 78, 691, 1984.

Gaunaurd, G. C., Elastic and acoustic resonance wave scattering, *Appl. Mech. Revs.*, 42, 143, 1989.

Givoli, D., Non-reflection boundary conditions, *J. Comput. Phys.*, 94(1), 1–29, 1991.

Goldberg, D. E., *Genetic Algorithms in Search, Optimization and Machine Learning*, Addison-Wesley, Cambridge, MA, 1989.

Graff, K. F., *Wave Motion in Elastic Solids*, Dover Publications, 1991.

Green, E. R., Transient impact response of a fiber composite laminate, *Acta Mechanica*, 86, 153, 1991.

Gulyaev, Y. V., Electroacoustic surface waves in solids, *JETP Lett.*, 9, 37, 1969.

Han, X., Elastic waves in functionally graded materials and its applications to material characterization, Ph.D. thesis, National University of Singapore, Singapore, 2001.

Han, X., Liu, G. R., Xi, Z. C., and Lam, K. Y., Transient waves in a functionally graded cylinder, *Int. J. Solids Struct.*, 38, 3021, 2001.

Hellbratt, S. E. and Gullberg, O., The development of the GRP-sandwich technique for large marine structures, in *Sandwich Constructions 1*, Olsson, K. A. and Reichard, R. P., Eds., 425, 1989.

Hikita, M., Koshiba, M., Tanifuji, T., and Suzuki, M., Methods of a transmission line model for microwave acoustic in piezoelectric media, *Trans. Inst. Electron., Info. Comm. Eng.*, J56-B, 56, 1973.

Huang, K. H., and Dong, S. B., Propagating waves and edge vibrations in anisotropic composite cylinders, *J. Sound Vibration*, 96, 363, 1988.

Jennings, A., A compact storage scheme for the solution of symmetric simultaneous equations, *Computer J.*, 9, 281–285, 1966.

Jones, R. M., *Mechanics of Composite Materials*, Scripta, Washington, D.C., 1975.

Junger, M. C., Sound scattering by thin elastic shells, *J. Acoust. Soc. Am.*, 24, 366, 1951.

Junger, M. C., Vibrations of elastic shells in a fluid medium and associated radiation of sound, *ASME J. Appl. Mech.*, 19, 439, 1952.

Karim, M. R., Awal, M. A., and Kundu, T., Elastic wave scattering by cracks and inclusions in plates: in-plane case, *Int. J. Solids Struct.*, 29, 2355, 1992.

Karunasena, W. M., Shah, A. H., and Datta, S. K., Plane-strain-wave scattering by cracks in laminated composite plates, *ASCE J. Eng. Mech.*, 117, 1738, 1991.

Kausel, E., An explicit solution for the Green's functions for dynamic loads in layered media, in *Technical Report R81-13*, Department of Civil Engineering, M.I.T., Cambridge, MA, 1981.

Kausel, E., Wave propagation in anisotropic layered media, *Int. J. Numer. Methods Eng.*, 23, 1567, 1986.

Kausel, E. and Roësset, J. M., Stiffness Matrices for Layered Soils, *Bull. Seismological Soci. Am.*, 71(6), 1743–1761, 1981.

Kausel, E. and Roësset, J. M., Frequency domain analysis of undamped systems, *ASCE J. Eng. Mech.*, 118(4), 721–734, 1992.

Kawai, T., Miyazaki, S., and Araragi, M., A new method for forming a piezo-electric FGM using a Dual Dispenser System, in *Proc. First Int. Symp. Functionally Gradient Mater.*, Yamanouchi et al., Eds., Sendai, Japan, 1990, 191.

Kerner, E. H., The elastic and thermo-elastic properties of composite media, *Proc. Phys. Soc.*, 63, 808, 1956.

Khdeir, A. A. and Reddy, J. N., Exact solution for the transient response of symmetric cross-ply laminates using a higher-order plate theory, *Composite Sci. Technol.*, 34, 205–224, 1989.

Kline, R. A., *Nondestructive Characterization of Composite Media*, Technomic, Lancaster, PA, 1992.

Koshiba, M. and Suzuki, M., A consideration on finite element analysis of piezoelectric elastic waveguides, *Trans. Inst. Electron., Info. Comm. Eng.*, J61-B, 689, 1978.

Koshiba, M. and Suzuki, M., Acoustic wave propogation in anisotropic rots of rectangular cross section, *Trans. Inst. Electron. Comm. Eng.*, 59-B, 513–520, 1976.

Koshiba, T., Tanifuji, T., and Suzuki, M., Acoustic wave propogation in rots of rectangular cross section, *Trans. Inst. Electron. Comm. Eng.*, 57-B, 734–741, 1974.

Krishnamumar, K., Micro-Genetic algorithms for stationary and non-stationary function optimization, *SPIE: Intelligent Control and Adaptive Systems*, Philadelphia, PA, 1989, 1196.

Kundu, T. and Boström, A., Axisymmetric scattering of a plane longitudinal wave by a circular crack in a transversely isotropic solid, *ASME J. Appl. Mech.*, 58, 695, 1991.

Kundu, T. and Mal, A. K., Elastic waves in a multilayered solid due to a dislocation source, *Wave Motion*, 7, 459, 1985.

Lagasse, P. E., Higher-order finite-element analysis of topograghic guides supporting elastic surface wave, *J. Acoust. Soc. Am.*, 53-4, 1116, 1973.

Lam, K. Y., Liu, G. R., and Wang, Y. Y., Characterization of vertical surface-breaking crack plates, *Computational Acoustics*, 3(4), 297, 1995.

Lam, K. Y., Liu, G. R., and Wang, Y. Y., Time-harmonic response of a vertical crack in plates, *Theoretical and Applied Fracture Mechanics*, 27, 21, 1997.

Lee, Y. and Reismann, H., Dynamics of rectangular plates, *Intl. J. Eng. Sci.*, 7, 93–113, 1969.

Lewis, M. F., On Rayleigh waves and related propagating acoustic waves, in *Rayleigh-Wave Theory and Application*, Ash, E. A. and Paige, E. G. S., Eds., Springer-Verlag, Berlin, 1985.

Liu, G. R., Wave propagation in an inhomogeneous anisotropic plate and its impact responses, Ph.D. thesis, Tohoku University, Japan, 1991.

Liu, G. R. and Achenbach, J. D., A strip element method for stress analysis of anisotropic linearly elastic solids, *ASME J. Appl. Mech.*, 61, 270, 1994.

Liu, G. R. and Achenbach, J. D., Strip element method to analyze wave scattering by cracks in anisotropic laminated plates, *ASME J. Appl. Mech.*, 62, 607, 1995.

Liu, G. R., Han, X., and Lam, K. Y., Stress waves in functionally gradient materials and its use for material characterization, *Composites Part B: Engineering*, 30(4), 383, 1999.

Liu, G. R., Han, X., and Lam, K. Y., An integration technique for evaluating confluent hypergeometric functions and its application to functionally graded materials. *Computers and Structures*, 79, 1039, 2001a.

Liu, G. R., Han, X., Xu, Y. G., and Lam, K. Y., Material characterization of functionally graded material by means of elastic waves and a progressive-learning neural network, *Composites Science and Technology*, 61, 1401, 2001b.

Liu, G. R. and Lam, K. Y., Characterization of a horizontal crack in anisotropic laminated plates, *Int. J. Solids Struct.*, 31, 2965, 1994.

Liu, G. R. and Lam, K. Y., Transient waves in a laminated composite plate to a point excitation, *Proc. 12th Intl. Conf. Composite Mater.*, Paris, 1999.

Liu, G. R. and Lam, K. Y., Scattering of SH waves by flaws in sandwich plates and its use in flaw detection, *Composites Structure*, 34, 251, 1996.

Liu, G. R., Lam, K. Y., and Chan, E. S., Stress waves in composites laminates excited by transverse plane shock waves, *Shock and Vibration*, 3(6), 419–433, 1996.

Liu, G. R., Lam, K. Y., and Shang, H. M., Scattering of waves by flaws in anisotropic laminated plates, *Composites Part B*, 27B, 431, 1996.

Liu, G. R., Lam, K. Y., and Ohyoshi, T., A technique for analyzing elastodynamic responses of anisotropic laminated plates to line loads, *Composites*, 28B, 667, 1997.

Liu, G. R., Lam, K. Y., and Shang, H. M., A new method for analysing wave field in laminated composite plates: two-dimensional cases, *Composites Eng.*, 5, 1489, 1995.

Liu, G. R. and Ma, W. B., An inverse procedure for identification of loads on composite laminates, *Composite Part B*, Accepted for publication, 2001.

Liu, G. R. and Tani, J., Surface waves in functionally gradient piezoelectric plates, *ASME J. Vibration Acoustics*, 116, 440, 1994.

Liu, G. R., Tani, J., Ohyoshi, T., and Watanabe, K., Transient waves in anisotropic laminated plates, part 1: theory, part 2: application, *J. Vibration Acoustics*, 113, 230, 1991a.

Liu, G. R., Tani, J., Ohyoshi, T., and Watanabe, K., Characteristic wave surfaces in anisotropic laminated plates, *ASME J. Vibration Acoustics,* 113, 279, 1991b.

Liu, G. R., Tani, J., Ohyoshi, T., and Watanabe, K., Characteristics of surface wave propagation along the edge of an anisotropic laminated semi-infinite plate, *Wave Motion,* 13, 243, 1991c.

Liu, G. R., Tani, J., Watanabe, K., and Ohyoshi, T., Lamb wave propagation in aniso-tropic laminates, *ASME J. Appl. Mech.,* 57, 923, 1990a.

Liu, G. R., Tani, J., Watanabe, K., and Ohyoshi, T., Harmonic wave propagation in anisotropic laminated strips, *J. Sound Vibration,* 139(2), 313, 1990b.

Liu, G. R., Tani, J., Watanabe, K., and Ohyoshi, T., A semi-exact method for the propagation of harmonic waves in anisotropic laminated bars of rectangular cross section, *Wave Motion,* 12, 361, 1990c.

Liu, G. R., Wang, X. J., and Xi, Z. C., Elastodynamic responses of an immersed composite laminate to a Gaussian beam pressure, *J. Sound Vib.,* 233, 813, 2000.

Liu, G. R., Xi, Z. C., Lam, K. Y., and Shang, H. M., A strip element method for analyzing wave scattering by a crack in an immersed composite laminate, *J. Appl. Mech.,* 66, 898, 1999.

Liu, S. W. and Datta, S. K., Scattering of ultrasonic wave by cracks in a plate, *ASME J. Appl. Mech.,* 60, 352, 1993.

Liu, S. W., Datta, S. K., and Ju, T. H., Transient scattering of Rayleing Lamb waves by a surface-breaking crack: comparison of numerical simulation and experi-ment, *J. Nondestructive Evaluation,* 10, 111, 1991.

Lu, C., Computational Modeling of Composite Structures and Low-Velocity Impact. Ph.D. Thesis, National University of Singapore, 1996.

Lu, I. T. and Felsen, L. B., Ray, mode and hybrid options for time-dependent source-excited propagation in an elastic layer, Geophys. *J. R. Astr. Soc.,* 86, 177, 1986.

Lysmer, J., Lumped mass method for Rayleigh waves, *Bull. Seismological Soc. Am.,* 60(1), 89–104, 1970.

Mal, A. K. and Lih, S. S., Elastodynamic response of a unidirectional composite laminate to concentrated surface loads, *ASME J. Appl. Mech.,* 59, 878, 1992.

Mal, A. K., Wave propagation in layered composite laminates under periodic surface loads, *Wave Motion,* 10, 257, 1988.

Mallikarjuna and Knat, T., Dynamics of laminated composite plates with a higher order theory and finite element discretization, *J. Sound Vibration,* 126, 463–475, 1988.

Markus, S. and Mead, D. J., Axisymmetric and asymmetric wave motion in ortho-tropic cylinders, *J. Sound Vibration,* 181, 127, 1995.

Markus, S. and Mead, D. J., Wave motion in a 3-layered, orthotropic isotropic ortho-tropic composite shell, *J. Sound Vibration,* 181, 149, 1995.

McCoy, J. J. and Mindlin, R. D., Extensional wave along the edge of an elastic plate, *J. Appl. Mech.,* 30, 75–78, 1963.

Meyer, C., Solution of equations; state-of-the-art, *J. Structural Division, ASCE,* 99, 1507–1526, 1973.

Meyer, C., special problems related to linear equation solvers, *J. Structural Division, ASCE,* 101, 869–890, 1975.

Michaels, J. E. and Pao, Y. H., The inverse source problem for an oblique force on an elastic plate, *J. Acoust. Soc. Am.,* 77, 2005, 1985.

Michaels, J. E. and Pao, Y. H., Determination of dynamic forces from wave motion measurements, *ASME J. Appl. Mech.,* 53, 61, 1986.

Miklowitz, J., *The Theory of Elastic Waves and Waveguides*, North-Holland, Amsterdam, 1978.

Mindlin, R. D. and Fox, E. A., Vibrations and waves in elastic bar of rectangular cross section, *J. Appl. Mech.*, 27, 152–158, 1960.

Moon, C. F., One-dimensional transient waves in anisotropic plates, *ASME J. Appl. Mech.*, 40, 485, 1973.

Moon, F. C., Wave surfaces due to impact on anisotropic plates, *J. Composite Mater.*, 6, 62, 1972.

Mota Soares, C. M., Moreira de Freitas, M., Araujo, A. L., and Pedersen, P., Identification of material properties of composite plate specimens, *Composite Structures*, 25, 277, 1993.

Murakami, H. and Akiyama, A., A mixture theory for wave propagation in angle-ply laminates, part 2: application, *ASME J. Appl. Mech.*, 52, 338, 1985.

Murakami, H., A mixture theory for wave propagation in angle-ply laminates, part 1: theory, *ASME J. Appl. Mech.*, 52, 331, 1985.

Nardin, M., Perger, W. F., and Bhalla, A., Numerical evaluation of the confluent hypergeometric function for complex arguments of large magnitudes, *J. Comput. Appl. Math.*, 39, 193–200, 1992a.

Nardin, M., Perger, W. F., and Bhalla, A., Algorithm 707 CONHYP: A numerical evaluator of the confluent hypergeometric function for complex arguments of large magnitudes, *ACM Trans. Math. Software*, 18(3), 345–349, 1992b.

Nayfeh, A. H., *Wave Propagation in Layered Anisotropic Media with Applications to Composites*, Elsevier, Amsterdam, 1995.

Nelson, R. B. and Dong, S. B., High frequency vibrations and waves in laminated othortropic plates, *J. Sound Vibration*, 30, 33, 1973.

Nelson, R. B., Dong, S. B., and Kalra, R. D., Vibrations and waves in laminated orthotropic circular cylinders, *J. Sound Vibration*, 18, 429, 1971.

Nigro, N. J., Steady-state wave propagation in infinite bar of noncircular cross section, *J. Acoust. Soc. Am.*, 40, 1501, 1966.

Nilsson, A. C., Wave propagation in sandwich plates: some dynamical and acoustical aspects, in *Sandwich Constructions 1*, Olsson, K. A. and Reichard, R. P., Eds., 195, 1989.

Ohyoshi, T., Energy propagation velocity of elastic waves in sandwich layer, *ASME J. Vibration, Stress, and Reliability in Design*, 107, 235–342, 1985.

Ohyoshi, T., Easy calculation for the response of an elastic plate to transverse impact, *Trans. Jpn. Soc. Mech. Eng.*, 55(A), 483, 1989.

Ohyoshi, T. and Miura, K., Experimental analysis of elastic wave propagation in rectangular plate by longitudinal end-impact, *Trans. Jpn. Soc. Mech. Eng.*, 52(A), 1169–1675, 1986.

Ono, S., Wasa, K., and Hayakawa, S., Surface-acoustic-wave properties in ZnO-SiO$_2$-Si layered structure, *Wave Electronics*, 3, 35, 1977.

Pao, Y. H., Elastic waves in solids, *ASME J. Appl. Mech.*, 50, 1152, 1983.

Pao, Y. H., The generalized ray theory and transient responses of layered elastic solids, *Physical Acoustics*, XIII, 184, 1977.

Payton, R. G., *Elastic Wave Propagation in Transversely Isotropic Media*, Martinus Nijhoff Publishers, 1983.

Polyanin, A. D. and Zaitsev, V. F., *Handbook of Exact Solutions for Ordinary Differential Equations*, CRC Press, Boca Raton, FL, 1995.

Pualey, K. E. and Dong, S. B., Analysis of plate waves in laminated piezoelectric plates, *Wave Electronics*, 1, 265, 1975.

Rattanwangcharoen, N., Shah, A., and Datta, S. K., Reflection of waves at the free edge of a laminated circular cylinder, *ASME J. Appl. Mech.*, 61, 323, 1994.

Rattanwangcharoen, N., Zhuang, W., Shah, A., and Datta, S. K., Axisymmetric guided waves in jointed laminated cylinders, *ASCE J. Eng. Mech.*, 123, 1020, 1997.

Reddy, J. N., Dynamic (transient) analysis of layered anisotropic composite-material plates, *Intl. J. Num. Meth. Eng.*, 19, 237–255, 1983.

Reddy, J. N., *Energy and Variational Methods in Applied Mechanics*, John Wiley & Sons, New York, 1984a.

Reddy, J. N., A simple higher-order theory for laminated composite plates, *ASME J. Appl. Mech.*, 51, 745–752, 1984b.

Reddy, J. N., A refined nonlinear theory of plates with transverse shear deformation, *Intl. J. Solid Structures*, 20, 881–896, 1984c.

Reddy, J. N., A review of the literature on finite-element modeling of laminated composite plates, *The Shock and Vibration Digest*, 17, 3–8, 1985.

Reddy, J. N., *Mechanics of Laminated Composite Plates: Theory and Analysis*, CRC Press, Boca Raton, FL, 1997.

Reismann, H., Forced motion of elastic plates, *J. Appl. Mech.*, 35, 510–515, 1968.

Reissner, E., A consistent treatment of transverse shear deformations in laminated anisotropic plates, *AIAA J.*, 10, 716–718, 1972.

Riche, R. L. and Haftka, R. T., Optimization of laminate stacking sequence for bucking load maximization by genetic algorithm, *AIAA J.*, 31, 951, 1993.

Rikards, R. and Chate, A., Identification of elastic properties of composites by method of planning of experiments, *Composite Structures*, 42, 257, 1998.

Rikards, R., Chate, A., Steinchen, W., Kessler, A., and Bledzki, A. K., Method of identification of elastic properties of laminates based on experiment design, *Composites*, 30B, 279, 1999.

Robson, B. L., The Royal Australian Navy Inshore Minehunter—Lessons learned, in *Sandwich Constructions 1*, Olsson, K. A. and Reichard, R. P., Eds., 1989, 395.

Rock, T. and Hinton, E., Free vibration and transient response of thick and thin plates using the finite element method, *Earthquake Engineering and Structural Dynamics*, 3, 51–63, 1974.

Rose, J. L., *Ultrasonic Waves in Solid Media*, Cambridge University, Cambridge, England, 1999.

Santosa, F. and Pao, Y. H., Transient axially asymmetric response of an elastic plate, *Wave Motion*, 11, 271, 1989.

Sasaki, M., Wang, Y., Hirano, T., and Hirai, T., Design of SiC/C functionally gradient material and its propagation by chemical vapor deposition, *J. Ceramic Soc. Jpn.*, 97, 539, 1989.

Scott, R. A. and Miklowitz, J., Transient elastic waves in anisotropic plates, *ASME J. Appl. Mech.*, 34, 104, 1967.

Seale, S. H. and Kausel, E., Point loads in cross-anisotropic layered halfspaces, *ASCE J. Eng. Mech.*, 115, 509, 1989.

Shiosaki, T., Mikamura, Y., Takeda, F., and Kawabata, A., High-coupling and high-velocity SAW using ZnO and AlN films on a glass substrate, *IEEE Trans. UFFC*, 33, 324, 1986.

Sinha, B. K., Some remarks on propagation characteristics of ridges for acoustic waves at low frequencies, *J. Acoust. Soc. Amer.*, 56, 16–18, 1974.

Slater, L. J., *Confluent Hypergeometric Functions*, Cambridge University Press, Cambridge, England, 1960.

Stanton, T. K., Sound scattering by cylinders of finite length: II. elastic cylinders, *J. Acoust. Soc. Am.*, 83, 64, 1988.

Sun, C. T., Propagation of shock waves in anisotropic composite plates, *J. Compos. Mater.*, 7, 366, 1973.

Sun, C. T., Achenbach, J. D., and Herrmann, G., Continuum theory for a laminated medium, *ASME J. Appl. Mech.*, 35, 467, 1968.

Sun, C. T., Achenbach, J. D., and Herrmann, G., Time-harmonic waves in a stratified medium propagating in the direction of the layering, *ASME J. Appl. Mech.*, 35, 408, 1968.

Sun, C. T. and Tan, T. M., Wave propagation in a graphite/epoxy laminate, *J. Astronautical Sci.*, 32, 269, 1984.

Takahashi, K., and Chou, T. W., Non-linear deformation and failure behavior of Carbon/Glass hybrid laminates, *J. Composite Mater.*, 21, 396, 1987.

Takahashi, M., Itoh, Y., and Kashiwaya, H., Fabrication and evaluation of W/Cu gradient material by sintering and infiltration technique, in *Proc. First Intl. Symp. Functionally Gradient Mater.*, Yamanouchi, et al., Eds., Sendai, Japan, 1990, 129.

Thompson, W. J., *Atlas for Computing Mathematical Functions: An Illustrated Guide for Practitioners with Programs in Fortran 90 and Mathematica*, 1997.

Thomson, W., Transmission of elastic waves through a stratified medium, *J. Appl. Phys.*, 21, 89, 1950.

Toda, K. and Mizutani, K., Propagation characteristics of plate waves in a Z-Cut X-propagation LiTaO$_3$ thin plate, *Trans. Inst. Electron., Info. Commun. Eng.*, J71-A-6, 1225, 1988.

Tolstoy, I., Resonant frequencies and high modes in layered wave guides, *J. Acoust. Soc. Am.*, 28, 1182–1192, 1965.

van der Hijden, J. H. M. T., *Propagation of Transient Elastic Waves in Stratified Anisotropic Media*, North-Holland, Amsterdam, 1987.

Vasudevan, N. and Mal, A. K., Response of an elastic plate to localized transient sources, *ASME J. Appl. Mech.*, 52, 356, 1985.

Vikstrom, M., Thermograhpic NDT of foam core materials, in *Sandwich Constructions 1*, Olsson, K. A. and Reichard, R. P., Eds., 331, 1989.

Vinson, J. R. and Sierakowski, R. L., *The Behavior of Structures Composed of Composite Materials*, Martinus Nijhoff, Dordrecht, 1987.

Waas, G., Linear two-dimensional analysis of soil dynamics problems in semi-infinite layer media, Ph.D. Thesis, University of California, Berkeley, CA, 1972.

Wang, Y. Y., Lam, K. Y., and Liu, G. R., Detection of flaws in sandwich plates, *Composite Structures*, 34, 409, 1996.

Wang, Y. Y., Lam, K. Y., Liu, G. R., Reddy, J. N., and Tani, J., A strip element method for bending analysis of orthotropic plates, *JSME Intl. J. (A)*, Japan, 40(4), 398, 1997.

Wang, Y. Y., A strip element method for the analysis of laminated composite plates, Ph.D. Thesis, National University of Singapore, Singapore, 1998.

Wang, Y. Y., Lam, K. Y., and Liu, G. R., Bending analysis of classical symmetric laminated composite plates by the strip element method, *Mech. Composite Mater. Struct.*, 7, 225, 2000a.

Wang, Y. Y., Lam, K. Y., and Liu, G. R., The effect of rotatory inertia on the dynamic response of laminated composite plates, *Composite Structures*, 48, 265, 2000b.

Wang, Y. Y., Lam, K. Y., and Liu, G. R., A strip element method for the transient analysis of symmetric laminated plates, *Solids and Structures*, 38, 241, 2001.

Watanabe, R. and Kawasaki, A., Overall view of the P/M fabrication of functionally gradient materials, in *Proc. First Intl. Symp. Functionally Gradient Mater.*, Yamanouchi et al., Eds., Sendai, Japan, 1990, 107.

Weaver, R. L. and Pao, Y. H., Axisymmetric elastic waves excited by a point source in a plate, *ASME J. Appl. Mech.*, 49, 821, 1982.

Whitney, J. M. and Leissa, A. W., Analysis of heterogeneous anisotropic plates, *ASME J. Appl. Mech.*, 36, 261–266, 1969.

Whitney, J. M. and Pagano, N. J., Shear deformation in heterogeneous anisotropic plates, *ASME J. Appl. Mech.*, 37, 1031–1036, 1970.

Wiggins, R. A. and Helmberger, D. V., Synthetic seismogram computation by expansion in generalized rays, *Geophys. J. R. Astr. Soc.*, 37, 73, 1974.

Wu, E., Tsai, T. D., and Yen, C. S., Two methods for determining impact-force history on elastic plates, *Experimental Mech.*, 35, 11, 1995.

Wu, T. T. and Liu, Y. H., Inverse determinations of thickness and elastic properties of a bonding layer using laser-generated surface waves, *Ultrasonics*, 37, 23, 1999.

Xi, Z. C., Liu, G. R., Lam, K. Y., and Shang, H. M., Dispersion and characteristic surfaces of waves in laminated composite circular cylindrical shells, *J. Acoust. Soc. Am.*, 108, 2179, 2000.

Xi, Z. C., Liu, G. R., Lam, K. Y., and Shang, H. M., Strip element method for analyzing wave scattering by a crack in an immersed laminated composite cylinder, *J. Acoust. Soc. Am.*, 108, 175, 2000a.

Xi, Z. C., Liu, G. R., Lam, K. Y., and Shang, H. M., Strip element method for analyzing wave scattering by a crack in an axisymmetric cross-ply laminated composite cylinder, *ASME J. Appl. Mech.*, 67, 427, 2000b.

Xi, Z. C., Liu, G. R., Lam, K. Y., and Shang, H. M., Strip element method for analyzing wave scattering by a crack in a fluid-filled laminated composite shell, *Compos. Sci. Technol.*, 60, 1985, 2000c.

Xu, P. C. and Mal, A. K., Calculation of the inplane Green's functions for a layered viscoelastic solid, *Bull. Seismological Soc. Am.*, 77, 1823, 1987.

Xu, Y. G., Liu, G. R., Wu, Z. P., and Hunag, X. M., Adaptive multiplayer perceptron networks for detection of cracks in anisotropic laminated plates, *Intl. J. Solids Structures*, 38, 5625, 2001.

Yen, C. S. and Wu, E., On the inverse problem of rectangular plates subjected to elastic impact, *ASME J. Appl. Mech.*, 62, 692, 1995.

Yuan, F. G. and Hsieh, C. C., Three-dimensional wave propagation in composite cylindrical shell, *Composite Structures*, 42, 153, 1998.

Zhuang, W., Shah, A., and Datta, S. K., Axisymmetric guided wave scattering by cracks in welded steel pipes, *J. Pressure Vessel Technol.*, 119, 401, 1997.

Zienkiewicz, O. C., *The Finite Element Method*, 4th ed., McGraw-Hill, London, 1989.

Index